W9-AYN-765

RAISING ELIJAH

ALSO BY SANDRA STEINGRABER

Having Faith: An Ecologist's Journey to Motherhood

Living Downstream: An Ecologist's Personal Investigation of Cancer and the Environment

Post-Diagnosis (poetry)

RAISING ELIJAH

*Protecting Our Children in
an Age of Environmental Crisis*

Sandra Steingraber

A MERLOYD LAWRENCE BOOK
DA CAPO PRESS
A Member of the Perseus Books Group

Many of the designations used by manufacturers and sellers to distinguish their products are claimed as trademarks. Where those designations appear in this book and Da Capo Press was aware of a trademark claim, the designations have been printed in initial capital letters.

Copyright © 2011 by Sandra Steingraber

All rights reserved. No part of this publication may be reproduced, stored in a retrieval system, or transmitted, in any form or by any means, electronic, mechanical, photocopying, recording, or otherwise, without the prior written permission of the publisher. Printed in the United States of America.

Designed by Jill Shaffer
Set in 11-point Giovanni by Eclipse Publishing Services

Cataloging-in-Publication data for this book is available from the Library of Congress.

First Da Capo Press edition 2011

ISBN: 978-0-7382-1399-6

Published as A Merloyd Lawrence Book by Da Capo Press
A Member of the Perseus Books Group
www.dacapopress.com

Da Capo Press books are available at special discounts for bulk purchases in the U.S. by corporations, institutions, and other organizations. For more information, please contact the Special Markets Department at the Perseus Books Group, 2300 Chestnut Street, Suite 200, Philadelphia, PA 19103, or call (800) 810-4145, ext. 5000, or e-mail special.markets@perseusbooks.com.

10 9 8 7 6 5 4 3 2

for Elijah

and for Jeremy Armstrong
(1979–2002)

"Ithaca is only a small island, yet it is my home and very dear to me. It is a rugged land, with narrow ways and no broad roads, and its pasture-land is best fitted for the sturdy sure-footed goats. But yet it breeds fine men, brave and bold, and I long once more to see it, because, my friends, it is my home."

—*The Odyssey of Homer,* retold by Barbara Leonie Picard
(New York: Oxford University Press, 2000)

"Go from here," she said. "You are in great danger if you stay. Our master is Kostchei the Deathless, the wizard of darkness. He turns to stone anyone who tries to rescue us. Only we are alive in all his realm."

She pointed to the garden and the stone statues.

Prince Ivan grew cold at her words. Only the feather over his heart remained warm.

"I am not afraid," he said, though he was.

—*The Firebird,* retold by Jane Yolen
(New York: HarperCollins, 2002)

Contents

Foreword xi

Author's Note xviii

ONE   Milk (and Terror).................................... 1

TWO   The Nursery School Playground
(and Well-Informed Futility) 27

THREE The Grocery List (and the Ozone Hole)............. 57

FOUR   Pizza (and Ecosystem Services)..................... 83

FIVE   The Kitchen Floor (and National Security) 111

SIX   Asthma (and Intergenerational Equity) 137

SEVEN The Big Talk (and Systems Theory)................ 167

EIGHT Homework (and Frontiers in Neurotoxicology) 197

NINE   Eggs (and Sperm) 229

TEN   Bicycles on Main Street (and High-Volume
Slickwater Hydraulic Fracturing) 263

Further Resources 287

Source Notes..................................... 291

Acknowledgments................................. 329

Index.. 333

About the Author 350

Foreword

IN ALTON, Illinois, downstream from the river town where I grew up, the abolitionist Elijah Lovejoy was pumped full of bullets on a dark November night by a mob intent on silencing the man once and for all.

On this evening, they succeeded.

By dawn, Elijah was dead, and his printing press—the means by which he distributed his radical ideas—lay at the bottom of the Mississippi River. The year was 1837. The Reverend Lovejoy was buried on his thirty-fifth birthday.

But the story doesn't end here.

Almost immediately, membership in antislavery societies across the nation swelled. Vowing to carry on the work of his fallen friend, Edward Beecher, president of Illinois College, threw himself into abolitionist efforts and, in so doing, inspired his sister, Harriet Beecher Stowe, who went on to write the most famous abolitionist treatise of all: *Uncle Tom's Cabin*. Meanwhile, Elijah's brother, Owen Lovejoy, turned his own house into a station along the Underground Railroad. Owen went on to win a seat in Congress, and, along the way, befriended a young Illinois politician by the name of Abraham Lincoln.

These facts impressed me as a child.

Reading Reverend Lovejoy's biography as a grown-up mother, I find other things impressive. Such as the fact that, at the time of his assassination, Elijah had a young family. And yet, in the weeks before his death—when it became clear that the terrorist mob pursuing him was growing bolder by the hour—he did not desist from speaking out against slavery. *While all around me is violence and tumult, all is peace within. . . . I sleep sweetly and undisturbed, except when awakened by the brickbats of the mob.* So declared Elijah in one of his final speeches.

Truly? With a pregnant wife in the bed next to him and a one-year-old son in the next room? He wasn't worried?

A letter to his mother in Maine tells a more nuanced story:

Still I cannot but feel that it is harder to 'fight valiantly for the truth' when I risk not only my own comfort, ease, and reputation, and even life, but also that of another beloved one.

And then there's this poignant aside:

I have a family who are dependent on me . . . And this is it that adds the bitterest ingredient to the cup of sorrow I am called to drink.

Here's something else that I've noticed while reading his words. To the slave owners and murderous thugs, Elijah spoke calmly. He reserved his fierce language for the members of the community who gladly lived in the free state of Illinois but wished to remain above the fray: the ones who added their signatures to a resolution asking him to cease publication of his newspaper and leave town but would not sign a resolution that urged protection of law against mob rule; the ones who agreed that slavery was a homicidal abomination but who feared that emancipation without recompense to slave owners for loss of property would be socially destabilizing; the ones who believed themselves upstandingly moral but who chose to remain silent about the great moral crisis of the day.

They included fellow clergymen.

In *Raising Elijah* I call for outspoken, full-throated heroism in the face of the great moral crisis of our own day: the environmental crisis. And, because the main victims of this unfolding calamity are our own children, this book speaks directly to parents.

In fact, the environmental crisis is actually two crises, although they share a common cause. You could view it as a tree with two main branches. One branch represents what is happening to our planet through the atmospheric accumulation of heat-trapping gasses (most notably, carbon dioxide and methane), and the other branch represents what is happening to us through the accumulation of inherently toxic chemical pollutants in our bodies. Follow the first branch along and you find droughts, floods, acidifying oceans, dissolving coral reefs, and faltering plankton stocks. (The oceans' plankton provides half of our atmospheric oxygen supply. More on this in Chapter 6.) Follow the second branch along and you find pesticides in children's urine, lungs stunted by air pollutants, abbreviated pregnancies, altered hormone levels, and lower scores on cognitive tests.

The trunk of this tree is an economic dependency on fossil fuels, primarily coal (plant fossils) and petroleum and natural gas (animal fossils). When we light them on fire, we threaten the global ecosystem. When we use them as feedstocks for making stuff, we create substances—pesticides, solvents, plastics—that can tinker with our subcellular machinery and the various signaling pathways that make it run.

Biologist Rachel Carson first called our attention to these manifold dangers a half century ago in her 1962 book, *Silent Spring*. In it, she posited that "future generations are unlikely to condone our lack of prudent concern for the integrity of the natural world that supports all life." Since then, the scientific evidence for its disintegration has become irrefutable, and members of the future generations to which she was referring are now occupying our homes.

They are our kids.

I mean this in the most basic of ways. When my son, at age four, asked to be a polar bear for Halloween, I sewed a polar bear costume—and I did so with the full knowledge that his costume might easily outlast the species. No other generation of mothers before mine has ever borne such knowledge—nor wondered if we should share this terrible news with our children. Or not. It's a novel situation. Indeed, according to the most recent assessment, one in every four mammal species (and one in three marine mammals) is now threatened with extinction, including that icon of Halloween itself: the little brown bat. Thus, animal costumes whose real-life correspondents have been wiped from the Earth may well become commonplace.

This leads me to wonder: What will we say when our grandchildren ask us the names of the departed? When bats, bees, butterflies, whales, polar bears, and elephants disappear, will children still read books about them? Will they want to dress up as vanished species? Or, by then, will the loss of favorite animals be the least of their worries? (Talking with children about environmental devastation is the topic of Chapters 7 and 8.)

At the same time, chronic childhood diseases linked to toxic chemical exposures are rising in prevalence. (These receive my close attention in Chapters 1, 3, 4, 6, and 9.) Here are a few of the current trends:

- 1 in 8 U.S. children is born premature. Preterm birth is the leading cause of death in the first months of life and the leading cause of disability. Its price tag is $26 billion per year in medical costs, special services, and lost productivity. Preterm birth has demonstrable links to air pollution, especially maternal exposure to fine particles and combustion byproducts of the type released from coal-burning power plants.
- 1 in 11 U.S. children has asthma, the most common chronic childhood disease and a leading cause of school absenteeism. Asthma symptoms have been linked to certain ingredients in plastic (phthalates) as well as outdoor air pollution, including traffic exhaust. The annual cost of childhood asthma is estimated at $18 billion. Its incidence has doubled since 1980.
- 1 in 10 U.S. children has a learning disability, and nearly 1 in 10 has attention deficit/hyperactivity disorder. All together, special educational services now consume 22 per-

cent of U.S. school spending—about $77.3 billion per year at last count. Neurodevelopmental disorders have significant associations with exposures to air pollution, organophosphate pesticides, and the heavy metals lead, mercury, and arsenic, among others.

- 1 in 110 children has autism or is on the autism spectrum. Annual costs are $35 billion. Causes are unknown, but exposure to chemical agents in early pregnancy is one of several suspected contributors.
- 1 in 10 U.S. white girls and 1 in 5 U.S. black girls begin breast development before the age of eight. On average, breast development begins nearly two years earlier (age 9) than it did in the early 1960s (age 11). A risk factor for breast cancer in adulthood, early puberty in girls is associated with increasing body fat as well as exposure to some hormonally active chemical agents. We have no cost estimates for the shortened childhoods of girls.

All together, asthma, behavioral problems, intellectual impairments, and preterm birth are among the "new morbidities of childhood." So concludes a federally funded investigation of pediatric environmental health. Ironically, by becoming so familiar a presence among children, these disorders now appear almost normal or inevitable. And yet, with an entirely different chemical regulatory system, farm bill, and energy policy, their prevalence might be much reduced. While environmental factors are not the *only* cause of the problems named above, they are unquestionably contributing to them, and they are preventable. The fact that we do not identify and abolish hormone-disrupting, brain-damaging chemicals to which children are routinely exposed raises profound ethical questions. As the authors of the pediatric environmental health investigation rightly point out:

> In the absence of toxicity testing, we are inadvertently employing pregnant women and children as uninformed subjects to warn us of new environmental toxicants. . . . Paradoxically because industry is not obligated to supply

the data on developmental neurotoxicity, the costs of human disease, research, and prevention are socialized whereas the profits are privatized.

In the absence of federal policies that are protective of child development and the ecology of the planet on which our children's lives depend, parents have to serve as our own regulatory agencies and departments of interior. Already maniacally busy, we are encouraged by popular media reports to read labels, consult Web sites, vet the contents of birthday party goody bags, shrink our carbon footprints, mix our own nontoxic cleaning products, challenge our school districts to embrace pesticide-free soccer fields, and limit the number of ounces of mercury-laced tuna fish consumed by each child per week.

Thoughtful but overwhelmed parents correctly perceive a disconnect between the enormity of the problem and the ability of individual acts of vigilance and self-sacrifice to fix it. Environmental awareness without corresponding political change leads to paralyzing despair. And so, eventually, we begin to discount or ignore the latest evidence for harm. We feel helpless in the face of our knowledge, and we're not sure we want any more knowledge. You could call this *well-informed futility syndrome.* (Chapter 2 does.)

And soon enough, we are retreating into silent resignation rather than standing up for *abolition now.*

In *Raising Elijah* I seek a path out of that despair. This book is not about shopping differently. Indeed, it rejects altogether the notion that toxicity should be a consumer choice. Instead, it seeks the higher ground of human rights in which to explore systemic solutions to the ongoing chemical contamination of our children and our biosphere. And because I believe that stories move us to action more than data alone, the scientific evidence is strapped to the hood of an autobiographical tale that begins with the birth of my son and spans the first nine years of my life as a biologist mother of two. Once I chronicled interspecies relationships in a central American rainforest; now I seek to understand the complex habitat of my own household. (Readers who seek more technical detail will find it in the Source Notes at the end.)

Throughout these chapters, I discover that the domestic routines of family life with young children—however isolated and detached from public life they seem—are inextricably bound to the most urgent public health issues of our time. Bedtime snacks are linked to global systems of agricultural subsidies. Sunburn at the beach is linked to the stability of the ozone layer, which, in turn, is threatened by particular pesticides used in the production of tomatoes and strawberries. Risks for asthma are related to transportation and energy policies. The highly explosive raw materials used for manufacturing my kitchen floor pose demonstrable threats to national security. The rabid bat I capture in the kids' bedroom reveals the precautionary principle in action as an enlightened public health policy. The proposal to shatter the shale bedrock of our rural county and extract from it natural gas reveals the abandonment of that same principle (Chapter 10).

And through these various explorations, two epiphanies emerge:

One: Current environmental policies must be realigned to safeguard the healthy development of children and sustain planetary life-support systems on which their lives depend. Only within a new regulatory framework can parents carry out our two most fundamental duties: to protect our children from harm and provide for their future.

Two: Such a realignment necessitates emancipation from our terrible enslavement to fossil fuels in all their toxic forms. In other words, as Elijah Lovejoy exhorted his fellow citizens when encouraging them to imagine a U.S. economy no longer dependent on the unpaid labor of people held as property: It's time to "show a spirit of freedom."

AUTHOR'S NOTE

The names of children in this book, other than my own two, have been changed to protect anonymity. Some details and circumstances have been likewise obscured.

S.S.

CHAPTER ONE
Milk (and Terror)

ALL THROUGH the unknowing summer of 2001, my unborn son lay horizontally within my body like a half-open umbrella, his head pressing hard against my left side and his bottom bulging out on the right. Midwives and obstetricians refer to this position as a transverse lie. It makes birth impossible. You can't come out of a woman sideways.

All sorts of alignments are common in early pregnancy, but, by seven months, most babies are head down. Of those still off-plumb at eight months, only a minority are able to get themselves vertically oriented in time for labor and delivery. A transverse lie once foretold death for both mother and child. It's what Caesarean sections are for.

Coming up on eight months, I wasn't worried yet. But I was soliciting suggestions. My midwife advised swimming. Buoyancy, she said, sometimes encourages side-lying babies to move. And so each evening, I waded into Cayuga Lake—sunlight folding itself into the water or rain pebbling its surface—and struck out for the opposite shore. *You swim, too,* I instructed my son.

He didn't budge.

A physician friend recommended crawling on all fours. This position stretches the cervical ligaments and so invites fetal rotation.

So I spent an afternoon on hands and knees, exploring the hollow of land behind our cabin. I told myself I was investigating the habits of wood frogs, with their eye masks and their freezable blood.

It's true, I told my crossways son as I scuttled along. *Wood frogs are freeze tolerant. They can't dig like toads or sink to the bottom of lakes like other frogs. So, in the winter, ice fills their abdominal cavities. Their eyes frost over. Breathing ceases. And in the spring, they thaw back to life and are the first to sing.*

I crawled over to the grassy slope that descends to a reed-filled swamp and rolled on my back, my head downhill from the bony funnel of my pelvis. I'd read that inverted poses also sometimes help turn babies.

I felt ridiculous but told myself I was surveying the local watershed.

The geological maps for Tompkins County, New York, name the swamp below my head the Thomas Road wetlands. Considered critical habitat, it filters and feeds rainwater into Sixmile Creek, a source of drinking water for the entire city of Ithaca, population 35,000. My feet pointed uphill to the steep, rocky ridges of Snyder Hill that block the afternoon sun far too soon during the short days of the long winters here. This hill also captures and directs the snowmelt, recharging our drinking water well and filling the swamp.

More than water flows down that hillside. Each spring, a great amphibian migration brings toads, frogs, and salamanders from the Snyder Hill highlands into the swampy hollow of the Thomas Road wetlands where they mate and lay eggs. Along the way, each migrant confronts the treacherously busy Thomas Road, and, at great peril, must crawl, hop, or slither across. Every spring, about 40 percent of them, including the rare and the endangered, are squashed in the passage. The odds are especially stacked against long-lived species like spotted newts, which spend the first two years of a fifteen-year lifespan as red efts—a special juvenile stage—before transforming into adults and so must cross the murderous asphalt year after year. Thus, during the nights of peak migration, local college students patrolled this road—environmental studies crossing guards—stopping drivers, lifting and carrying across as many animals as possible.

Nothing within me moved.

Lying on the hillside, I was aligned east–west. My son, north–south. Our two bodies resembled the compass rose on the corner of the map. Locked together, we were perpendicular arrows aimed at the four directions. At odds. Crossed. Still, the atmosphere rose above us both. The groundwater flowed below us both. And below that water lay a continuous sheet of shale— the old Devonian sea floor, a graveyard of squids and sea lilies that had lived here long before the earth knew wood frogs or newts. We rested together on the same bedrock.

On what seemed to me the hottest day of the summer, I spoke at a press conference in lower Manhattan. The topic was PCB—polychlorinated biphenyl—contamination in the Hudson River and what should be done about it. It was like being asked to play a bit part in a long-running Shakespearean tragedy. Or, more to the point, a badly directed, over-budget summer blockbuster.

Everything about this story was an epic catastrophe. PCBs are the most vile collection of chemicals ever birthed from coal tar. All together, PCBs possess links to the most feared health problems—cancer, infertility, learning disabilities, and premature birth. By turning eel and striped bass into carcinogenic threats, they had destroyed commercial fishing in the Hudson River.

And they possessed a ghoulish ability to resist ecological degradation. From their burial grounds in the sediments of the Hudson, decades after their 1976 abolition, PCBs are still seeding themselves into the atmosphere. They are still insinuating themselves into the food chain. They are found in trace amounts in everyone's blood. And they are not going to go away on their own.

Matching the stubborn resistance of PCBs was General Electric (GE). For thirty years GE dumped these oily chemicals into the river from its capacitor plant in Hudson Falls. For another thirty years—through protracted legal challenges—it had stiff-armed calls to clean them up. The U.S. Environmental Protection Agency (EPA) investigation dragged on for ten years. Whole careers were devoted to the study of the situation. And the situation was maddening. PCBs had not only been discharged into the river through pipes, they had also seeped through the factory floor and *into the bedrock*. As the dense oils migrated through fractures and fissures in the shale, they traveled to the river. The PCB-saturated bedrock itself was a source of ongoing contamination.

But citizen advocacy was equally persistent. As a result, in the summer of 2001, the EPA was about to announce a plan. My task was to describe the prenatal effects of PCB exposure. And so I did—from atop an outdoor dais in Battery City Park, with the state's attorney general on one side of me and Robert Kennedy Jr. on the other. At our backs, the Hudson and the Statue of Liberty. Before us, the World Trade Center.

As I spoke about the mechanisms by which PCBs prevent thyroid hormone from reaching fetal brain cells, I was aware that my body was as much a part of the presentation as my words. Behind me, PCBs were evaporating from river sediments. Within me, an amphibious baby was swallowing amniotic fluid. And knitting together a brain.

It was hard to catch my breath. I felt as though a baton was wedged between my ribs. Or a truncheon. Could a baby feel like a truncheon? Yes, it could.

For all my talk of water, I had foolishly forgotten to bring any. I started to worry about fainting. During the speeches that followed mine and while the cameras whirred and clicked, I imagined myself atop the World Trade Center towers. It would be cool up there. And quiet. I focused on thinking of a name for my son.

I had a new idea about that. Earlier in the day, I'd visited the Metropolitan Museum of Art with my artist husband, Jeff, and our two-year-old. Wandering among the contemporary collection, I discovered a watercolor by Faith Ringgold. My daughter's name is Faith, so I thought I'd show it to her. The painting was a simple American flag, but written on the red stripes were the words of the First Amendment, and written on the white stripes were names that referenced acts of censorship. In between the words "peaceably" and "redress" appeared the name *Elijah Lovejoy.*

I became so entranced, I forgot to point out the letters of Faith's name on the artist's plaque. Elijah Lovejoy. The name sounded so familiar. Who was he?

Now, just as the press conference was wrapping up, I remembered: Elijah Lovejoy, abolitionist writer from my home state of Illinois who had walked from Boston to Alton. Whose printing press was destroyed multiple times by proslavery mobs from St. Louis. Who had refused to stop writing. Whose speech on freedom of the press and appeal for protection was ignored by an Illinois court. Who was killed. Whose death captured the attention of Abraham Lincoln. This was a story once known to every Illinois school child. Elijah Lovejoy was a persistent and uncompromising man.

Elijah. It sounded good.

If I'd thought that naming him would make the baby turn around, I was wrong. Back at the Thomas Road cabin, the wildflowers set seed, carpenter bees excavated holes in the kitchen wall, and my daughter explained to me a complicated career plan. She wished to be a midwife and a train engineer. She would run her trains express while delivering babies. She also announced she was done with diapers now. The baby could have them.

Everyone seemed to be moving on but the two of us. Elijah and I remained in opposition.

And, except for fitful, dream-filled dozing, sleep eluded me. The nocturnal sounds of summer, I reported to Jeff, mostly consisted of the distress calls of foxes. I had no idea what might cause a fox to feel distress in the middle of July, but there it was, every damn night, making unholy sounds out in the woods—shrieks, squalls—that most closely resembled, well, a baby.

Jeff suggested that I was dreaming about the baby and had confused it with a fox.

So the next night I listened hard. There it was. Definitely a fox. Was I asleep? No. There it was again.

Finally came the sound I'd been waiting hours to hear: Far down Thomas Road, a muffler-challenged car was fast approaching. In another minute, it would bomb up to my mailbox, pause, and then roar off into the night again. The paperboy. At least, I thought of him as a boy. In any case, the distant rumble signaled the arrival of another awake human being, whoever it was.

It also signaled the beginning of the end of the night. Since I was too nearsighted to see the hands of the clock and too unwilling to move from whatever least-uncomfortable position I had worked out for myself, the hours spent in bed had become weirdly amorphous. But now I knew that the time had to be, more or less, 4 a.m. And if 4 a.m. comes, can dawn be far behind?

As I entered the final sleepless month of my pregnancy, I began to imagine night as a black body of water I had to cross. The newspaper courier was the lighthouse keeper on the distant shore.

I wondered if he—or she—had any idea how much the unfailing appearance of the newspaper in the predawn darkness comforted me. I wondered if he—or she—could sense, while rolling and stuffing the paper into the plastic box, that I was lying awake inside the darkened house just behind the trees, with various pillows rolled and stuffed around me, listening hard. I told myself that, after the baby was born, I'd write a note and leave it in the newspaper box. *Thank you for being my lighthouse.* But immediately the idea sounded so hokey that I forgot about it.

Then one sticky night, the baby wiggled and squirmed more than usual. Hour after hour, he persisted. He was working hard. This, too, marked a long-awaited arrival. When at last I heard the paperboy roar off, I got dressed. I walked out to the road in the misty darkness and fetched the newspaper. I made tea, poured granola into a bowl, and sat by the east window. The air turned ink blue and separated itself from the earth. For the first time in a month, I could breathe. Later on in the day, the midwife would confirm what I already knew: We were in alignment, my son and I. He was head down, ready to go.

Upstairs, my husband and daughter went on sleeping. I matched my breaths to theirs. Everything I loved was here.

Outside, a fox returned to its reclaimed woodchuck burrow. A red eft crawled from the swamp, inhaled oxygen through its new lungs, and headed up the hill to higher ground.

On August 6, 2001, President George W. Bush received a now-famous intelligence briefing informing him that Osama Bin Laden was determined to strike within the United States—and, more specifically, that he wanted to hijack aircraft. And, even more specifically, that patterns of activity "in this country consistent with preparations for hijackings" had been noted.

On the same day, I began to notice signs of labor.

Given that I had given birth only once before, I was surprised how familiar the sensation was. Vague crampiness. A band of achy pressure that came and went. *Oh, yeah. I remember this.* By the time I saw the midwife later that afternoon, I was four centimeters dilated but not yet experiencing progressive contractions. She sent me home to get some rest while I could. "Stay on high alert, though."

August 6 was blisteringly hot.

So was the next day.

The day after that broke a record high. A persistent haze silted the air. Each morning and evening, the episodes of achy pressure intensified—only to subside again.

What did it mean to go *into* labor anyway? The phrase made having a baby sound like a location rather than an endeavor. I'm in church. I'm in school. I'm in labor.

So was I truly *in* labor? Or just approaching it? Well, I was a mother. I knew how to be patient. If indeed childbirth were a room that one entered—crawling on hands and knees, beseeching, naked—it was a place I'd seen before. Wherever I was now, I was sure to recognize the furnishings when I arrived.

Here was an interesting truth about entering motherhood for the second time: I was just having a baby; I wasn't taking on a new identity. Furthermore, I now carried with me a skill set. I already knew how to change a diaper in the dark. I already knew how to nurse a baby with one hand and type with the other. I already knew how to talk my way past the pediatrician's surly after-hours answering service and get a doctor on the phone.

It turns out that field biology also offers a number of transferable credits to the college of parenting. In motherhood, as in ecology, data analysis is complicated, confounded by multiple variables, and boundlessly fascinating. Both activities require humility and a high tolerance for dashed plans and unanticipated outcomes. I realized, soon after the birth of my first child, that I was more prepared for parenthood than I thought I was.

Nevertheless, I was now starting to suspect that having two children was going to require of me something that having only one didn't. Of course, it went without saying that another baby was going to eat whatever was left of my solitude. That was an easy prediction.

I'd been a mother for fewer years than three. Being a nonparent still felt like my normal state. It was still a surprise to wake up and find a child in my house. But I wasn't nostalgic. In my experience, a happy single life was like an exotic orchid with particular moisture, light, and temperature requirements. It required a lot of tending. You had to make weekend plans, have a therapist or a pet, and keep track of how many days had passed since you had last dined with someone. Small disappointments had to be monitored lest they balloon into despair.

When you are by yourself, surreal things can happen: You could get lost in a rainforest while hiking to a study site and then step on a snake at the precise moment you realize that you had neglected to tell anyone where you were going. Or an infected salivary gland could send you to the ER on Thanksgiving morning. Five hours later, you might find yourself shivering with fever in a check-out line at a Walgreens, a scrip for antibiotics in one hand, a can of condensed mushroom soup in the other, and a homeless guy in front of you kindly offering to give you, the girl with the puffed-out face, his can of tuna fish. Which makes a nice story but you have no one to tell it to. (Until now.)

Still, in the way it aligns you with all the other nonaffiliated people of the world, there is something democratic and outwardly focused about solitary living. By comparison, family life had seemed to me, when I was an outsider to it, oblivious and tribal. There is something darkly thrilling about watching a crisis unfold and realizing, *nobody even knows I'm here.*

But from now on, my mental health would be besieged by too much symbiosis rather than too much isolation. My whereabouts would be planned, publicized, monitored, and made a topic of constant negotiation. Including trips to the bathroom. Fine. It was an adventure of another sort.

Two other things worried me more.

First was the growing awareness of the fact that I had no saltshaker. Nor any chest of drawers. I just shook the salt right out of the cardboard cylinder it came in and stored my clothes in bins. And my bedspread was a sleeping bag. And the cabin we lived in was rented. These seemed to me signs of a provisional, ad hoc approach to setting up house. As though I were camping out in my own life.

In this, Jeff and I were well matched—he kept his toothbrush and razor in a travel bag on the bathroom shelf, as if any moment he might be summoned to the airport—but I was beginning to wonder if we both still imagined ourselves as work-driven gypsies who could fold up our tents in the night and be gone. That sort of make-do approach to our surroundings might be compatible with one child—you and me and Baby McGee—but it seemed unsuited, in ways I couldn't fully explain yet, to life with two children.

One of the lessons of ecology is that scale matters. And when you add another organism to the ecosystem, the complexity of interactions increases exponentially. At the very least, moving into a four-square family from a two-parents-and-a-kid triangle meant changes to the food chain, the laundry cycle, the infectious disease rate, and the ambient noise level. Matter, energy, and chaos were going to start flowing faster and in multiple directions. It seemed worth thinking about the efficiencies of the household—and more profoundly, what kind of family we wanted to create and in what community we wanted to embed it. Less whimsy, more stability.

My second concern was straightforward: I was having a boy, and I didn't know anything about boys. In fact, those were my exact words to Jeff six months ago. Biologist or not, I was shocked by the news from the ultrasound technician—how was it possible that I had a *male* inside of me?—and so I, the brotherless mother of a daughter, had turned to my husband in the hospital parking lot and cried, *I don't know anything about boys!*

To his abiding credit, he waited a moment before saying the obvious. *But I do.*

No one knows what sets in motion the onset of uterine contractions known as labor. A 2010 review of the evidence in the *Archives of Gynecology and Obstetrics* concludes, "Despite impressive progress in the science and technology of reproduction, the mechanism by which labor is initiated in humans remains obscure."

Theories abound, and they all focus on intrinsic factors: the cascade of biochemical changes that soften the cervix of the uterus and coax its muscle fibers into synchronized action. Some of these changes are guided by hormones from the endocrine system, some by signals from the nervous system and some by inflammatory reactions involving the immune system. But the ultimate source of them remains unelucidated.

The idea that the physical environment surrounding a pregnant woman may play a part in influencing the onset and tempo of her labor—perhaps by modulating neuroendocrine and immune responses—is mostly not part of the discussion within the gynecology and obstetrical communities. But there is growing evidence for such a role, and it comes from two sources: epidemiologists and midwives.

Epidemiological studies show that women exposed to certain chemical pollutants are at higher risk for preterm birth. Chemicals with the power to shorten human pregnancies include hormone-disrupting chemicals such as PCBs. They also include benzene, lead, and certain pesticides. And they include DEHP, a member of a widely used but little known family of chemicals called phthalates. DEHP is short for di(2-ethylhexyl)phthalate.

One billion pounds of phthalates are manufactured each year. DEHP is used to soften vinyl plastic and, along with other varieties, is added to perfumes, hairsprays, and nail polish. Indeed, perfume use is linked to higher levels of phthalates in the urine of pregnant women. As described in a study published in 2009, researchers at the Columbia Center for Children's Environmental Health persuaded 311 very pregnant women living in Harlem and the Bronx to wear backpacks containing personal air monitors. These monitors measured actual levels of phthalates in the air that the women breathed during their last months of pregnancy. Researchers found phthalates in 100 percent of these personal air samples. DEHP was associated with shorter pregnancies: Women exposed to higher levels went into labor sooner.

So how might a chemical that imbues shower curtains and dashboards with that instantly recognizable new-car smell abbreviate human pregnancy? Evidence from animal studies suggests that DEHP exposure can induce inflammatory responses in the lining of the uterus. Inflammation is a known risk factor for premature birth.

A second possible route to preterm labor involves DEHP interference with an obscure set of proteins called peroxisome proliferator-activated receptors. These subcellular receptors are the shushing librarians of late pregnancy: they enforce what's called *uterine quiescence*. By regulating the expression of certain genes, they keep the taut, stretched muscle fibers of the uterus quietly focused on remaining taut and stretched. As the time of birth nears, the receptors—stationed along cell membranes—are quietly stripped of their authority. Released from the receptors' stern oversight, the uterine muscle fibers are now free to start twanging. And they do. Soon enough, there's a riot in the library.

In the presence of phthalates, however, the story changes. Phthalates retire from duty the peroxisome proliferator-activated receptors before their work is finished. In other words, phthalates have the power to shush the shushers. The result may be earlier onset of contractions.

Phthalates are not the only air pollutant that can disturb uterine quiescence. So can second-hand tobacco smoke. So can car exhaust. So can smog. In general, the more fine particulate matter a pregnant woman breathes, the greater her risk for an early birth.

Air pollutants act through multiple pathways to induce labor. Some trigger inflammatory reactions. Others disrupt hormonal messages. Diesel fumes and heavy metals such as cadmium, for example, can interfere with the production of progesterone, which is a key player in regulating the timing of birth. Epidemiological studies have identified links between timing of birth and traffic intensity: The closer a mother lives to a busy highway, the higher her risk for preterm labor. Other studies have identified links between earlier onset of labor and proximity to certain industries. When the Utah Valley Steel Mill closed in 1986, for example, rates of premature birth among pregnant mothers living in the valley dropped. When the mill reopened a year later, rates went back up.

In short, the atmosphere itself (through pathways we do not entirely understand) interacts with a complex, exquisitely calibrated biological mechanism (whose engineering we do not entirely understand) to influence the initiating events of human birth.

From the midwifery community comes an observation about weather events and the onset of labor. Part of the oral tradition among midwives is that human births occur more often during times of decreasing barometric pressure, as during a storm or arrival of a cold front. Such a relationship is not without precedent—surgeons have long noted an association between low atmospheric pressure and increased rates of abdominal aneurysm ruptures—but it has new backing from studies in biometeorology. Although the data are not completely consistent, the results show that the rupture of amniotic membranes and/or onset of labor tend to increase during frontal changes, especially those involving falling air pressure.

Both the epidemiologists' findings on air pollution and the midwives' observations about storm fronts have relevance for mothers living in a world altered by climate change. In a warming world, heat waves become longer and more frequent. The higher temperatures speed up chemical reactions between air pollutants—such as car exhaust and evaporating paint fumes. As a result, smog thickens, ozone levels rise, and air quality deteriorates. At the same time, storm systems intensify and barometric pressures fluctuate more wildly.

The twin impact of air pollution and climate instability on the duration of pregnancy is not yet part of biomedical discourse. It should be. In the United States, preterm birth is already the leading cause of death within the first month of life. It's also the leading cause of disability. In addition to the suffering inflicted on families because of the death or lifelong disability of a beloved child, prematurity is expensive. Each year, preterm births cost the nation more than $26 billion in medical expenses, special education services, and lost productivity.

Much remains to be learned about the ways in which the environment influences the timing of human birth. "The vast numbers of pollutants to which a woman may be exposed have never been considered in an investigation of preterm birth," concluded the Institute of Medicine in a recent report. Nevertheless, we know two things already: Pregnant women need clean air and a stable climate.

On August 9, 2001, the mid-afternoon high hit 98 degrees, and another temperature record fell. Faith was desperate for a nap but couldn't settle down. We tried various sleeping spots, various storybooks, various lullabies. Her fretfulness went on and on. During the third encore of "Daniel in the Lion's Den," I noticed that my come-and-go contractions seemed to be recurring more purposefully, so, as I sang, I began surreptitiously to time them. They became impressively regular. As did, gradually, Faith's breathing.

One child was entering sleep just as the other one was about to make an entrance.

A sweaty arm was flung across my chest, and a hand twirled in my hair. Faith was exhausted. The next contraction was hard and long. I knew that if I got up now, she would rouse and start crying again. And that would be the end of naptime. And without a nap, she would be miserable. Not everyone's needs were going to get met here. This was a new thought for me. My usual mantra—*family first, work second, and everything else after that*—didn't really offer any insights. And so, not for the last time in my life as the mother of two children, I looked at one of my kids and said, *I'm sorry, but I can't always be fair.* And got up.

The barometer fell all afternoon. The big storms skirted around Ithaca—thundering in the distance—but the cold front they brought with them arrived later that night. The next day's high was only 84. By then, I was carrying my son outside my body rather than inside.

He was shy. That was my first opinion of him. And peaceful. The prophetic name *Elijah* seemed too fierce for him. He was also blond. And in each of those three ways he was unlike his sister. Faith had looked around the delivery room, so her father claimed, before she was even all the way delivered. Elijah kept his eyes firmly shut. *Decidedly* shut, as if he had not yet completely arrived in this world. Faith's birth had been like a reunion with someone I'd known my whole life. I'd loved her instantly, hugely, involuntarily. This new child was someone I didn't recognize.

Back at home, we received a steady stream of friends bearing food, and I recounted the details of the previous night. The story, it seemed to me, contained some notable observations about the influence of the environment on the timing and tempo of labor.

Getting from our cabin to the September Hill Birth Center—a renovated farmhouse next to the county hospital in Montour Falls—had required a 45-minute drive. At the 35-minute mark was Watkins Glen. The road through this village passes by vineyards, dairy farms, bait shops, and offers views of a waterfall-laced gorge. It's a tranquil place—except on a few nights of the year when it isn't. Watkins Glen is a NASCAR town. Its fast and fearsome speedway is hailed as America's Road-Racing Mecca, and qualifying races for two different series were scheduled the night of Elijah's birth.

When we'd pulled into town, the entire village was a car-stereo-thudding tailgate party. And of all the people navigating the clogged streets that night, I was the one least interested in the news that the best lap time in the qualifyings was an awesome 138 mph.

Here were the results of a study I conducted entitled "The Impact of NASCAR Racing on the Timing of Human Birth" [sample size of 1]: Labor contractions ceased. Entirely. As though they had never existed.

With the approval of the midwife on call, we decided to find a nearby hotel for the night rather than return home. Thus began the Joseph-and-Mary search for a room, except that instead of a donkey we were accompanied by a sleep-deprived toddler and, an hour later, had secured something slightly cleaner than a stable. For the second time that day, I lay down with a frantic child and tried to coax her to sleep. And for the second time, just as she drifted off, a hard contraction took my breath away. It was immediately followed by another and then another.

Jeff's qualifying times were impressive.

The midwife met us at the door and led me, gratefully, to a whirlpool tub. Immediately, in the way the water floated me, something on the verge of unbearable became bearable again. Something on the verge of terrifying receded. Yes, I was now *in* labor. Truly in it. Labor was a room with a ticking clock, a tiled wall, and swirls of light. Labor was the shale floor of a briny sea. Magma. Pressure. I descended by Braille. I rose through the squeezing passageway of my own breathing. Labor was a sparkling lake. I could swim a long way. *And nobody even knows I'm here.*

When it came time to push, I pushed hard. I needed to push. I wanted to push. And whatever other verbs exist in the space between *needed* and *wanted*, I did those, too. And then he was in my arms, and the rain arrived, and somewhere in a nearby pond, frogs, like loose banjo strings, started twanging.

I don't recognize you. But I'm going to love you anyway.

With the first baby, you realize that you would sacrifice everything for your child. With the second baby, the impulse is toward self-preservation. The realization is that the whole lurching family enterprise, which now includes two small children, depends on one's own ability to hold up and carry on.

Two weeks after the birth of my son, I turned 42. I already knew that my own capacity for perseverance depended on accumulating, within each 24-hour period, six hours of sleep in at least three two-hour chunks. Anything less—or more fragmented—for more than a few days in a row—unhinged me. As co-breadwinner for the family, I was back at my desk within days of Elijah's birth. I needed my hinges intact. So all decisions were now made with an eye toward saving time. Extra minutes were squirreled away and devoted to sleeping. Convenience mattered.

With this new perspective, I looked again at the literature on breastfeeding.

With my first child, what had impressed me about breast milk was its clear superiority over infant formula. Breastfed babies have greater protection against infectious diseases. They hit developmental milestones sooner. They grow into children less likely to contract leukemia or diabetes or asthma. They have higher verbal IQs. They are less likely to become obese. Ergo, breast was best. And I wanted what was best for my daughter, so I gladly nursed her.

With my second child, what jumped out at me were studies that focused on the impact of breastfeeding on maternal health—

For mothers, failure to breastfeed is associated with an increased incidence of premenopausal breast cancer, ovarian cancer, retained gestational weight gain, type 2 diabetes, myocardial infarction [heart attack] and the metabolic syndrome.

With lists like these, I felt my motivation shift from *I'll breastfeed if it kills me* to *I'll breastfeed to stay alive*. Items on the lists of risks incurred by mothers by *not* breastfeeding were, to say the least, examples of supremely inconvenient events. And to say the most . . . well, my biggest fears were on those lists. If there was anything I could do to lower, however modestly, the odds that my children would grow up with a cancer patient as a mother, I would do it.

As for heart disease and osteoporosis, I was already willing to engage in time-consuming attempts to mitigate my risks for those problems (jogging in the middle of the night). Improving my cardiovascular health and bone density by sitting on a couch and nursing sounded like an ingenious form of multitasking to me. I mean, I had to feed him anyway.

I began to look at the data on health outcomes of infants from this new vantage point as well. Feeding babies formula increases the chances that they will get sick. When babies are sick, mothers don't sleep. Households descend into chaos. The needs of other children go unmet. Pneumonia. Diarrhea. Gastroenteritis. Ear infections. Contagious diseases could sweep through the household. Who had time for that? But to say that breastfeeding offers benefits—as though nursing were a job with stock options and a retirement plan—doesn't really provide those insights. The notion of "benefits" doesn't help a new mother know that not breastfeeding likely means more 2 a.m. phone calls to the pediatrician, more trips to the pharmacy, more time spent reading the fine print on the back of the Tylenol bottle, more missed days of work. Or that not breastfeeding may jeopardize her own future health.

Critics who complain that breastfeeding advocacy creates guilt in mothers who choose not to nurse are missing the point. The choice is not between a gold-plated but sometimes tricky, painful, and inconvenient way to feed a baby (breastfeeding) and the perfectly adequate standard model that offers ease and convenience (formula)—even though the sly language we use to talk about infant feeding practices implies as much. According to a 2010 study published in the journal *Pediatrics*, low breastfeeding rates in the United States kill 911 infants per year and cost $13 billion. That's the choice.

But pediatric cost analyses of the burden of suboptimal breastfeeding are of minimal use to new mothers. So here are two personal observations from my own untidy, working-mother trenches: 1. Breastfeeding allows you to make crying children fall asleep on demand. Anytime. Anywhere. Nothing to sterilize. No floors to walk. 2. Bottle-feeding takes two hands. Breastfeeding only one. With your free hand you can—read a story to a toddler, analyze data, make dinner, give interviews over the phone, draft a grocery list, write a book.

As everyone knows, the weather in New York State on the morning of Tuesday, September 11, 2001, was sunny and warm.

We were up early. *Look, Elijah. Sun. Trees. Crows. Look. Blue sky. Blue asters. Goldenrod.* Elijah peeked out through one eye. I laughed. He nursed, and I worked on a lecture for an upcoming talk in Manhattan. When he fell asleep, I took him outside—the morning so lovely—and laid him on a blanket in the grass. He looked like he would nap for a while, so I did a few laps around the backyard with a push mower.

The leopard frog jumped just as the blade reel spun over him. I jerked the mower back but too late. Cursing, I extricated from the blades what was left of the frog and laid the pieces under the blackberry canes at the edge of woods. Leopard frogs are intensely sensitive to chemical pollution. Their populations are in decline. In graduate school, I dissected dozens of leopard frogs. That fact didn't make me feel any better. Through tears, I decided I would atone for this stupid accident by making a donation to an amphibian society of some kind. In fact, I would do it right now. I carried the baby back inside.

The kitchen radio was on. I never wrote the check.

Elijah began a 72-hour jag of crying on Wednesday. He didn't seem hungry, and, in fact, he cried harder after a feeding than before. On the other hand, he wasn't eating much. He wasn't sleeping much either. He was so distraught his breaths came in great, ragged gulps. He wasn't running a fever. Jeff and I took turns holding him. He cried. He cried in the baby sling. He cried in the baby swing. He cried in the car. Whatever grief and shock Jeff and I were experiencing along with the rest of the nation was soon subsumed by the task of attempting to console an unconsolable baby.

By Friday, we were all exhausted, and Elijah had received a diagnosis of colic from the pediatrician's office.

Back at home, I glanced at the week's postings on an international listserv for lactation consultants to which I had subscribed a year earlier for a research project. It turned out that I was not alone: Ever since Tuesday, breastfeeding mothers across the United States were complaining they were losing their milk, suffering dips in production, coping with nipple pain, or waking up to newly fussy babies. The most reliable reports came from mothers in neonatal intensive care units who had been regularly pumping and measuring their milk for their preterm infants. Hospital-based lactation consultants confirmed that, for many of these mothers, milk volume had plummeted with no other explanation except stress brought on by the terrorist attacks. A lactation consultant from Tel Aviv wrote to say that she was unsurprised. She assured her U.S. counterparts that the effects were temporary.

On September 11, terrorism reached nursing mothers across continents. Entered their milk. Entered their babies. Terror was inside us now, Elijah and me. The connections between the political events in the outside world and the interior ones within the ducts and lobes of my own body were more profound and intricate than I had ever imagined.

By the following Monday, the interminable crying had ebbed. The Israeli lactation consultant was right. That much, anyway, was temporary.

The lecture I had been invited to deliver on pediatric environmental health at the New York Academy of Medicine had been planned months earlier. It wasn't at all clear, in the days leading up to it, whether the event would happen nor not. My host said, frankly, he could not guarantee an audience. I had misgivings of my own. The drive to Manhattan took five hours even before the George Washington Bridge was outfitted with security checkpoints, and I had a one-month-old who would be riding with me across that bridge.

In the end, we decided, as so many people did in those first dazed weeks, that since all possible actions felt wrong anyway, we should just get on with it. And so Jeff paced marble corridors with Elijah on his shoulder—and Faith in a stroller—while I addressed a half-filled auditorium.

In the audience were a number of pregnant women and, as I was getting ready to leave, they approached me as a group. They wanted to know about that most toxic of all synthetic chemicals, dioxin, which, at vanishingly small concentrations, can cause developmental problems as well as cancer. Had the incineration and collapse of the World Trade Center sent a dioxin-filled cloud over Manhattan? They had heard that the towers were filled with PVC plastic and that PVC makes dioxin when burned. Is that right? Were their babies in danger?

I tried to keep my voice calm. Yes, I said, PVC—or polyvinyl chloride, or vinyl—makes dioxin when it burns and yes, the Trade Center was surely full of PVC. It's used in electrical cables, flooring, wallpaper, and office furniture. I said I didn't know what health threats the cloud created for the people breathing it—or for the fetuses they might be carrying. Colleagues of mine were researching those very questions. Unfortunately, the answers would be years away. Science takes time, especially when actual exposures are unknown and the outcomes, like subtle developmental deficits, can take years to manifest.

The pregnant women watched Elijah nurse. I looked down at their various-sized bellies. We all fell into silence.

Early next morning, I left everyone in the hotel and took a nearly empty subway downtown. The trains under lower Manhattan were not running, so at some point, I got out and started walking. There were no towers to navigate by, a fact that seemed as surprising as it was obvious. How many weeks had gone by since I was here last, Elijah lying sideways within me, the Towers rising coolly above the summer heat?

I figured I was getting closer when the faces of the missing began appearing on every wall and pole.

And then I rounded a corner near a boarded-up barbershop, and there it was.

I felt as if I were looking back in time at the ruins of some ancient civilization. There was the broken pillar. There was the curl of smoke before the crumbled façade. There was the scrap of cloth fluttering from the blasted window. There was the rubble. There was the gaping hole.

After a while, I noticed that my shirt was wet with milk. Somewhere uptown a baby was wondering where I was. Fumbling with the buttons to my jacket, I dropped my book bag onto the sewer grate. Just then, in a surreal transaction, a cab appeared, a sobbing woman got out, and I got in.

Back in the cabin later that evening, I saw that my green book bag had turned gray with Ground Zero ashes. Lying on my kitchen floor, it was still holding my breast pump and a couple of diapers along with my lecture notes. I lifted the bag gingerly and carried it out to the porch. I'd deal with it tomorrow. For now, I focused on the floor. Around and around I went with the broom, until there was not a speck of dust or ash anywhere. The floor's vinyl beige squares and bland blue flowers seemed innocent and reassuring. The radio said we were going to war. And I was the mother of a son.

A decade later, here is what we know about the Ground Zero cloud that resulted from the pulverization of two of the world's tallest buildings and from the fires that burned for three months: It contained cement dust, glass fibers, heavy metals, pesticides, PCBs, asbestos, benzene, soot, and the combustion byproducts of jet fuel. It contained dioxin.

Here is what we know about the impact on children: In the months following the attack, the number of new asthma cases among children younger than five who lived in lower Manhattan jumped. The greater the exposure to the dust cloud, the greater the risk of developing asthma. Among children living in China-town near the World Trade Center, rates of asthma increased and preexisting asthma cases worsened. These effects were not temporary. Asthma rates among Chinatown's children are still elevated.

Here is what we know about the impact on pregnant moth-ers and their babies: Compared with those living farther away, women who resided near Ground Zero gave birth to shorter, lighter babies. Women who were in the first trimester at the time of the event had shorter pregnancies, and their babies had smaller heads. Newborns whose mothers lived within one mile of Ground Zero had higher levels of DNA damage in the blood cells of their umbilical cords than babies whose mothers lived farther away. The closer a pregnant woman was to the Towers, the greater the damage in the cells of her blood and in her baby's blood. At age three, children with elevated levels of DNA damage—if they had also been exposed to environmental tobacco smoke—showed modest deficits in cogni-tive development.

Here is what we know about the boy babies of women preg-nant during the 9/11 attacks: Some of them disappeared. That is to say, they were never born at all. And they vanished not just among women living in New York City but throughout the United States. Three to four months after 9/11, significantly fewer boys were born, and the death rate of male fetuses (those more than twenty weeks gestational age) increased by 12 percent.

This gender-selective loss and consequent reduction in the male birth rate is not without precedent. The male birth rate has long been known to decline after "natural disasters, pollution events, and economic collapse." No one understands the bio-logical underpinnings for this phenomenon. Why should male fetuses be disproportionally sensitive to environmental and social upheavals in the world outside their mothers' bodies? As the beginning of an explanation for male fetal loss in the case of the World Trade Center disaster, researchers have posited *communal bereavement*, which is defined as widespread distress after events in which institutions, such as governments, fail to maintain safety and security.

"To know me is to travel with me," says the frequent flying anti-hero of Walter Kirn's 2001 novel, *Up in the Air*, which does not anticipate what air travel became in the months that followed 9/11. Pre-dawn cab rides. Three-hour security lines. Armed soldiers by the X-ray machine. Passengers pulled aside at boarding gates for further questioning. Tape-looped announcements exhorting vigilance. This was the world into which Elijah, finally, opened both eyes. And this was the world in which I learned to know him—as my traveling companion during a fall 2001 book tour.

Our fellow travelers were mostly tight-lipped men on business trips. Elijah was almost always the only baby on the plane, and I was the only passenger hauling an infant car seat. That didn't exempt us from suspicion. In a Texas airport, my diaper bag was seized and searched between flights, and we missed our connection. We were often culled from the shuffling line through the metal detector and shunted into another for additional security procedures. Elijah learned to hold out his arms when the security guard approached with a beeping wand. We nursed under the watch of National Guard troops—22-year-old boys with bayonet-tipped rifles. Elijah studied them closely.

But mostly we studied each other. We talked to each other. We smiled at each other. We sought refuge together from too much noise and too much light. We learned to sleep and wake in unison in whatever hotel we landed. And when the world became too much, he turned away, burrowing down in the cloth cave of the baby sling in which I carried him. *Nobody knows I'm here.* And in this impulse toward privacy, toward solitude— I recognized him.

When we arrived home, fall was stiffening into winter. Jeff was eager to reconnect with the new, fully conscious Elijah and

I with Faith, who had so, so, so many stories to tell me. "Switch kids" became the phrase Jeff and I used to reorient our parenting so each of us could spend time with the child who had most recently occupied the periphery of our attention. And so I became the daytime parent to Faith and the nighttime parent to Elijah.

The nights in the sleeping loft were much longer now than those I endured in the Summer of the Transverse Lie, but they were no less timeless. I still couldn't read the clock without my glasses, and, while nursing or changing a diaper, half-sleeping in the darkness, I still wondered what time it was from inside an exhausted unwillingness to go get the answer I was looking for. Not knowing the hour of the night was bewildering somehow. If it was midnight, that was one thing. If it was 4 a.m., that was another. I felt vaguely lost, as though hiking without a compass.

And then, like discovering a cairn on the side of the mountain face, a marker of stones in a wilderness, came the distant, ragged sound of an old Chevy. The paper boy at last. Okay, the *newspaper courier* at last—somebody who, in my mind, would always be a boy. That meant 4 a.m., with 5 a.m. in view. The night was almost over; dawn was on the way. Soon, I could hand Elijah off to Jeff and sleep for another blessed hour.

In February, the engine sound stopped.

And, for a few mornings, no newspapers appeared in the box. And when they did again, there was an obituary.

Jeremy Armstrong, age 22, beloved son and brother. Whose Oldsmobile sedan had been found overturned in the creek. Who was a pianist, a maple syrup producer, a sailor, a builder of wooden boats. Who had plans for racing stock cars. Who supported his interests by working in the family newspaper business, delivering the *Ithaca Journal* throughout Ellis Hollow early every morning.

He was a boy.

At the funeral, I saw no one I knew. But I guessed that the gentle-looking man in the ill-fitting corduroy suit was his father. He introduced himself as Bob. The skin beneath his eyes was gray. He smiled and looked at me curiously. He took my hand. He thanked me for coming.

It was my turn to speak. I hadn't come with a script. I wondered how I was going describe to him the relationship I had with his son. How I had depended on Jeremy for far more than the local news. How I had listened for Jeremy's daily arrivals through the whole long summer of late pregnancy. How his hovering presence at the end of my driveway had welcomed me home from the birth center and from post-9/11 travels. How he had brought, in the dead of every night, a stroke of order to a nearsighted, disorderly, work-driven parent. And, more even than that, how, in a time of terror and war, he had brought fidelity and, somehow, peace. Ultimately, Jeremy's daily arrival at my house had filled me with a sense of peace. The peace that came with the knowledge that we had all made it through another night. Yes, that was really it. That, and how I had meant to write him a letter last summer and express my gratitude, but I never did.

Jeremy delivered my paper, I began.

Bob smiled again. *Come,* he said. And he steered me to open casket where Jeremy lay, unadorned. Not in a suit. Not on a satin pillow. Thin. Small shoulders. Dark feathery hair. An attempt at a beard—the kind that 22-year-olds try to grow. *Jeremy's brother made this coffin. Both of our boys are sawyers and woodworkers. We homeschooled them.* Bob was going to serve as my guide through Jeremy's life. *When the police found him, his hand was still on the gearshift.*

Bob showed me photos of his son as a child, running naked through a field, a swamp in the background. I recognized the landscape. We were practically neighbors. Now he wanted to introduce me to his wife Betty, Jeremy's mother. She was the woman with long black hair, fiercely beautiful, standing motionless at the head of the casket, rooted to the floor.

She took me in. And she, too, spoke before I could find the words I wanted to say. Gesturing at Elijah, sleeping on my shoulder, she said, *You never stop feeling that way about them, you know.*

And I knew, in that moment, that she was right. And I knew, too, that I would remember her words in the years to come.

Bob smiled again. *Is this your son?*

Yes. His name is Elijah.

Elijah?

Yes.

Elijah. Bob reached out a hand and stroked his hair. *It's nice to meet you.*

CHAPTER TWO

The Nursery School Playground (and Well-Informed Futility)

CHILDREN ARE NOT like you and me, every authority on pediatric environmental health wants you to know.

Whether published in the peer-reviewed medical literature, a government report, or a popular parenting magazine, whether spoken at the podium in a Ph.D.-filled conference hall, into the microphone at a Congressional hearing, or as part of a Power-Point presentation in a church basement, the message is the same: Ecologically speaking, children occupy another world, separated from us by dietary habit, behavioral pattern, and body size. Any speaker or author on the topic of children and the environment is practically required to preface his or her remarks with some version of the following:

> Pound for pound of body weight, children drink more water, eat more food, and breathe more air than adults.

One wonders who this memo is for. It's not news to parents that four-year-olds dine on fewer foods in proportionally larger quantities than the grown-ups at the table. (It was not an adult member of my household who decided to subsist for a month solely on bananas, eggs, and avocados.) And compared with those of us who share our homes with children, non-parents seem even more attuned to their unique attributes. The adjectives used by one of my childless friends to describe babies—*loud*, *leaky*, and *rude*—are all points of contrast with the full-grown sophisticates who occupy his world. I'm quite sure that he, along with other members of the children-are-aliens school of thought, would be unsurprised to hear the news that preschoolers, as compared with adults, have a greater tendency to mouth breathe.

In fact, we are not the intended audience for this message. The reason that the experts keep harping on the myriad ways in which children are different is not to surprise members of the public but to highlight the ways in which our environmental policies pretend that children—who make up 40 percent of the world's population—do not exist. Entire regulatory systems are premised on the assumption that all members of the population basically act, biologically, like middle-aged men. The laws and rules so generated by those systems are thus blinded to the unique characteristics of children that should be obvious to everyone.

Disregard of juvenile differences begins in the laboratory with the methods used to test chemicals for toxicity. Traditionally, these tests are performed on post-pubertal rodents who are exposed to a given dose of a potential toxicant and then examined for health effects when their corresponding age in human years is between sixty and sixty-five. The results of these toxicology studies are then fed into risk analyses that attempt to model routes of human exposure and estimate their effects. Reference doses are set—with arbitrary safety factors added in—and decisions are made as to how much of an inherently toxic chemical is permissible in, say, air or water. Needless to say, the presumptions used in these models have not always been representative of children. Until 1990, for example, the reference dose for radiation exposure was based on a hypothetical 5'7" tall white man who weighed 157 pounds.

Reference Man and his adult lab rat friends illustrate how environmental laws have historically ignored children in all their splendid, vulnerable Otherness. Given that our environmental laws are, at this point, mostly historical—with the big federal statutes on air, water, and toxic substances control dating back to the 1970s—it's fair to say that we adults basically do not—and, absent sweeping changes to those laws, cannot—adequately protect children from environmental dangers. Consider, too, that only 200 of the more than 80,000 synthetic chemicals used in the United States have been tested under the Toxic Substances Control Act of 1976, and exactly none of them are regulated on the basis of their potential to affect infant or child development.

Our chemical regulatory system has essentially ground to a halt. It is not only locked in an old Reference Man mentality, it is unresponsive to the development of new chemicals and emerging evidence about previously unknown types of danger, such as cumulative impacts and the additive effects of chemical mixtures. It's as though the government were choosing to ignore electronic identity theft because our laws about fraud predate the invention of the Internet.

Recently, some well-intentioned attempts have been launched to redress the problem. In 2000, the U.S. Congress authorized the prospective National Children's Study, which will measure the environmental exposures of 100,000 U.S. children from conception through age 21 and will examine their relationship to various health outcomes, including pediatric cancers, learning disabilities, early puberty, asthma, and autism—all of which show signs of increasing prevalence. But recruitment of subjects for the National Children's Study began only in 2009. The final results will come long after the children of today are no longer children. In 2010, the Toxic Chemicals Safety Act was introduced. It seeks to reform the Toxic Substances Control Act with an eye toward more stringently regulating chemicals to which children have unique vulnerabilities. At this writing, its successful implementation is far from guaranteed.

Parents trying to protect their children need reform of our antiquated regulations that reflects how infants and children are unlike us. Here are some additional differences to contemplate:

Children have alternative metabolic pathways and are thus slower to detoxify and excrete pollutants. They have a more porous barrier between blood and brain and are thus more prone to neurological toxicity. They eat and inhale more house dust—which is why children typically have higher blood levels of lead and flame retardants than the adults in the household. And they practice what researchers call "frequent hand-to-mouth interactions." To be specific, young children insert their hands into their mouths an average of 9.5 times per hour with the average amount of hand entering the mouth equivalent to three fingers worth.

Keep this last statistic in mind. I'll come back to it.

In spring 2001, Jeff and I set out to find a nursery school for Faith to attend the following fall. By then, she would be a big sister, and I would be leaving on an extended book tour with the individual responsible for her new status as sibling. Jeff was facing grant deadlines and needed more child-free hours. The choice seemed particularly fraught. Or maybe all parents approach the day-care decision for their first-born with a similar sense of peril.

I looked at large parent-run cooperatives and visited small home-based operations. Jeff studied the pink towers and chiming bells at the Montessori school on the hill, and he considered the wonder balls and woolen fairies at the Waldorf school in the valley. We reviewed snack policies, peered into dress-up boxes, examined art supplies, and, at the end of each day, compared notes as though we were a team of day-care inspectors. We pondered tuition schedules. We met with various teachers. They all seemed nice. They all asked me the same question, "So, what's your daughter like?"

I never knew how to answer this. Faith was the only child I'd ever really known, and I wasn't sure which adjectives applied to her alone and which to her whole age cohort. And yet, my lack of vocabulary about her personality seemed amateurish. I felt like a freshman. *So, what's your major?* Blurting out *undeclared* was sure to invite awkward silence.

Faith liked to sing, dance, listen to stories, and squish soap. She was inquisitive. Beyond that, the only particularity that I could dredge up about my undeclared daughter that seemed somehow relevant was that she was fond of pretending that she was a character in a story about herself. So, she sometimes talked about herself using third-person dialogue and incorporating lines from her favorite story books. Our conversations went something like this:

> Mother: "Faith, let's go down and have breakfast."
> Child: "Okay, said Faith, *as she descended the stairs.*"
> Mother: "Do you want jam on your toast?"
> Child: "No! . . . *came the unexpected reply.*"

To my recitation of that script, one teacher had said, "Oh, so she likes structure!" And another one had said, "Oh, so she probably wouldn't do well with structure."

In the end, Jeff and I chose a nursery school that operated out of a community center at the end of a quiet road. There was a frog pond out front and a play structure out back. The trees were full of chickadees and nuthatches. It was the least expensive. It was close to home. The snack policy extended to blue doughnuts, but, all other things being equal, keeping our daughter here in our own community seemed like the right decision. Moreover, when Faith herself visited the school, she had immediately joined the other kids sitting on the rug for story time and refused to leave when the story was done.

"This, *Faith said,* is my home," said Faith. "I live here now, *Faith said.*"

Having made some unsurprising statements about children, let me now make a few about arsenic.

Arsenic is a poison. One ounce can kill 250 adults. It's ranked number one on the federal list of the top 275 most hazardous substances found at toxic waste sites. Arsenic is an element (number 33 on the periodic chart), which means it can't break down and go away. At levels far below what's needed to kill a rat or a cheating spouse, arsenic causes cancer of the lung, bladder, and skin. It's also linked to kidney, nasal, liver, and prostate cancers. Both the U.S. Environmental Protection Agency and the World Health Organization classify arsenic as a known human carcinogen. Low-level arsenic exposure also carries risk for stroke and diabetes.

Arsenic is soluble in water. This property means it remains in the body for only three days and can be peed out (which is why it is associated with kidney and bladder cancer) and sweated out (which is why it is associated with skin tumors). It causes lung cancer when inhaled as dust. Arsenic can be absorbed through the skin but not easily. Eating and drinking are thought to be the main routes of exposure.

Arsenic is found in bedrock. From here it can leach into groundwater or rise to the surface when coal or metal ores are mined. Elemental arsenic can be scraped off the inside of the stacks after copper is smelted. Indeed, smelter dust is the source of most of the arsenic sold on the world market. Most commercial-grade arsenic comes from China.

Here are some less known attributes of arsenic. It is deadlier when all by itself than when combined with carbon. That is to say, inorganic arsenic is more powerful poison than organic arsenic. (Curiously, the opposite is true for mercury, element number 80 on the periodic chart. Mercury is far more poisonous in its organic form, methylmercury, than as an inorganic element.) In its elemental form, arsenic impersonates phosphate ions. By replacing phosphate inside the cellular machinery that extracts energy from sugar, it can grind respiration to a halt. In a sufficient dose, it destroys the heart, the brain, the nerves, and the lining of the intestine. Death follows.

As a carcinogen, arsenic is a far more shadowy assassin, and its modus operandi not well described. In trace amounts, arsenic appears to interfere with glucocorticoids, a family of hormones with two very different functions. Glucocorticoids regulate blood sugar. That's their Clark Kent day job. They also serve to suppress the growth of tumors. That's their Superman job. According to results from recent studies, arsenic atoms can bind with glucocorticoids, which may explain why arsenic raises the risks for diabetes as well as cancer.

The villain named arsenic can also take out another super-hero: our tumor-suppressor genes. According to a 2010 review of the evidence, arsenic silences these genes by attaching carbon groups to them. This subtle change, known as methylation, is the genetic equivalent of a chloroform-soaked handkerchief to the nostrils. When tumor-suppressor genes are knocked out, nascent malignancies are allowed to bloom.

Inconsistencies remain in the studies of arsenic carcino-genesis, but the main point is this: Even though much remains to be learned about how arsenic causes cancer, there is no doubt that it does. If indeed, as now seems true, arsenic is an endocrine disruptor as well as a silencer of cancer-protecting genes, then no safe level of exposure may exist.

The absence of safe threshold levels for arsenic exposure holds particular dangers for children. Their alternative metabolic path-ways are less able than ours to convert the highly poisonous inorganic form of arsenic into the less toxic organic form. Accordingly, children receive higher doses of arsenic at equiva-lent exposures, and these doses remain in their bodies for longer periods of time before being peed out.

Arsenic and children don't belong in the same sentence, says the reader.

Yes, I know (came the unexpected reply). But, for a series of peculiar historical reasons, arsenic and children have ended up together. And the resulting predicament, which isn't easily remedied through parental acts of vigilance and self-sacrifice, raises some fundamental questions:

• If it turns out that we, as parents, can't easily protect our children from exposure to environmental toxicants like arsenic, is it better to know about the evidence of harm? Or to not know?

- If we decide to know, can we consider the evidence without rationalizing it away? (*Well, I can't protect my kids from everything.*)
- And if we decide we are better off not knowing about problems like arsenic—because down that road lies only despair and futility—what else are we willing to close our eyes to?

The close encounter between children and arsenic began in 1933, when an Indian engineer, Sonti Kamesam, made a discovery that saved the lives of countless coal miners: Namely, copper and arsenic, when injected into wood, prevent timber beams from rotting. (Copper kills fungus. Arsenic kills insects.) Both metals were well-known pesticides at the time. Kamesam's special technique was to add chromium to the mix, thereby binding the toxic metals to the wood fibers. The result was stronger roofs in the damp underground tunnels through which coal is extracted. A patent was granted in 1938.

Researchers in Mississippi pounded wooden stakes treated with Kamesam's formula into fields that swarmed with termites. Months later, they were still standing. In 1950, a highly impressed Bell Telephone applied for permission to use chromated copper arsenate—CCA—on utility poles. In 1950, arsenic was acknowledged to be an acute poison. It was also suspected to cause skin tumors, but its ability to cause bladder and lung cancers was not known.

For the next two decades, CCA-treated wood remained a specialty product. Porches, fences, docks, and boardwalks were still constructed out of tree species that were naturally rot- and insect-resistant. Cedar. Redwood. Cypress. Fir. Then, in the 1970s, the price of wood soared. Cheap, plantation-grown southern pine became the construction material of choice. But it rotted and drew insects and fungus. It could be made to repel these with a formulation originally intended to prevent mines from caving in. So treated, it became known, euphemistically, as pressure-treated wood and, with this new name, entered the home building and construction trades. CCA-treated pine was promoted as an ideal framing material in spaces where dampness was a problem.

The word *pressure* refers to the way the mix of pesticides is applied to the wood. Pressure-treated lumber is made by placing a freshly milled board inside a vacuum chamber and sucking from its fibers all water and air. The pressure is then reversed so that copper, chromium, and arsenic are forced into the now empty cells. Pressure treatment is essentially a form of embalming. Chromated copper arsenate is 22 percent pure arsenic by weight.

Would this story have a different ending, I ask myself, if CCA-injected lumber had informally become known as *pesticide*-treated wood instead of *pressure*-treated wood?

Early concerns surfaced during the 1970s as faux-redwood, California-style decks became popular additions to suburban homes in the East and Midwest. Pressure-treated wood was an affordable material for their construction, and it was advertised to homeowners as such. This second marketing leap—from the building trades to weekend do-it-yourselfers—brought arsenic into the backyards of countless homes. It was as though a controlled substance, previously obtainable only through prescription, had become an over-the-counter drug recommended for all kinds of off-label purposes. A CCA deck right off the kitchen was a long way from telephone poles and coal mines. No one eats, barbeques, or sunbathes on the pesticide-soaked rafters of mine tunnels, and yet the backyard deck was specifically designed with such activities in mind.

In 1978, an increasingly concerned EPA began a special review of all arsenic-containing pesticides with the possible intent of revoking their registration. It was increasingly clear that CCA posed a cancer risk greater than the EPA's risk criteria. The EPA's review took ten years. During that time of regulatory inaction, the manufacture of pressure-treated wood increased by 400 percent, and its use expanded to other types of construction: including picnic tables, gazebos, and . . . children's playgrounds.

During the 1980s, wooden, castle-style play structures, complete with towers and suspension bridges, became the rage. They were erected in school playgrounds, city parks, and day care centers. Almost without exception, they were constructed from pressure-treated wood.

In 1988, the EPA reached a decision: CCA was re-registered as a pesticide for wood; its use by the wood preservation industry would be allowed to continue. (However, by 1993, all other arsenic-containing pesticides were banned.) To address concerns about carcinogenic exposures from contact with the wood, the EPA recommended that pressure-treated wood sold directly to consumers via lumberyards and home improvement centers bear warning labels. The timber industry objected, proposing instead that retail stores distribute fact sheets to educate buyers about the wood's potential hazards. The government agreed. And so, lumber departments of home improvement stores were provisioned with pads of tear-off sheets whose warnings often read like infomercials—touting the virtues of the product rather than describing its dangers to children who might play or eat on its surfaces. Moreover, many buyers were not even told about the availability of these consumer information sheets. Thus, almost no one in the general public was aware that pressure-treated wood contained pesticides. Nor that those sawing or sanding the wood should wear goggles and gloves. Nor that clothing that comes in contact with the wood should be washed separately. Nor that frequent, prolonged exposure to skin should be avoided. Nor that food should never touch it.

Signs of harm began to trickle in during the 1980s. Workers in wood treatment plants were found to have elevated levels of arsenic in their urine. A government employee became completely disabled after building picnic tables in an unventilated shop. Eight members of a rural Wisconsin family fell ill with a mysterious neurological disease that turned out to be arsenic poisoning caused by burning pressure-treated lumber in the wood stove.

Would this story have a different ending, I ask myself, had the following convoluted sentence not appeared in the 1992 Code of Federal Regulations?

The following solid wastes are not hazardous wastes. . . . Solid waste which consists of discarded arsenical-treated wood or wood products which fails the test for the Toxicity Characteristics for Hazardous Waste.

In other words, CCA was granted a special exemption to hazardous waste rules. This decision allowed pressure-treated wood to be dumped in ordinary, unlined landfills rather than in hazardous waste landfills—even though the chemicals in the wood constitute hazardous waste (and did so even by the standards of the day). Had this sleight-of-hand exemption not been granted, the market for CCA lumber would have undoubtedly contracted. How many dads would have lined up to buy pressure-treated wood at the hardware store if they knew that they would have to pay hundreds of dollars in hazardous waste tipping fees once the kids outgrew the play fort?

Between 1970 and 1995, the production of CCA wood increased by fourteenfold. Meanwhile, the scientific case against arsenic became more damning. In 1998, the National Research Council reported that arsenic exposure through drinking water was linked to lung and bladder cancers and could exert its carcinogenic powers at much lower levels of exposure than previously believed. Children were shown to be at particular risk. Other discoveries followed, including the finding that arsenic interferes with glucocorticoid hormones.

Far from the lab bench, a Connecticut chemist, David Stilwell, began crawling around backyard decks throughout New England. In 1997, he reported that the soil under and around pressure-treated structures contained concentrations of arsenic far in excess of background levels, and, in some cases, far in excess of the clean-up standard for toxic waste sites. The amount of arsenic in the soil under the decks increased with the age of the wood: the older and more weathered the deck, the more arsenic leached out. He also found that he could wipe arsenic from his hands after running them along the vertical poles of children's playground equipment.

In short, more than sixty years after Kamesam's invention, Stilwell discovered that chromium does not serve as such an effective binding agent after all. Eventually, the arsenic and copper leach out. Especially if the wood is rained on.

In spring 2001, an investigative journalist in Florida, Julie Hauserman, followed up on Stilwell's findings and published the results of her own investigation in an exposé headlined "The Poison in Your Backyard." She collected soil beneath playgrounds in a five-county area and sent it to labs for testing. "Arsenic," she wrote, "is leaking out of huge wooden playgrounds. . . . It's leaking beneath decks and state park boardwalks at levels that are dozens of times—even hundreds of times—higher than the state considers safe. And discarded pressure-treated lumber is leaking arsenic out of unlined landfills . . . posing a threat to drinking water."

Hauserman's reportage was a cultural tipping point. Newspapers and television stations throughout the country followed up with investigative stories of their own. Two environmental organizations demonstrated that arsenic wiped off easily from the surface of wood purchased in retail home improvement stores. Levels ranged from 20 to 1,000 micrograms per 100 square centimeters, which is about the size of a four-year-old's handprint. This was considerably more arsenic than the EPA allowed in drinking water. Moreover, the dislodgable film of arsenic that coated new wood also coated old wood. Investigations revealed that, as pressure-treated wood weathers and rain penetrates the cracks, arsenic dissolves in the raindrops and moves toward the surface.

In September 2001, the National Research Council announced that cancer risks from arsenic in drinking water were even greater than it had estimated two years earlier. Only 10 micrograms per day of long-term arsenic exposure in an adult were required to raise the lifetime risk of lung or bladder cancer to 1 in 300. It was easily possible that children climbing on arsenic-treated play structures were accumulating on their hands more than this amount.

At the beginning of January 2002, I hadn't heard about CCA wood. By the end of the month, I knew quite a lot.

Home from travel, I'd resumed daily life as a working mother, shuttling between my office at home and my office at Cornell. Jeff took the morning shift, dropping Faith off at school, and I took care of the afternoon pick-up. Glancing through the newspaper in the parking lot while waiting for Faith on a snowy Thursday, I noticed a story on toxic playgrounds. A school district near Rochester, New York, had decided to rip out its play structures after tests had found startlingly high levels of arsenic in the soil under and around them. The EPA was expected to make some kind of announcement within the next month about arsenic in the wood of these kinds of structures and the cancer risks posed by them. But the superintendent of this district wasn't willing to wait. He wanted precautionary action.

And thus I was introduced to chromated copper arsenate. Except that it didn't feel like an introduction so much as it did opening a door and finding an old and unwelcome face staring back. It felt like an unexpected visit from a bad ex-boyfriend. *What are you doing here?*

I am a bladder cancer survivor. Arsenic is a bladder carcinogen. That much I knew. Indeed, what had led me into the field of environmental health in the first place was my intimate experience with this particular disease and all the ongoing medical surveillance that it required. (Of all cancers, bladder cancer is one most likely to recur.) I had written an entire book about my life as a cancer patient and about the demonstrable evidence linking bladder cancer to environmental exposures. Having been diagnosed when I was barely out of childhood myself meant that childhood cancer risks were, for me, neither remote nor hypothetical.

I looked out at the nursery school playground where recess was in full swing. Three- and four-year-olds, including my own daughter, were charging back and forth along a wooden gangway, crawling through wooden tunnels, calling to each other from top of wooden towers, hiding behind wooden posts, climbing into a wooden boat, and rolling wood mulch and half-melted snow into slushy balls. Right next to the wooden bridge was the wellhead for the nursery school's drinking water.

Was it possible that the play structure was impregnated with known bladder carcinogens? It was difficult to believe. Lovingly erected by parent volunteers in 1987, the play structure was a jewel of community pride in little Ellis Hollow. And it was also a memorial site: a brass plaque dedicated the playground to the memory of Ed Rosenberg, a local father who had died unexpectedly. His own children had attended this school. Nestled in a grove of maples, the playground attracted kids in all seasons of the year. The previous summer I had watched a boy dressed only in swim trunks stick his chewing gum onto the banister of that play structure, eat a sandwich he had lain down on a stair rail, and, when finished, put the gum back in his mouth. I had laughed at his resourcefulness.

Back in my office, I dug into the scientific literature. The problem was that very little was known about the behavior of arsenic in the bodies of children. Nearly all the toxicology studies and occupational health studies had been conducted on adult animals or adult humans. Routes of exposure were also ill defined. In the case of children at play, transfers of arsenic first from wood to hand and then from hand to mouth were presumed the primary pathway of exposure, but quantifying these potential exposures—and then estimating cancer risk based on these numbers—was vexing work. Researchers at the University of Miami were currently investigating the issue.

How could we estimate the arsenic exposure of a child on a play structure? The answer seemed to hinge on how often the child put her hands into her mouth. This line of inquiry brought me to the arresting jargon about pediatric rates of hand-to-mouth interactions: 9.5 times per hour. And yet, this is just an average. It's clear that very young children put their hands in their mouths far more frequently. The National Exposure Research Laboratory estimates eighty-one mouthing events per hour for children under the age of two (a statistic that, admittedly, included mouth and tongue contact with body parts other than hands). But, then again, the hands of older children had larger surface areas. But, then again, younger children were more likely to lick or chew on the wood. But, then again, older children had longer playtimes. But, then again, younger children did more "soil play," and thus were more likely to inhale arsenic-contaminated dust. And what about children who eat dirt?

According to one estimate, the average five-year-old playing on a CCA play structure could exceed within two weeks the lifetime cancer risk considered to be acceptable under federal pesticide laws. The EPA had identified children in day-care settings with CCA decks and play equipment as high risk, especially if they also lived in homes with pressure-treated wooden decks.

There was such a deck on the back of our cabin.

Four-year-old Becca was Faith's special friend in the nursery school. Becca's mom was a research biologist. At the next afternoon's pick-up, I shared with her the newspaper story I had read and the results of my research, and we immediately became co-principal investigators. Within a week, we confirmed that our nursery school playground was indeed constructed from pressure-treated wood; we gained the unanimous support of parents—and the wholehearted permission from the community center's board of directors—to test the wood for arsenic leaching; we identified a lab certified to conduct chemical analyses; we gathered and submitted samples.

Becca's mom did the fieldwork. Following standard protocols, she collected three swipes from the play structure itself as well as four soil and mulch samples from underneath and beside the structure. Just a preliminary screening.

The results: All of the swipes of the playground equipment came back positive for arsenic, as did all of the soil samples. In all cases, the levels of arsenic greatly exceeded the naturally occurring background level for arsenic in New York soils—as well as the clean-up standard to which industrial sites in the state of New York are expected to attain. One soil sample collected near the slide contained 101 parts per million of arsenic. The clean-up standard for arsenic in the soil in the state of New York is 7.5. The mean arsenic level on the swipes collected from the surface wood, while somewhat lower than that reported for freshly milled boards, still exceeded 15 micrograms/100 square centimeters.

In February 2002, the EPA announced that backyard decking, picnic tables, and playgrounds could no longer contain arsenic: CCA's registration for use in residential settings was hereby cancelled. In making this announcement—which was a reversal of its 1988 decision—the Agency noted that arsenic was a known human carcinogen. *However* . . . the ban on CCA would not start until 2004—two years hence—to allow the wood treatment industry to use up old inventory and find chemical alternatives. And, under the terms of the phase-out, the millions of preexisting wooden structures made with arsenic-treated lumber, like ours in the nursery school, would remain in service. No recall. No replacement. No testing. No recommended actions for removing the threat. Until it could complete an assessment of pediatric health risks, the agency simply advised that children should wash any exposed skin after contacting CCA play equipment.

Until the numbers came back from the lab, a sense of concerned unity bound us nursery school parents together. A Playground Committee was formed. Lines of communication were opened with the community center's board. We learned that the company that had designed our playground happened to be located right here in Ithaca. Indeed, one of the company's employees was a regular customer in the motorcycle shop owned by one of our fathers. From him we heard that the wood on our play structure had been sealed. Although no real data existed yet on whether or not painting CCA wood with sealant prevented it from sweating arsenic, we felt reassured. To be on the safe side, we consulted with an agent from the county cooperative extension office who suggested that we not only insist on handwashing after trips out to the playground but also eliminate playground snacks and wipe children's shoes thoroughly before re-entry to the school. And so we did.

As soon as the lab report we had commissioned came back, everything changed. As one father said later, "As soon as you know, you can't not know."

The playground was abandoned. Whether or not it had been previously sealed, it was clearly leaking arsenic. Until we could devise a more permanent solution, we moved the children out to the unfenced front field for outdoor play. This created crowd-control challenges for the teachers, so extra parent volunteers were recruited to help oversee recess, and those of us who could, signed up for tours of duty. But beyond this collective action, parents were at odds about what course of action to pursue. About half thought we should say nothing publicly because we risked shutting down our nursery school, putting two very good (and very underpaid) teachers out of work. The rest thought that speaking out was the only way to get the problem redressed and that knowledge carries with it the obligation to do something.

I volunteered to represent the Playground Committee before the community center's board of directors. The meeting was a disaster. When I passed around copies of the lab results, they were quickly shuffled under stacks of other papers. No one even pretended to look at them. Instead, my remarks prompted board members to compare stories about their own backyard decks and reminisce about the erecting of swing sets. More than one person mentioned that his own kids had grown up on the nursery school playground. One board member expressed concern that the community center would be deemed a toxic site, as though the problem were not arsenic but people talking about arsenic. Replacing the play structure—one possible choice—was rejected out of hand as cost-prohibitive. The reverse-engineered conclusion seemed to be *if we don't have the money to remediate the problem, then it is not a problem.*

A decision was made to run an appeal in the next issue of the newsletter calling for community volunteers to recoat the playground with sealant during the coming summer—how about June 9?—and replace the wood chips. Meanwhile, the board would await the release of the forthcoming EPA report, which promised to address the question of how much cancer risk pressure-treated wooden playgrounds posed to children. Who knows? Maybe the results would even prompt state grants to defray the cost of equipment replacement. In a subsequent newspaper story about our playground imbroglio, the president of the board said, "We don't want to get too far ahead of the curve."

In fact, the report entitled *A Probablistic Risk Assessment for Children Who Contact CCA-Treated Playsets and Decks, Final Report* would finally be released by the EPA six years later. When Faith was nine. The curve was moving very slowly.

As winter melted into spring and no good solution to our problem emerged, rancor among parents swelled. The lab results were a Rorschach test. The arsenic data—65.0 micrograms; 13.5 micrograms; 30.4 micrograms—were numbers onto which an individual could project a worldview. Some saw in these figures low exposures and negligible risks. Compared to the lab results from some of the other arsenic play structures that were in the news, ours were not the worst. These parents argued for playground restitution. They wanted their children to have daily play adventures, for which, after all, they'd paid good tuition money. They reminded the rest of us that daily physical exercise was important for child health—probably more important than worrying about arsenic. And anyway, it was just time to move on. The school was starting to receive inquiries from community members who had noticed that our children were no longer using the play structure. How would the resulting rumors affect recruitment of students for next year? "More attention should be paid to the consequences our decisions will have on the overall functioning of the preschool," said Toby's mom in a memo to the rest of us.

The only safe level of arsenic is no arsenic, countered those who saw in the numbers unnecessary danger. They didn't want their children anywhere near the playground. To the accusation that they were alarmist, these parents pointed out that other communities had decided to replace their playgrounds on the basis of numbers similar to ours. Moreover, CCA wood was already banned for use in playgrounds in much of Europe, as well as Japan and Australia. They asked the others to remember how some of our children liked to splash in the water that pooled on the playground's wooden decks, wipe dew off the railings, and race their small hands up and down the boards over and over. "I do know that our children are playing with carcinogens and I cannot get beyond this," said Derrick's dad in a memo to the rest of us. "I would really like to refocus the community on removing and working on replacement of the structure."

At home, parents were consulting various Web sites and bringing in statements and studies that bolstered their position. By the end of March, no one appreciated receiving any more information from alternative points of view. After reporting that the San Diego and Bronx Zoos had initiated prohibitions on arsenic-treated wood, I was accused by one mother of comparing our children to zoo animals.

Through all this, the school's two teachers remained neutral. Finally in April, they broke their silence and expressed their strong belief that children need unstructured, imaginative outdoor play and the opportunity to negotiate among their peers with little adult restriction. The fenced-off playground in its shady grove of trees had provided that. It was an integral part of the school's built environment. The unfenced, unsheltered, wind-swept field, where the children were now exiled for an hour each day, did not. To play in a treeless expanse of grass required adults to supervise games and maintain boundaries. Moreover, with one group of parents insisting on playground right of return and another insisting on arsenic liberation—hey, why not just organize more field trips to the frog pond?—our nursery school, said its teachers, was becoming a house divided.

The president of the nursery school—Howie's mom—called an emergency meeting. Parents were told to come ready to vote. Should we return to the playground or not? But when we were all gathered, the first item on the agenda was an unannounced Presentation of Research on the Potential Risk of CCA-Treated

Wood. Behind the scenes, Ethan's dad and Connor's mom had quietly teamed up in an attempt to determine how much arsenic exposure our kids might be getting from one hour of daily play on the playground and then to evaluate the danger of that exposure. In the absence of any sort of government risk assessment, they had leaped into the breach and created their own. And now they were going to present their results.

Using parameters from an EPA document, the two co-authors plugged in some numbers for hand-to-mouth frequencies, calculated estimates for utilized amounts based on ingestion rates, and compared arsenic loading of hand swipes versus tissue swipes (taking into consideration potential nonlinearity in loading as a function of surface area). They then compared the number they derived for ingestion, corrected for bioavailability, with the EPA's lowest observed adverse effect level and with estimated intakes from other sources.

From their equations and spreadsheets, they concluded that the health risks to our children from playground-derived arsenic were minimal. According to their calculations and the assumptions contained therein, exposures from the playground were less than exposures from other sources, such as food and water.

And this is how the parents of a rural nursery school, in the absence of any direction at all from their government, acted as their own regulatory agency. And, shortly thereafter, the vote was to return to the playground. Except for the families who refused. One of which was mine.

Well-informed futility refers to a particular kind of learned helplessness. It's a term that was coined in 1973 by psychologist Gerhart Wiebe who was writing in an age when television had brought war into the living rooms of Americans for the first time. Wiebe noticed that a steady onslaught of information about a problem over which people feel little sense of personal agency gives rise to futility. Ironically, the more knowledgeable we are about such a problem, the more we are filled with paralyzing futility. Futility, in turn, forestalls action. But action is exactly what is necessary to overcome futility.

Just down the street from well-informed futility resides denial. According to contemporary risk communication expert Peter Sandman, we all instinctively avoid information that triggers intolerable emotions—such as intolerable fear or intolerable guilt. In the face of knowledge too upsetting to bear, there is nothing to do but look away. Well-informed futility and its inattentive neighbor, denial, especially flourish, says Sandman, when there are discontinuities in the messages we receive, as when we are told that a problem is dire (mass extinctions, melting icecaps) but the proposed solution (buy new lightbulbs) seems trivial. If the problem were really so dire, wouldn't we all be asked to respond with actions of equivalent magnitude? So . . . maybe the problem isn't so dire. Discontinuity provides an exit door.

Given this, I could understand why the initial solidarity among nursery school parents had fractured, why many parents just wanted to put the issue behind them, and why a homemade risk assessment that made our playground seem less threatening was so attractive. After all, if pressure-treated playgrounds were really so dangerous—I could imagine parents and teachers thinking—wouldn't the EPA or the Consumer Product Safety Commission have demanded an immediate recall?

The discontinuities of arsenic were multiple. On the one hand, the federal government had placed arsenic on the top of its toxic Most Wanted List. Lethal, indestructible, and water soluble, arsenic was the Osama Bin Laden of hazardous waste. On the other hand, between 1964 and 2001, the government had allowed 550 million pounds of arsenic to be injected into lumber, much of which was used to build things that children would have intimate contact with—and it did so even after good evidence showed that the wood leaks and serves as a reservoir for toxic releases.

On the one hand, the government set allowable limits for arsenic in the soil of industrial and hazardous waste sites. These limits represented thresholds above which the risk for cancer was calculated to be more than negligible. Exceeding them triggered regulatory action. On the other hand, when they occurred in playgrounds and backyards where children reside, these same releases were ignored. Arsenic released from pressure-treated wood in residential spaces routinely exceeded allowable limits for industrial or hazardous waste sites—sometimes by one or two orders of magnitude. Yet the EPA had no allowable loss rates for arsenical chemicals from pressure-treated wood.

On the one hand, the government of Norway was so concerned about arsenic-treated wood that it had decided to remove it, along with the soil it had contaminated, from 40,000 schools and parks and from 6,000 day-care centers. On the other hand, my own government was leaving the problem for individual citizens to solve by themselves, and the unofficial recommendation for excessively worried people was to coat the wood with sealants. Indeed, that was the solution the community center was pursuing. Yet, there were no studies to demonstrate the effectiveness of applying sealants, and layers of paint did nothing to stop migration of arsenic into soil from underground posts.

I thought about that a lot. I also thought a lot about our nursery school's well. The water our children drank and used to mix their finger paints lay beneath a leaking arsenic-filled play structure. Becca's mom and I hadn't thought to bring a tap water sample to the lab for analysis as part of our first-pass sampling, and the community board and the majority of parents were now adamant that further chemical testing was unnecessary. I could only hope that the high clay content of the soil was preventing the arsenic from migrating into the water table. There was a lot of wishful thinking going on.

And then there was the mother of all discontinuities: On the one hand, chromated copper arsenate was universally acknowledged to be wickedly hazardous. On the other hand, thanks to an administrative exemption that reclassified wood pumped full of chromated copper arsenate as non-hazardous, piles of old pressure-treated lumber enjoyed unlimited access to unlined landfills as though they were paper bags full of coffee grounds, broken crayons, and empty toothpaste tubes. And yet, if all the arsenic contained in the wood were extracted and placed in a bag, its disposal would require a call to a hazardous materials hauler.

Here, then, lay the real conundrum: What should happen to all the CCA wood out there when it finally reached the end of its usable lifespan? Pressure-treated lumber lasts a long time—twenty to fifty years—but it's not immortal. Sooner or later, somebody was going to start falling through the floorboards of their deck, play structures were going to chip, crack, and warp, and the word *demolition* was going to be spoken out loud. Seventy percent of single-family households in the United States have pressure-treated structures of some kind. All together, those facts had to mean that, during the lifespan of my two children, hundreds of millions of cubic feet of pressure-treated lumber would be dumped in landfills, carrying within their splintery fibers thousands of tons of arsenic. Decks, fences, swing sets, picnic tables, treehouses—and all of the 1,000 or so large, community-built, pressure-treated wooden playgrounds now in service across the United States—will go there. So buried, they will pose unending threats to groundwater.

What would the alternative be? CCA wood certainly shouldn't be burned in incinerators: the arsenic evaporates and enters the air and also creates highly poisonous ash. Mulching pressure-treated wood dramatically increases the leaching rate and is actually illegal (although it happens anyway because CCA lumber gets into the woodstream that finds its way to the chippers). And because of potential liability, there is little market for reuse. CCA wood, it seemed, violated every precept of sustainability. It was unrecyclable, unburnable, and uncompostable. It was dangerous to bury, dangerous to reuse, and dangerous just standing there, quietly shedding arsenic into the environment—although this last option might actually pose the least threat for the short-term future.

Was it possible that the EPA's odd decision to cancel CCA for new residential construction while doing nothing about CCA in preexisting residential construction was an unspoken acknowledgement of this exact predicament? Without a coordinated program to remove and safely encapsulate old decks and play structures, a recall would just trigger a rush to landfills and bring about a massive transfer of arsenic from the surface of the earth to a hole underneath the earth. Which could, some scientists were alleging, create even bigger threats to public health if aquifers became contaminated.

I was now futilely well informed. And here is where the growing fractures within our nursery school community extended into my own heart. I honestly didn't know what should happen to our playground, and I couldn't solve the problem on my own. I was not the nation state of Norway. No amount of fundraising would pay for its transport and disposal into a lined hazardous waste landfill. (Nor could I ever convince the neighbors that their beloved memorial to Ed belonged in a haz mat dump.) And it was possible that dismantling the playground and sending it off, board by board, to the regular landfill ultimately posed a bigger threat to children than letting it stand. Arsenic-contaminated drinking water is a far more efficient carcinogen delivery system than bridges and turrets—whatever one thought about all those micrograms of arsenic that Becca's mom had so easily wiped from their wooden surfaces.

Burying the toxic castle was not the answer.

And yet, I could not watch my three-year-old narrate stories about herself while climbing around on a structure that contained carcinogens. Known carcinogens. Bladder carcinogens. It was my job to keep my children safe. Whatever I could do to prevent my daughter from entering the world of biopsies, ultrasounds, and phone calls from the pathology lab, I would do. It wasn't even a choice. If I couldn't remove the play structure from the community, then I would have to remove Faith from the community.

I did. And so did three other families, who, at the end of May, pulled their children from the school and enrolled them elsewhere.

This is my home, said Faith.
I will keep you safe, said Mama.

And in so doing, two sentences that belonged together became discontinuous.

We avoid information that elicits in us feelings of intolerable fear or intolerable guilt, says Peter Sandman, who is surely right about this. But I'd like to submit that intolerable rage can, by contrast, serve to blaze new pathways of inquiry.

While reading through a mind-numbingly dense technical document on the regulatory history of chromated copper arsenate, I came across a curious detail: When the very last remaining use of arsenical pesticides was voluntarily ended—arsenic acid was prohibited from cotton agriculture because of unacceptable cancer risks to workers—its cancellation allowed for the sale of existing arsenic stock to the wood preservative industry for reformulation into "registered wood preservative products." Otherwise known as CCA. The year was 1993, a boom time for CCA-treated playgrounds and backyard decks.

In other words, a chemical deemed too carcinogenic to be handled by adults was sold off to an industry for use in products that children would handle. Adults: large people with protective clothing and access to Material Safety Data Sheets. Children: small people who stick gum onto playground equipment and then put it back in their mouths.

At this point, I felt capable of reaching into the pages of the Federal Register, grabbing someone by his 1993 collar, and hauling him out to my daughter's nursery school.

So that was the deal? The sellers of arsenic acid lose their agricultural market but get to dump their poison into playgrounds like this one? Is that how it went down? Go ahead, stand over there on the mulch by the slide—101 parts per million arsenic, by the way—and tell me, the mother of a three-year-old, how I've got it all wrong.

I would say that I was experiencing an episode of intolerable rage. And the problem was not that it led me to inattentive despair but rather—obviously enough, I guess—that the person with whom I needed to have an intolerably enraged chat was not identifiable to me. I think this is the place where a lot of parents find themselves. It's not that we're not paying attention to the environmental threats surrounding our children, it's just that the web of causation and responsibility is so complicated that we don't know how to navigate it or where to focus our actions. Or it becomes navigable only in hindsight after the damage is done.

By contrast, when chains of custody are clear—and public safety agencies are responsive—knowledge about potential dangers to our children can be a source of power for parents rather than a source of impotence. For example, even as I was getting ready to wave the white surrender flag from the top of a pressure-treated castle tower, I was successfully addressing another menace, about which I was equally well informed but feeling far less futile. It involved a dog.

A quarter mile from us along our road was an A-frame back in the woods where a revolving collection of young adults lived with a revolving collection of dogs, all of them unneutered males with Nazi-sounding names. Periodically, there were drug busts. Periodically, their dogs came into our woods. I once mistook one of their Rottweilers for a black bear.

And one afternoon that same dog chased me up onto my own porch and then lunged at me through the screen door. So I invited one of the young men in the A-frame over for a conversation. We stood on the porch, and I explained my problem, baby in arms, Faith clinging to the back of my skirt. And when he smirked and said something flip, I found myself saying, in a voice I had never heard myself use before, *If your dog so much as touches my children, I . . . WILL . . . TAKE . . . YOU . . . APART.*

Intolerable rage caused a drug dealer to topple backwards off my steps. And between me, the sheriff, and the SPCA, the problem got solved. No risk assessments required. It was a triumph for the precautionary principle.

The first draft of the EPA's promised risk assessments was released in late 2003. It provisionally concluded—in contrast to the homemade one presented at our parents' meeting—that children who frequently play on pressure-treated wood experience, over their lifetimes, elevated cancer risks. The extra deaths could be as many as five per 10,000 children exposed, but there was debate about the validity of setting threshold doses. The

Consumer Product Safety Commission reached a similar conclusion in a 387-page report.

Refinement of the mathematical model used to calculate the number of children who would eventually get cancer from contact with CCA wood continued in 2004. The problem was the "considerable uncertainty associated with quantitative estimates of children's arsenic exposure from CCA-treated wood." Specifically, there was a need to refine estimates of hand-to-mouth transfer, skin absorption, and gastrointestinal absorption. To assess children's exposure to arsenic from wood preservatives, the EPA developed the Stochastic Human Exposure and Dose Simulation model. A complicated formula was developed to estimate the number of micrograms of arsenic per square centimeter of a child's palm. This parameter was called *handloading*.

In 2005, the EPA reported the results of its sealant study. The good news was that sealants could—by a factor of ten—reduce levels of dislodgeable arsenic on the surface of a deck. The bad news was that arsenic rebounded to its original level after a period of several months. Moreover, the tests were conducted in Maryland and North Carolina; the results were not applicable to regions with freeze-thaw cycles. The following year, a science advisory panel concluded that "many uncertainties exist with regard to using surface coating or sealant studies to estimate reduction of dislodgeable arsenic."

In 2007, the simulation model for handloading was challenged by a study funded by a wood preservative trade association.

In 2008, the EPA's final risk assessment mostly confirmed the findings of its 2003 draft. It also reported that children who have regular contact with CCA playgrounds as well as CCA decks have, not surprisingly, double the absorbed doses.

In a 2010 survey of playgrounds in New Orleans, researchers found arsenic concentration in soil under pressure-treated wood play structures at 57 parts per million—nearly 40 times higher than background levels. Quoted in an editorial, lead author Howard Mielke said, "If we want to protect our children, don't expose them to toxics. If we find arsenic in the environment, we should deal with it directly."

There is little doubt that some number of children will die from cancers acquired from playing on arsenic-treated playgrounds and backyard decks. But after a decade of study, we still don't have a good estimate of what that number of cancers is, except that it could be as high as one in 10,000 for highly exposed subgroups and is almost certainly greater than one in a million. One in 10,000 does not include cancers other than bladder and lung, cancers caused by exposures to mixtures of chemicals of which arsenic is just one contributor, or other diseases like diabetes and stroke in which arsenic also plays a role.

One in 10,000 is a number that exceeds the annual child drowning rate.

The difference between death by drowning and death by low-level arsenic exposure is timing. As any water safety expert will tell you, child drownings are silent and swift. One minute you're talking to a toddler in the kitchen; the next minute he's face down in the pool. By contrast, arsenic requires 20 to 45 years to cause cancer. By then, nobody remembers daily recess in the nursery school castle or the paper doll parties out on the deck. The long lag time between exposure and onset of disease— together with the fact that lots of things other than arsenic cause lung and bladder cancers—means that the cancer patients created by pressure-treated wood are anonymous. We can't name the dead, and we can't name those responsible for the dead. But the suffering will be real.

It's worth thinking about what we are willing to do to protect our children from potential harm and why. When do we take precautionary action and when do we just shrug our shoulders and hope for the best? And how are these actions judged? When I enrolled Faith in Red Cross water safety classes and spent my Saturday mornings sitting in the bleachers at the pool, I was seen as a good mom. When I brought data on arsenic contamination into the community center, I was seen as alarmist. (Actually, I believe the adjective was *hysterical*.)

Finally, it's worth revisiting that fundamental parameter around which working parents orbit: *convenience*. In this fast-paced world of ours, so goes the dominant narrative, trace chemical exposures are the price we pay for convenience—whereas the items in the sustainability bin trend toward time-consuming and inconvenient. But CCA wood offers no such trade-off. It manages to be both unsustainable and terribly inconvenient.

Here, from *Pediatrics for Parents*, is a partial list of safety tips for homeowners with pressure-treated decks and play structures:

- Wash hands thoroughly with soap and water immediately after touching the wood.
- Keep a towel by the door for children to wipe their feet on after they have played on the equipment or walked on the deck.
- Wear gloves and a mask when sawing or sanding the wood. Safely dispose of all sawdust.
- Avoid bringing sawdust indoors.
- After working with the wood, thoroughly wash exposed areas of the body.
- Don't allow children or pets under the deck.
- Till the soil near the play equipment and deck and cover with clean topsoil or virgin mulch.
- Seal the wood annually.
- Launder clothes separately.

As soon as you know, you can't not know.

After reading through the list, I looked out the sliding glass window at the arsenic-treated planks of our deck, with all the irksome inconveniences they generated. Beyond them stood living trees, whose cells contained sap, not poison. They would not rot in the rain. They would not necessitate disposal in a hazardous waste landfill. They required none of my time. No towels. No handwashing. No gloves or masks. And one of them looked like a pretty good climbing tree.

CHAPTER THREE

The Grocery List (and the Ozone Hole)

WHEN ONE DOOR CLOSES, ANOTHER ONE OPENS. That was the lesson from my pre-parenting life. I had become fairly skilled at recognizing a closing door when I saw one—or slamming it shut myself—and finding some other intriguing portal to wander through. But that was not supposed to be the lesson anymore. I was no longer interested in doorways. I was trying to push roots through clay and wrap them around boulders, sink wells into groundwater, dig a foundation. I wanted the little tribe I'd created to originate from *this* place, their bodies constructed from *these* good molecules of food, water, and air.

Ithaca, New York, was admittedly an arbitrary choice for this project. Jeff and I had no family ties to this place at all. We had chosen it because it had offered me a research library and Jeff cheap studio space. And because when I had visited here once from Boston, I had stopped by the food co-op on my way back to the airport, stood in the produce aisle, picked up a locally grown peach, and thought, *Hey, I could live here.*

It all seemed provisional. And yet after only three years, I was already feeling like Odysseus. Whenever I traveled—which was a lot—I yearned to be back. My favorite moment of every trip was scanning the departures screen at the terminal for the single word *ITHACA*. *I have slain the airport Cyclops and return now to Ithaca, end of all my wanderings.*

And thus, I was taking our break-up with the nursery school pretty hard. Ultimately, our playground had ended up on national television. Long before any of us knew about its arsenic problem, I'd been filmed on the play structure for a PBS documentary called "Kids and Chemicals." The swings and slide had innocently offered a backdrop for an interview about the special vulnerability of children to toxic exposures. Once the results came back from the lab, I shared them with the producer, and they were incorporated into the program's script. If I had somehow thought that hearing about our playground hazards from Bill Moyers was going to change the hearts and minds of the other parents, I'd been sorely mistaken.

The nursery school had been my first attempt to shepherd my children into the larger world outside our cabin. It was also my first attempt at enfolding them in a community larger than the cabin's inhabitants: me, Jeff, a cat left over from Jeff's first marriage, and a now elderly dog that had been a gift from a past boyfriend of mine.

But, when Elijah was born and I was about to leave on a book tour with him, my beloved mother-in-law—herself an adventurous soul who lived in England with her second husband—had offered me a tip about motherhood: *be mobile.* This, she said, had served as her own mantra when she had four small children at home. So maybe that was the grown-up version of all those opening and closing doors.

I found a new nursery school for Faith. It was located in a suburban neighborhood by a strip mall, but appearances were deceiving. Behind the house where the wife of a retired surgeon lived and ran her little school were thirty acres of as-yet undeveloped woods and fields. The five children who were her small charges hiked there every day. There were two grumpy sheep to check on and a handful of chickens to feed. Instead of blue doughnuts for the birthday child, the children collected eggs from the henhouse and made a cake. When it was sunny, they gathered branches to build forts and castles in the backyard. When it rained, they gathered sheets to build forts and castles inside the house. In October, they gathered walnuts. In March they made maple syrup. In all the ways that it nurtured creative play, the school seemed like a laboratory for the Campaign for a Commercial-Free Childhood. Except that it was very un-lab-like. And there was no arsenic-treated wood. And without the thirty-acre patch of wild land for the children to roam and zoning that allowed for farm animals, the whole operation would not even be possible.

While driving home from the nursery school drop-off one morning, Elijah and I discovered a back road that led from the nursery school up to the summit of Snyder Hill. Thus, although there was no direct path between our cabin in the woods and the school in the suburbs, it turned out that we were, in fact, connected by a glacial ridge that created one side of the bowl of Ellis Hollow. Snyder Hill was the frog- and newt-filled slope that rose to the west of us. Faith and her nursery school friends were playing in the woods on the other side of that hill. And on top was a flat, wide expanse of good drainage and good views that was also the location of an organic agricultural operation called Narrow Bridge Farm.

We already belonged to this farm as shareholders of sorts. Organized along the principles of community supported agriculture, it sold its produce directly to its consumer members who bought a share of the season's harvest in early spring. As members, Jeff and I helped pay for seeds, machinery, and labor, thereby guaranteeing the farmers' production expenses. This investment was returned to us throughout the months that followed, in the form of boxes of fresh berries and vegetables. It also gave us flowers, honey, herbs, and eggs. Joining had been Jeff's idea. We both were grateful to Narrow Bridge not only for the food but because the snow and rain that fell on its fields eventually descended the west slope of Snyder Hill and seeped into the hollow where we lived, recharging our drinking water well and filling up our froggy swamp. Our home shared a watershed with this farm. We were glad it did not use pesticides.

CSA farms originated in Japan in the 1960s. They typically enjoy extraordinary crop production from tiny parcels of land precisely because they are required to produce a large diversity of crops for their members. Each species and variety absorbs and leaves behind in the soil a different suite of nutrients. As the seasons go by, these intricate patterns of cropping enrich the soil rather than deplete it. CSA fields resemble patchwork quilts. The secret to their high output is this botanical complexity—which extends beyond the crops for harvesting. During the off-season and the fallow times, CSA farmers typically plant an array of cover crops, like rye and vetch. These hold the soil in place and allow it to regenerate the elaborate bacterial and fungal ecosystems that plowing and planting disturb.

At the same time, all this botanical variety fools the pests—so few or no chemicals are needed. A farmer can plant strawberries, for example, in the patch where last year's broccoli grew, and the naturally occurring fungus-inhibitor exuded by the broccoli roots will protect the mold-prone strawberries. The Ghost of Broccoli Past stands guard over the strawberry patch.

These are the sorts of relationships that my colleagues in agroecology wrote dissertations about and commanded large grants to explicate. Figuring all this out on the fly while also dealing with hailstorms and tractor repairs made CSA farmers seem to me inspired geniuses. CSA farms also proved that there was no conflict between our desire as eaters for novelty and variety and the farmer's desire for high, pest-free yields. Indeed, each made the other possible. It was a beautiful and harmonious system. It made me happy just thinking about it.

When I realized that our nursery school and our cabin were ecologically connected by Narrow Bridge, I had a new thought. We were already receiving, in exchange for our $350 annual membership, weekly boxes of produce. But the farm also offered opportunities to its member-eaters for deeper kinds of involvement. Perhaps the path to the community that I was seeking for my children ran through food. We as a family could reorient ourselves away from the community center and toward this farm. At the same time, Elijah, at age one, was beginning to take his place on the human food chain, my body no longer his sole source of sustenance. I was interested in nurturing in him food preferences that leaned toward the novel and the varied. A deeper engagement with the farm might help me accomplish that. That would be convenient.

To be sure, the idea of me being a working member of any enterprise called Narrow Bridge was ironic. I had once turned down an opportunity to participate in an ecological research project in a rainforest because I wasn't willing to traverse suspension bridges over gorges and through tree canopies. I sometimes panic when confronted with ordinary pedestrian bridges. And here was a farm whose very name, according to the promotional pamphlet, was inspired by a directive from Hasidic scholar Reb Nachman of Breslov.

Know! The whole world is a very narrow bridge. The important thing is not to be afraid.

And so, as an adopted woman from the Protestant Midwest with no ancestral ties to this place—with no known ancestral ties at all—I set about to commit my children to a farm based on Japanese principles where Hebrew was heard as often as English and which was named after something that terrified me. It was going to be our farm. And I was going to improve my cooking skills and learn to put up food.

All that summer and into the fall, Faith picked green beans and basil. Elijah picked cherry tomatoes. They both learned the joy of raw fresh sugar snap peas. Meanwhile, I gathered recipes from among the farm's hundred or so members. Jeff joined with other fathers to construct a children's play area out by the bean fields. We attended all the monthly potlucks. And we initiated some rituals that would serve forevermore as our family's seasonal benchmarks: on summer solstice, pick strawberries; on Elijah's birthday, eat peaches; on Faith's birthday, serve plums.

Along the way, I began to think of the CSA relationship as a kind of goofily unbalanced marriage. The farmer is the superachieving mom—who manages the household, oversees scheduling, balances the books, and is tasked with all the monotonous, unpopular drudge work—while the CSA members are the clueless dad—the guy who waltzes in periodically, washes a few dishes, and feels like a hero. Equally shared parenting, this was not. The farmer filled out spreadsheets, consulted field maps, planned the fall cover crops, and decided when to chisel-plow the oats. All we adoring CSA members had to do was pull a few weeds now and then and wax ecstatic about the leeks. It was glorious that way.

At the harvest festival in the fall, while a band played in the barnyard and a yellow moon rose over the fields, I danced with the guy with whom I do equally share parenting. Around us a stream of sticky-mouthed children, including two of our own, ran and scattered in the tall grass.

Organic farming prohibits the use of synthetic pesticides. And here is the good news. Organic food really does contain fewer pesticide residues than conventional food. The first systematic comparison of pesticide residues in organic and nonorganic foods was carried out in 2002. Examining the data from more than 90,000 samples of produce, the authors of this study found that nearly three-quarters of conventionally grown foods had detectable pesticide residues. Three-quarters of organic crops had none. And among the one-quarter of organic samples that did test positive, levels of pesticide contamination were far lower. (Organic farmers can't control rain and wind, which can carry pesticide residues into chemical-free fields.) Conventionally grown foods were also more likely to test positive for multiple pesticides than were their organic counterparts.

Children fed organic food have in their bodies lower residues of certain pesticides—organophosphates—than children fed conventionally grown food. Frequently used in fruit and vegetable farming, organophosphate insecticides kill by attacking the nervous systems of insect pests. They have the same effect in humans: Organophosphates block the action of an enzyme that regulates a neurotransmitter and are thus brain poisons. (More on their effect on children's brain development in Chapter 8.)

A study that measured levels of these chemicals in the urine of pre-school children living in Seattle found marked differences between two groups of children. Children who ate mostly organic food had levels of organophosphate insecticides in their urine that were one-sixth of those who ate conventionally grown food.

A follow-up study pursued this difference in a more exacting way. Instead of comparing two groups of children with two kinds of diets, these researchers changed the diet within one group of children and then looked to see what happened to pesticide levels: That is to say, children who typically ate conventional diets were given one-to-one substitutes of organic food for five consecutive days. (Organic juice was swapped in for conventionally produced juice; organic corn chips for the usual brand, and so forth.) Within the five-day organic time span, levels of pesticides in urine dropped to below detectable levels. When children were returned to a nonorganic diet, pesticides once again showed up in their urine.

So we know that most organic produce is free from pesticide residues and most conventionally grown produce is not. And we know that children fed organic produce have significantly lower pesticide residues in their bodies than children fed conventional produce. We also know one more thing with certainty: Compared to those of adults, children's food preferences skew toward the more contaminated food items on the menu. They prefer to eat proportionally more of the foods on the high end of the pesticide residue continuum—such as apples, peaches, and strawberries. When produced conventionally, for example, 94 percent of strawberries have detectable residues of pesticides and most bear residues of more than one pesticide. (And, no, they don't wash off.) What we don't know is what levels of pesticide exposure are sufficient to endanger the health of children or alter their pathways of development. Maybe there are safe levels of exposure for children. But maybe there are not.

The EPA, tasked by the 1996 Food Quality Protection Act, is working to figure this out. The law itself arose out of concerns highlighted in a 1993 report from the National Research Council about the special vulnerabilities of children to pesticides. This report documented that the major source of pesticide exposure for kids was food. It also uncovered regulatory lapses, such as the fact that the maximum allowable level of pesticides in apple juice had been set with adults in mind even though—prepare not to be surprised—the average one-year-old drinks, in a year, fourteen times more apple juice than the average adult.

Thus, among other things, the Food Quality Protection Act directed the EPA to review all the pesticide residues allowable in food with an eye toward quantifying children's exposures. This part of the work, which took ten years, is now finished. Along the way, the EPA cancelled the registration for several organophosphate pesticides—but, in a 2006 decision that disappointed even some EPA scientists, it re-registered several others that had been red-flagged.

The law also directs the EPA to apply a tenfold safety factor when assessing risks of exposure to infants and children. And these assessments must now consider all possible exposure pathways. A child might, for example, be exposed to a pesticide that is used in his soccer field at school, in his backyard at home, and in the fields where his breakfast cereal is grown. All those potential exposures must now be estimated, added up, and incorporated in the risk assessment model.

The Food Quality Protection Act also requires that the government screen pesticides for hormone-disrupting effects. This is the task perhaps most crucial to the question of how adequately our children are protected from the harmful effects of pesticides. Hormones direct biological processes such as metabolic rate and glucose absorption and also prompt the unfolding of big events within the body, such as the onset of sexual maturation. In this, hormones are the veritable conductors of the symphony called child development. Just as the flick of a small baton can cue the crashing of cymbals, hormones can elicit dramatic effects at parts per billion concentrations. Pesticides with the ability to mimic or interfere with human hormones can thus create risks that are disproportionate to dose and dependent on the timing of exposure. (In Chapter 9, we'll look more closely at the ability of hormone-disrupting chemicals to influence the timing and tempo of puberty.)

Assessing the power of pesticides to influence children's hormones is the part of the job mandated by the 1996 law. It is not yet done, even though the EPA was originally given a 1999 deadline. The problem is that the screening assays to test for the ability of pesticides to act like hormones have to be invented, and then they have to be validated. There are also political pressures. The program's advisory panel has been dissolved at least four times. At this writing, not a single pesticide has been tested under the EPA Endocrine Disruptor Screening Program.

So to the question of whether or not organic foods are healthier for our kids, I have two answers. As a biologist, I say I don't know. (Nutritionists also say they don't know.) Accumulating evidence does seem to point in that direction, but more study is required to know for sure. For example, in an investigation published in 2010, researchers measured levels of organophosphate pesticides in the urine of more than one thousand eight-to-fifteen-year olds, selected to be representative of the U.S. population. They found that children with higher levels of pesticides were more likely to have symptoms of attention-deficit/hyperactivity disorder (ADHD). These results are consistent with others that have also reported connections between organophosphate exposure and difficulties in memory, attention, and behavior. Do these correlations point to a cause-and-effect relationship between pesticides and ADHD? Maybe. The ongoing National Children's Study, which will measure pesticides in umbilical cord blood and follow children throughout their development, will certainly shed more light on the issue. (More in Chapter 8.)

As a mother, my position is less equivocal. When the results of the National Children's study are finally published in twenty years, I won't have any children living in my house anymore. As long as I still do, my job is to avoid situations that seem inherently dangerous. All pesticides are inherently poisons, and all organophosphate pesticides are, inherently, brain poisons. So I don't feed my children food grown with pesticides. Period.

Now that Elijah had picked up a spoon and joined us at the table, and we all sat down together for dinner, I had to figure out an approach to family mealtime. Yes, I did some research.

From Cornell University researcher Wendy Wolfe, expert on child nutrition and obesity prevention, I learned that healthy eating habits tend to flourish when parents exert absolute control over the provisioning of food (no junk food) and almost none over consumption (how many pea-loaded spoons make it into the mouth). If healthful foods are brought into the house and artfully presented, some subset of them will get eaten. Bribery or cajoling, says Wolfe, is counterproductive. That sounded right. So, anything we didn't want the kids to eat, we didn't buy. And although we casually encouraged tastings, we didn't say, "Eat your peas."

From nutrition educator Antonia Demas, who designs food-based curricula for grade schools, cooks with children, and is working on transforming the School Lunch Program, I learned about the No Yuck Rule: You don't have to eat it, but you can't call it yucky. Noted and implemented.

I also made up some of my own rules. Everybody eats off the same menu. (No separate meals for kids.) When it's time to eat, everybody helps. (A one-year-old can ring the dinner bell and carry napkins.) Whoever sets the table gets to decide where everybody sits. (And whoever doesn't, takes out the compost.) Thursday night is ice cream night. (It's not Thursday; don't ask.) Finally, I appropriated a time-honored rule from my own mother's household: When the kitchen is closed, it's closed.

Since it was fun to talk about food, I also followed the advice of nutritionist Laurine Brown, who believes that children should recognize three food groups: go foods (whole grains and complex carbohydrates for energy), grow foods (protein for building body parts), and glow foods (brightly colored fruits and vegetables, rich in vitamins and minerals for health). So, when I wanted to explain what was on our plates tonight, that was the language I used. The kids and I also played lots of imaginary games at the table. Sometimes we were the Flopsy bunnies eating lettuces in Mr. McGregor's garden. ("Highly soporific!") And sometimes we were chicks pecking up chickpeas.

Reader, the results were pleasing. Faith developed the more adventurous palate—tempted by the garlic-y and the ginger-y—whereas Elijah preferred the tried and true: eggs, beans, squash, and any kind of soup. Both of them were fond of red bell peppers, broccoli, cucumbers, beets, and tomatoes and were willing to try any sort of vegetable as long as it was colorful and served with "special sauce" for dipping. Rice, tofu, and corn were always dependable side dishes. Faith loved spinach; Elijah preferred carrots. Watermelon, raspberries, peaches, and cherries were ambrosia. Yogurt and apples were staples. As were, mysteriously, green olives. Raisins went inexplicably in and out of favor at random intervals. By the time Faith was five and Elijah two, the most requested bedtime snack was baked sweet potatoes. I had two good eaters.

Some of the credit for all this goes to Narrow Bridge. Knowing where their food came from invested my kids in the idea of eating it. (When a cloudburst cut short a family outing to the lake one afternoon, Faith cheerily announced, "I think our carrots are drinking now.")

I'd also like to acknowledge the guy who stole the TV out of the moving van.

When Jeff and I had moved from Boston, our balky television set—one of the last objects to be trundled out of the empty apartment—didn't arrive with us in Ithaca. Somehow we never got around to replacing it, and the VCR/DVD player went into storage. Consequently, our kids had never been advertised to. For the most part, the images, jingles, and pitches of the food industry hadn't reach them, and their food preferences were thus mostly shaped by their direct experience with the food itself and the farmers who grow it. And for the most part, they made healthy choices, enjoyed trying new things, and shunned very few foods (although Elijah did once inform a guest, "I am vegetarian to radishes.")

Consider the results of a recent experiment on the influence of fast-food branding on the taste preferences of young children.

In this study, three-, four-, and five-year-olds were given pairs of identical food items and were asked which one tasted better. One item of the pair was wrapped in McDonald's packaging and one was unbranded. Overwhelmingly, the children reported that they liked the food and drinks better if they thought they were from McDonald's. And those children with more television sets in their homes showed a greater effect of branding than those with fewer.

Look, I realize that my kids are a sample size of only two. I acknowledge that most kids are not raised in log cabins without televisions or extended family members to shape their eating habits. I'm not advising that we all become mom and dad versions of Henry David Thoreau and raise our children in isolated Waldens in order that they swoon over vegetables. I'm not offering any advice here at all. But because a commercially unmediated relationship to food is so exceedingly rare, it seems worth reporting the following story.

When Faith was six and Elijah three, we were all delayed in an airport for several hours, and we ran out of snacks. Forbidden from leaving the gate area—the problem was alleged to be a computer glitch that could be resolved any moment—I looked around for something to eat. The only vendor within sight was McDonald's. *Well, here is a watershed moment in parenting.* I handed each of my children a red and yellow sack, warm with food.

They hated it.

"Too spicy," said Elijah.

I urged him to eat it anyway; we wouldn't be home for another four hours. But he refused. In addition to being too spicy, it was *too brown*. That was hard to argue with. I looked inside the sack—brown bun, brown meat, brown fries. Only the packaging was colorful. Even the drink was brown.

"Look, Mama," Faith shot back. "Look at their sign."

She pointed at the big yellow M.

"So?"

She whispered, "Their name is made of French fries."

I got it. She didn't see the world-famous logo as golden arches. No one had ever told her that's what it was supposed to be. To her the *M* in McDonald's looked like a pair of soggy, bent-over fries.

I drew from this experience the following conclusion: Fast food is neither innately nor universally attractive. Without aggressive marketing, it's, well, oversalted and brown.

Soon after, Faith changed her career plan from midwife-on-a-train to gourmet chef. She would open a restaurant called the Flowing River Café. It would be located on an island. Her brother would be the farmer, and he would deliver the ingredients to her kitchen on his boat. A *speedboat*, said Elijah the farmer.

Feeding my family an all-organic diet, with foods drawn as much as possible from local farms, brought a lot of stars into alignment. Protected from chemical exposures, my children chose from a banquet of healthy, delicious food. They developed a sense of community and place. And, at the same time, my grocery list helped support the integrity of three other things I cared about: farm workers, frogs, and antibiotics.

Among the people most heavily exposed to pesticides are the children of farm workers. Not only do these children receive doses of pesticides in their daily diets, they are also exposed to chemicals carried home from the fields by their parents. In Washington State, for example, farm workers in apple and pear orchards have elevated levels of pesticides in their urine. These pesticides also turn up in their cars and in the dust of their homes—and in the urine of their children. Similarly, in the Salinas Valley of California, researchers found farm pesticides in house dust of homes near the fields. They found the same pesticides in the socks and union suits of toddlers who live in those homes. These are the results of a pair of new studies that, along with others, demonstrate how pesticides used in fields, vineyards, and orchards find their way to children.

Two routes of exposure have been documented. In some cases, parents inadvertently carry farm chemicals home in their clothes or on their shoes and skin. In other cases—or in addition—chemicals drift in the air and, like second-hand smoke, seep into the home as vaporous fumes. Studies of farm families and farm worker families consistently find that the pesticide levels in children's urine correlate with that of adults in the household. Thus, the occupational risks borne by the parents who work in fields are also borne by their children who never set foot there. And, as we'll explore further in Chapter 8, a growing body of evidence finds links between pesticides in children's blood and risks for neurodevelopmental disorders.

The apples and pears I want to feed my own children are those grown in orchards that do not poison other mothers' children.

Like children, amphibians are also uniquely sensitive to pesticides. Their eggs lack shells. Their skin is thin. They live in creek beds and swampy, pond-filled bottomlands, often downhill from chemically treated farm fields. And their pubertal transformation from limbless gill-breather to walking, hopping, lung-equipped adult is accomplished by the actions of hormones.

Hormone-disrupting pesticides are contributing to the ongoing decline of North America's frog populations. In recent studies, trace exposure to the weed killer atrazine, for example, emasculated male tadpoles. It did so by stimulating an enzyme that converts male hormone into female hormone. Thus altered, male tadpoles metamorphosed into hermaphroditic adults. Similarly, nitrates from synthetic fertilizers can trigger deformities in developing tadpoles or kill them outright—at levels well below their legal limit in drinking water. Adult frogs are also vulnerable to the sex change operations of farm chemicals. In one recent study, male frogs exposed to atrazine turned into fully functional females that mated with males and produced eggs.

On April nights I am sometimes kept awake by the shrill *EEP EEP* of spring peepers. In upstate New York, peepers are soon joined by the quieter *ZZZIPPP . . . ZZZIPPP* of chorus frogs, whose call resembles a finger drawn over the tines of a comb. Later in the season comes the loudspeaker *JUG-O-RUM* of the bullfrog. Because I want my own children to enjoy frog-induced insomnia someday, I try to support farm practices unthreatening to the annual spring amphibian festival.

Agriculture is not only the largest user of pesticides; it is also the largest user of antibiotics. Encouraging this practice is the rise of concentrated animal feeding operations in which large numbers of animals are raised indoors—often far from the fields where forage crops are raised. One of the most striking trends in contemporary U.S. agriculture is the growing geographic separation between farm animals and the sources of food that farm animals eat. In factory farms, animals can be quartered time zones away from farm fields, estranged not only from their food sources but from sunlight and the green earth itself. In environments of lifelong confinement, without access to pasture or fresh air, animals are more prone to stress and are kept on antibiotics both to speed their growth and to prevent disease outbreaks.

Routine use of antibiotics in conventional dairy and livestock operations is one of the factors driving the emergence of drug-resistant bacteria. Humans are exposed both through the eating of animal products as well as through the release of these antibiotics into the environment—and children are more vulnerable to the risks so created. Children are more likely than adults to develop infections with those drug-resistant bacteria that have been linked directly to the use of antibiotics in agriculture. "Experts continue to call for reevaluation of and changes in these practices," concludes one recent analysis, "but in the United States, little action has been taken."

During the winter that Elijah and I both succumbed to pneumonia, I didn't have to wonder if, come spring, Jeff was going to be burying one of us. Thanks to azithromycin, we both recovered. When Faith's abdominal pain was diagnosed as strep, I didn't have to worry that she—like Beth in *Little Women*—would perish from that rare but frightening complication of a strep infection, rheumatic fever. We had penicillin.

Jeff and I use antibiotics very judiciously in our household because we need to know that they are going to continue working when we really need them. We don't want our food dollars supporting systems of agriculture that undermine the miracle of antibiotics. It's not in our interest to do so.

The presence of Narrow Bridge and the absence of television were two possible explanations for the joyful eating habits of our children. There was also a third. They spent no time in supermarkets.

Whatever food we didn't get from the farm, Jeff bought at the natural foods co-op. Not only did it stock organic teething biscuits, it had a play area near the deli where little kids could assemble puzzles or host tea parties while their parents could read, say, the arts section of the *New York Times* while drinking much-needed cups of coffee.

Because the store was organized as a cooperative, we could join it as working members. Food co-ops are, essentially, retail CSAs. If Jeff worked two hours a week, we could get a 17.5 percent discount on groceries. That discount meant that prices approached those in regular supermarkets. And this meant that he didn't have to drive anywhere else for toilet paper, soap, and toothpaste. Running errands with small children, Jeff pointed out, took a lot longer than just the driving time. There were minutes lost to the buckling and unbuckling of car seat straps, the zipping and unzipping of jackets, and the diaper changes in the men's room. There were hours of consequences brought on by disrupted nap schedules. So even with his two-hour weekly shift, the result of exclusive co-op shopping was a net gain of time.

As we also discovered, shopping at the food co-op was simply more convenient. No matter where we bought our weekly groceries, pausing in the aisles to ponder ingredient labels was inadvisable. With two preschoolers in the cart, speed was a requirement. To hunt down the organic options in a supermarket involved too much navigating, too much reading. But in the co-op, Jeff or I could, more or less, mindlessly grab the foods off the shelf that matched the words on the grocery list. The co-op's board of directors—elected by members—made the decisions about what products could appear on the shelves, and organically produced, locally grown foods were privileged. They were clearly labeled as such, and those labels typically included the name of the farm where the item originated. The institutional commitment of our co-op to local, organic farming freed Jeff and me from having to think too much.

One fortuitous consequence of co-op shopping was that no cartoon characters stared out at our children from boxes of presweetened cereals displayed at pediatric eye level in the aisles. No candy bars waited at the cash register to spark a parent–child battle of wills. No one suffered meltdowns in the check-out lane, although one of us did experience a noteworthy fit of pique involving kale.

On the way home late one afternoon with the kids in the backseat, I remembered we were out of shampoo, so I pulled into the co-op to pick some up. I let both kids know that no one should ask me for treats. Elijah, nevertheless, shot right over to the deli, where, behind the glass was his new favorite food item: steamed kale with sesame seeds and tamari sauce. He pleaded with me for some. I reminded him of our singular task. He asked again. I said no. Swamped by frustration, he threw himself, wailing, onto the floor. *Kale! Kale! I want kale!* Shoppers left their carts and walked over to investigate.

I . . . WANT . . . KALE!

A toddler was throwing a full-volume, operatic tantrum over a dark green leafy vegetable. It was a show worth watching.

Strawberries and tomatoes became the red bookends of our summers. In June, when days were long, we brought home baskets of strawberries. The ones not eaten immediately were frozen. Ten gallons of strawberries in the chest freezer was my annual goal. In September, when days and nights hung in equal balance, we brought home a parade of tomatoes. The ones not eaten immediately were frozen or canned. Sixty quarts of tomatoes on the pantry shelf was my annual goal.

All during the winter months, a stream of berries trickled back out of the freezer and were tossed onto cereal, stirred into oatmeal, or whirred into smoothies. Two quarts were reserved for Valentine's Day when I made strawberry shortcake, thus ensuring my position as the family's sweetheart. And all during the long winter, quart jars of tomatoes were cracked open and dumped into the Crock-Pot. Chili. Soup. Spaghetti sauce. In all these forms, we ate the color of summer all winter long.

More than any other food, Jeff and I decided, strawberries and tomatoes were connected to memory, but it was the *smell* of the tomatoes and the *taste* of the strawberries that were the transporting qualities. Viewed through the glass of a Ball jar, home-canned tomatoes were fleshy and animalistic, floating in a pale yellow fluid that looked like blood plasma. But as soon as the lid was pried off, the tomato-y aroma returned me immediately to the botanical—to fuzzy vines and sticky, serrated leaves. Within that smell was the whole scene: a blue September sky, slant of sun on the blond feathers of Elijah's hair, the heft of the bushel basket as we carried it together, purple asters in the ditch, fiddling of crickets, the retained heat inside each just-picked globe as we spread the tomatoes gently on the kitchen counter. A home movie inside every jar.

The flashback of strawberries came when the thawed berries hit the inside of our mouths. One bite, and suddenly, there was June: drifting lint of cottonwood, cry of a killdeer, slide of straw under our knees and palms as we crawled through the rows, Faith's fingers pushing aside the glossy, zigzagged leaves, Elijah's recurring complaint that everyone's basket was fuller than his. Meanwhile, in the brown-and-gray world outside the breakfast window, snow might be piling up.

On one such dark February morning, while serving themselves from a bowl of half-thawed berries, the kids announced they had found a *particular strawberry that they recognized.* Bickering ensued as each claimed credit for having harvested this misshapen, and unusually large, strawberry. I left off washing dishes to referee a nonsensical argument. But when I looked down at the disputed berry, I recognized it, too. It was the one that, we had all agreed eight months ago, looked like an enormous nose. We had gathered around to admire it, dangling so comically over its straw nest at the end of a row. *You're both wrong. I was the one who saw it first.*

Commercially produced strawberries and tomatoes tell a different story. It's not a story about memory or seasons. It's about the destruction of Earth's ozone, that blessed layer of sunscreen in the sky.

Fresh-market tomatoes, which mostly come from Florida, and commercial strawberries, which mostly come from California, are both produced using a pesticide called methyl bromide, which dates back to the nineteenth century. Methyl bromide is an all-purpose sterilant that kills everything its vapors touch—insects, weeds, fungus, earthworms, microbes, disease pathogens. In commercial strawberry beds, the soil is mounded up, covered in black plastic sheeting, and then fumigated with methyl bromide before nursery plants are plugged into holes punched through the tarp. Commercial tomato fields in Florida employ similar methods.

Methyl bromide is a wicked neurotoxicant. It is also a powerful destroyer of stratospheric ozone—more powerful than chlorofluorocarbons and aerosols. As a volatile gas, methyl bromide rises out of the tarped soil of strawberry beds and tomato fields and enters the stratosphere—six to thirty miles above our heads—where it picks apart the fragile triangles of oxygen known as ozone. These triple-atom molecules filter out harmful ultraviolet B radiation from the sun. (By contrast, ozone at ground level, created from the interaction of sunlight and car exhaust, is a noxious, smog-creating air pollutant linked to asthma. More on this in Chapter 6.)

The shredding of the stratospheric ozone layer allows more harmful ultraviolet B rays to reach us down here at the earth's surface, raising risks—especially among children—for deadly skin cancers, such as melanoma. And it's not just humans who are affected. Whales in the Gulf of California, who sleep and breastfeed at the ocean's surface, are also beginning to suffer blistering sunburns. A 2010 study has linked skin blisters in whales to rising levels of ultraviolet radiation, the result of ozone depletion.

Methyl bromide's stealthy ability to corrode our ozone layer is not a new discovery. Its ozone-eating properties landed it, in 1992, on the list of chemicals slated for worldwide phase-out under the Montreal Protocol on Substances that Deplete the Ozone Layer, an international treaty to which the United States is a signatory. By 2005, methyl bromide was supposed to exist no more.

But the United States petitioned the Ozone Secretariat of the United Nations for critical-use exemptions for California strawberry growers and Florida tomato growers, arguing that, for them, no economically feasible alternatives existed. These were granted. And this is how our strawberry- and tomato-buying habits can subsidize an industry that desperately lobbies for the right to use a chemical known to destroy the vital part of the atmosphere that protects us from sunburn and skin cancer. (In 2008, 2,708,710 pounds of methyl bromide were used on commercial strawberries in California.)

Given a choice, I would like the strawberries and tomatoes I feed my children to align with my efforts to keep their hats on at the beach—and with the empathy I feel for nursing mother whales napping in the ocean's rocking waves, I don't want any of us to suffer blistering sunburns.

There is a chemical alternative to methyl bromide—but it represents a leap from frying pan to fire. It's a chemical called methyl iodide used in laboratories. Like methyl bromide, methyl iodide is a drift-prone fumigating gas that kills on contact. Unlike methyl bromide, methyl iodide does not destroy the ozone layer. It's so chemically unstable that it can't travel very far without reacting with something. But this eagerness to combine with other chemicals also makes it one of the most toxic chemicals commonly used in the lab. With the ability to mutate DNA, it is classified as a human carcinogen by the state of California. Indeed, methyl iodide is sometimes used to induce cancer in lab rats. And because it targets the thyroid gland—whose hormones guide brain development—children are at special risk. In November 2010, methyl iodide was approved by the state of California as a substitute for methyl bromide. Workers who fumigate the fields will have to wear respirators.

Given a choice, I'd like the strawberries and tomatoes I feed my children to be grown by people who do not require chemical weapons protection.

Almost twenty years have passed since the clock started ticking on methyl bromide's eventual prohibition. And yet, after two decades of research and development, the only viable alternative to a known ozone destroyer is a classified carcinogen and a known thyroid destroyer? How can this be? I feel as though I discovered a partial answer from the late Charles Wilber, an Alabama tomato grower and holder of the 1987 *Guinness Book of World Records* record for the Biggest Tomato Yield on a Single Tomato Plant. (That would be 342 pounds of tomatoes.) Wilber was an organic grower. His secrets, he said, were mulch and compost. Also birds. Wilber built birdhouses around his fields and provided birdseed in the winter. The birds, in turn, ate the hornworms on his tomatoes. He used cardboard collars to keep cutworms off the stalks and planted Austrian peas around his fields to collect aphids. In his book, *How to Grow World Record Tomatoes*, Wilber asserts, "I can testify that by and large the university professors will not listen. . . . There is way too much money tied up in the chemical companies for America to change unless we have to."

It's up to parents of young children—and lovers of strawberries and tomatoes everywhere—to say firmly, loudly, and all together, *we have to.*

And in this *we have to* is the recognition that the food preferences of children are indivisible from the ecological world in which both the children and the food are grown. In this *we have to* is the acknowledgement that most U.S. children, like most farm animals, have become separated geographically from origins of their food. In this *we have to* is the admission that only 2 percent of U.S. children succeed in eating each day the five to nine daily servings of fruits and vegetables that is the official dietary recommendation of our federal government—and that in 30 states, the percentage of obese or overweight children now stands at or above 30 percent.

These problems cannot be solved by interstate fleets of diesel trucks full of lettuce and carrots (and tomatoes and strawberries) from California and Florida where cancer-causing, ozone-destroying fumigants are used to kill every living creature in the soil. If we want U.S. children to eat fruits and vegetables, then surround them with gardens, berry bushes, fruit trees, and urban farms. Plant vegetables in schoolyards and rooftops. Establish a CSA in every community. Reorient the food purchasing practices of institutions like hospitals, colleges, and nursing homes toward support for local, organic farms and thus provide growth opportunities for local agriculture. If we want children to eat healthy food, then provide their parents markets full of healthy food. And make it affordable.

As the family cook, I had a lot going for me. On top of the hill, I had a CSA farm. In town I had a well-stocked organic grocery store that sourced from family farms scattered throughout the local area. A Ph.D. nutritionist was willing to come to my house and teach me how to can tomatoes. Indeed, I was surrounded by people who shared my ideas about the primacy of healthy, chemical-free, locally produced food. I had a boy who was crazy for kale because, in part, my message to him about food was reinforced by the larger culture he lived in. There were few discontinuities between the food signals he received from his mother and the food signals beamed at him from the landscape itself. Because Ithaca had a thriving farm economy, our rural landscape was not filled with fast-food joints and billboards advertising Big Gulps. That made my life easier.

Also, organic, local food tastes better, and all those tempting colors and flavors seemed to inspire food curiosity. Thus, I didn't have to cater to finicky eaters. That definitely made life easier.

But I was still a working mother. I needed to learn to cook meals for four people from scratch, using the seasonal ingredients that surrounded me, and get them on the table fast. Whatever skills I brought with me from eighth grade home economics—I dimly remember learning that you shouldn't make meringue on rainy days—were not sufficient. So I began to talk with other parents of young children whose cooking skills

I admired and who seemed equally devoted to putting fresh healthy food on the table every night.

As you might expect, there is no special secret to making it all work, and when both parents hold full-time jobs, family dinners sourced from local farms require a kind of resolve worthy of military campaigns. One mother arose at 4 a.m. each Sunday and cooked meals for the whole week. One father cooked at midnight after everyone was asleep. Another made good culinary use out of naptime. Some cooks involved their children in the meal preparation, and some decidedly did not. There were two habits, though, practiced by every highly effective family cook I talked with. The first was a belief that family meals were a priority and not a chore or an afterthought. Food was the independent variable, the organizing parameter around which other variables circled. The other was meal planning. As in, you had do it.

And this is a true thing. You can't walk in the door at 6 p.m., confront a pile of unwashed kale, a jar of dried beans, and a sack of heirloom potatoes, and feel anything other than despair and panic. For Jeff and me, meal planning was the big difference between cooking for ourselves and cooking for us all. It's not that our children's food preferences were poles apart from our own; it's that we now had to beat an 8 p.m. bedtime. This meant that the menu had to be set in advance and ingredients pre-prepped. Both Jeff and I liked food—preparing meals for and with each other had been a notable element of our courtship— but our pre-children approach to the kitchen ran toward wild-hair improvisation. Those days were over.

Happily, I grew fond of the weekly ritual of inventorying the contents of freezer, pantry, cupboards, and crisper while consulting recipe cards and cookbooks, pondering the possibilities. How could I incorporate all that I had on hand into seven dinners without serving the same one twice? It was like food Sudoku. I'd sketch out a rough draft of the week's dinner menu, check the vegetable forecast from next week's CSA share, revise the plan, and finally, generate a grocery list based on the ingredients needed. And hand it to Jeff.

We've had our share of minor disasters—meals that were aborted when someone frantically had to wave a towel at a screaming smoke detector and at least one that Elijah described as "cryful." But, as with running and driving, the more you cook, the better you do it, with or without formal coaching.

The lynchpin to my own personal system is a Crock-Pot. It means that I can cook while I sleep and cook while I work. I also cook while flying through turbulence two time zones away. In my household, many dinners are assembled before I leave for the 6 a.m. flight out—even if that only means shaking frozen lumps of beans and vegetable stock into the Crock-Pot and pushing the *on* button. The food is local, even if the cook is not. Tellingly—in a somewhat pathetic and Freudian way—my children's pet name for the Crock-Pot was *Mama*. (The two cookbooks I rely on to fill it: Robin Robertson's *Fresh from the Vegetarian Slow Cooker* and Marian Morash's *The Victory Garden Cookbook*.)

And where was this parenting-sharing husband of yours? the reader asks. He can't manage the soup pot on his own?

In fact, Jeff is the breakfast and lunch chef. Anything involving batter, maple syrup, or sandwiches happens on his watch. My shift begins with the dinner hour, and I prefer single authorship—even more so when I am on the road. While sitting in the F terminal waiting for a delayed flight or at the bedside desk in some unmemorable hotel room, I like to envision my three beloveds gathered at our table back at home, eating the food I have cooked for them. I like believing that the weekly menu, pinned by magnets to the refrigerator door, is somehow holding us all together. By imagining how the table is set—the color of my bean soup in the blue bowls, the bread knife next to the loaf, the apple slices atop the cabbage salad that I'd assembled the evening before—I can imagine myself at that table. I can imagine the water pouring from the pitcher, the words spoken, the dying light in the trees outside the window. Elijah should be carrying the compost out right about now, and, hey, it's ice cream night.

It turns out that more than the smell and taste of food is transporting. If you need it to, even the idea of food can bind a family together.

Jeff and I don't dine at midnight any more. But we do sometimes find ourselves together in the kitchen at that hour, food passing through our hands. While he and I disassembled peaches late one August night—Jeff blanched them and stripped their fuzzy jackets, and I sliced and stuffed skinned crescents into freezer bags—I recalled the single peach that had prompted the whimsical decision to raise a family here. It was the feeling of it in the palm of my hand that had made me think, *Hey, I could live here.* Now a bushel of peaches was spread across my kitchen table. *And now, I do.*

CHAPTER FOUR

Pizza (and Ecosystem Services)

THE EARLY YEARS after Elijah's birth were lean ones. Jeff was making a transition from studio-based artist to state-certified art teacher, which necessitated night classes and tuition payments and child-care costs. At the same time, both of us, as second-time parents, were acutely aware of the brevity of infancy and childhood. We didn't want to miss anything. Money was a means to buy time with Faith and Elijah. We competed to be the primary parent, and the farther we could stretch a dollar, the more time we had with our kids.

To that end, Jeff and I sought out the advice of a financial counselor. We thought a third pair of eyes looking over our household budget might identify places where we could take up even more slack. And Becky Bilderbeck—who ran a bed and breakfast and sewed her own curtains—seemed to possess the ideal eyes for the job. Becky wasted no time scanning down the list of our monthly expenses. But, to both our relief and disappointment, she couldn't find much room for improvement. We owned one car, bought clothes at consignment shops, paid off our credit card in full each month. And there wasn't much that could be done about the health insurance premiums and all my various medical co-pays. Finally, she tapped her finger on one of our line items.

"Here," said Becky. "Right here. This seems high to me."

We leaned over the table. It was our groceries: $140 per week for food for a family of four. She thought, with careful coupon clipping, we could whittle that down.

Resisting the urge to issue an organic proclamation, I responded only with a polite, nodding smile. But the rest of the conversation continued without my full attention. I felt defensive and somehow ashamed. I had avoided supermarkets for so long that I had no idea what the prices of conventionally produced foods were like. I knew that organic food carried premium prices, but was it possible that even with Jeff's co-op discount and our CSA farm, we were paying significantly more for food? I cooked beans in the Crock-Pot while sleeping. Jeff cooked oatmeal on the stove while awake. Dried beans and bulk oatmeal—even organic dried beans and organic bulk oatmeal—were surely cheaper than, I don't know, Hamburger Helper and a pound of ground chuck. Right? I ran through a mental grocery list. Maybe it was the fair-trade coffee. Maybe the once-a-week organic chicken.

Back at home, I checked the Official USDA Food Plan, which consists of four different budgets—complete with grocery lists—all of which can meet minimal daily nutritional requirements. Developed during the Great Depression with an eye toward preventing low-income families from spiraling into malnutrition, the Food Plan is divided into quartiles of food spending—liberal, moderate-cost, low-cost, and thrifty—each corresponding to quartiles of household income. (These standards are now used for the basis of food stamp allotments as well as in setting alimony payments in divorce courts.) In 2003, for a family of four with young children, the average weekly food outlay was $145 for those who followed the moderate-cost plan.

So, we were essentially eating at the low end of the third quartile, although our income did not really qualify us to do so. We were clearly not as thrifty as those prudent souls following the thrifty plan, who were, in fact, able to limit their grocery budgets to about $100/week. Nor did we come in under $120/week, as those following the low-cost plan were managing to do. Still, we certainly looked like the paragons of parsimony compared to families following the USDA's liberal plan. Those eaters spent, on average, nearly $180 every week for their groceries.

Why *does* organic food cost more than conventionally grown food? I spent some days in the basement of the agriculture library at Cornell University trying to figure this out. There seemed to be three basic reasons. First, organic prices are higher because of retail mark-ups. Organic farms are often small, local, and seasonal. Retailers have to source with more suppliers who provide less predictable quantities. This takes more work. By contrast, conventional growers can keep prices down by sheer volume. Second, organic food provides higher profit margins for those who produce it. Increased farm income, of course, is not a bad thing. It provides jobs, prevents bankruptcies and foreclosures, and strengthens the fabric of rural communities.

But the principal reason that organic food costs more than conventional food is that organic food costs more to produce. It is more tightly regulated, and it requires more labor. And, in the United States, labor is more expensive than chemicals.

Encouraging job creation while decreasing the demand for toxic chemicals seemed like worthy use of my food dollars. In a sense, my generous food budget was subsidized by my lack of charitable giving—the financial counselor had noticed the absence of a line item for this expense category—because I had long ago decided that I was more interested in preventing social problems than trying to cure them. By directing my food dollars toward organic farmers, I felt that I was helping to prevent cancer, preterm birth, and rural unemployment, for example, while also investing in a healthy environment for my own children.

As a mother of modest means who cleaves to an organic diet, I'm hardly alone. Consumer demand for organic food has risen swiftly over the past two decades, at some points racing past domestic supply to the point where shortages constrained further growth. U.S. sales of organic food hit $24.8 billion in 2009—up from a mere $1 billion in 1990. Even during the

economic downturn, the growth of organic sales—up 15.8 percent in 2008 and another 5.1 percent in 2009—outpaced total
sales of conventional food. (Organic dairy was an exception.
More on this momentarily.)

Nevertheless, even with retail growth figures like these—
which *Food Business Week* called "whopping"—the amount of
organic food in the U.S. marketplace is paltry. To quantify paltry:
In 2009, the most recent year for which there is data, organic
food sales reached—drumroll, please—3.5 percent of all food
product sales in the United States.

Measured by the number of acres in production, the statistics
on organic food production are even more unimpressive. Yes,
certified organic cropland increased 41 percent between 2001
and 2005 and another 51 percent between 2005 and 2008. But
even with the collective enthusiasm for chemical-free farming,
these organic acres still only account for 0.7 percent of total U.S.
crop acreage. Of U.S. cows, 2.7 percent are raised organically. Of
U.S. egg-laying hens, 1.5 percent are raised organically. These
numbers are dramatically lower than those in many other
nations. (Carrots are a bright spot, though: the organic kind now
takes up 25 percent of total U.S. carrot acreage.)

The U.S. Department of Agriculture provides some clues
about the tempest-in-a-teapot progress of organic agriculture in
its reports. In spite of high consumer demand, obstacles still
stand in the way of farmers who are considering making the
move from conventional to organic. Organic farmers need to
know they have access to credit, markets, trade assistance, and
crop insurance. They need infrastructure: meat lockers, processing facilities, warehouses, and mills. They need university
research dollars directed down lines of inquiry relevant to them.
They need knowledgeable assistance from cooperative extension
services. Without this support, farmers are on their own when
problems arise, and, no matter how enthusiastic consumer
demand, the decision to embrace organic represents a perilous
and possibly foolish career move.

Institutional neglect of organic farming should make parents
of young children sit up and take notice. It means that the kind

of agriculture that does *not* rely on hormones, antibiotics, and neurological poisons receives less public investment than the kind that does. It means that even when we grocery shoppers create such demand for pesticide-free food that we blow past supply, systematic bottlenecks prevent supply from catching up. It means that organic farming will remain a niche market rather than a transformational force in the lives of families everywhere. It means that retail prices will remain high. It's what prevents organic agriculture from becoming . . . well, agriculture.

To quantify the depth of my financial commitment to organic food, I decided to conduct an experiment. I would make two pizzas using the same recipe. The first would be assembled from conventionally grown ingredients purchased at the supermarket near Faith's nursery school and the second from organically grown ones purchased at our food co-op. I would find out what I could about the agricultural origins of the ingredients. I would conduct a taste test. And I would compare the costs of the individual ingredients, based on the price I paid per unit amount.

Pizza seemed like a happy metric. It was the food that had fueled all my childhood birthday parties. It was notably present the first time a boy put his arm around me. In college, pizza was a recurring motif in my first serious love affair. Brian—linebacker, poet, and eldest of four brothers—had taught me the fine art of rolling up a pizza slice and eating it like sushi. The night he left his fraternity and moved in with me—setting in motion a small scandal that I had fully intended to enjoy—we had ordered pizza. Now I was married to a man who gladly made pizza because the dough reminded him of plaster, and I had wolfishly consumed the results throughout both my pregnancies. Both of our children, the then almost five-year-old as well as the almost-two-year-old, identified pizza as their favorite entrée. In this, they were hardly exceptional. Fully 70 percent of U.S. schoolchildren make the same claim.

In the fall of 2003, I published the results of my pizza study. Essentially, they were these: With one-to-one substitutes of organic ingredients for conventional ones, an organic pizza cost over 40 percent more than one assembled with conventional ingredients purchased at a supermarket. However, thanks to Jeff's discount at the co-op, the real additional cost to us was closer to 25 percent (conventional pizza: $4.50; organic pizza: $5.65). That 25 percent premium was roughly the difference between the Moderate-Cost Plan and the Low-Cost Plan on the USDA's chart of official food plans. The pizza experiment seemed to illustrate that my organic stubbornness was indeed the explanation for why we were eating with the modest but well-heeled third-quartile families rather than with the cost-cutting second-quartile tribe.

In a blind taste test, no one could tell the two pizzas apart.

In fall 2010, I repeated the study but tweaked the methodology a bit. In the seven years between the first pizza experiment and the second, my organic ingredient options had expanded considerably. For example, thanks to a local grain collaborative, I now had available to me a source of organic flour produced from heritage wheat varieties and milled in my own county. (By contrast, the organic dough of 2003 was kneaded with flour that had originated in the wheat fields of Montana.) In addition to possessing a CSA membership, I was now within walking distance of a weekly farmers' market whose vendors included at least one artisanal cheese maker. Alternatively, an Amish grocery a few miles up the road had expanded operations to include locally made cheese. I could compare prices.

Along with my shopping options, my culinary skills had also improved. Over the years, I had become, for example, an experienced home-canner of tomatoes. Along the way, I discovered that I could forego the addition of a can of tomato paste to my pizza sauce, as the recipe called for, because a quart jar of my own tomatoes—when crushed and drained—was zippy enough in flavor and substantial enough in texture to stand in for a can of paste and two watery, chopped tomatoes.

In fact, between 2003 and 2010, so many more organic ingredients and methods had become available to me that I felt confused. What was the purpose of my experiment now? Was I trying to prove that I could make an all-organic pizza for the same price as a one assembled from conventional supermarket ingredients? If so, yes, I could do that. Although the locally milled, stone-ground, organic flour cost twice as much as its conventional supermarket counterpart, I could more than compensate for that expense with my home-canned tomatoes. Two tomatoes and a can of tomato paste at the supermarket cost $3.70, whereas a jar of my own tomatoes priced out at $2.14.

Or was the point of my new experiment to assemble an entirely local pizza out of the most delicious ingredients I could find, while still remaining within the 25 percent organic price differential? If so, I could do that, too (not withstanding the essential but decidedly unlocal quarter-cup of olive oil). Since it seemed somehow unfair to use tomatoes and garlic from my CSA farm—the overall cost of my weekly share of produce from the farm came in at only a dollar a pound, as near as I could figure—I decided to buy everything. The choices were still tantalizing. Not only did I have sources for locally produced artisanal organic flour and cheese, my new BFF was a farmer named Margaret whose stall at the weekly market looked like the Museum of Garlic. She had rare varieties of every sort—some hot, some sweet, some with indescribably complicated flavors. At one point, thanks to Margaret, I had thirteen different kinds of garlic bulbs piled in a big wooden bowl on my kitchen counter. They had names like Music, Lukak, German Red, and Shang Tung Purple. They made me feel rich. And they stored better than conventional bulbs.

Full disclosure: The second time around, the blind taste test wasn't blind. Everybody recognized the scent of heirloom garlic and the nutty taste of our local flour. And that's the one we all said we liked best.

Driven by concerns about childhood obesity, the high price of cheap food is currently receiving well-deserved attention. And therein lies growing public acknowledgement that the money we hand to supermarket cashiers is only part of the price we pay for a form of agriculture that makes a twelve-pack of Ding Dongs cheaper than a bag of apples. Not appearing on the cash register receipt that flutters from a bag of groceries are the costs of treating obesity-related cancers, heart disease, stroke, and diabetes.

Right behind this critique lies another one: This same system of agriculture that fills store shelves with Ding Dongs requires pesticides and synthetic fertilizers to function, and this dependency, too, carries hidden economic price tags. These include higher utility bills triggered by the need to filter farm chemicals out of tap water; lost productivity caused by the pesticide poisonings of farm workers; higher taxes to pay for elaborate systems to monitor pesticides; loss of revenues prompted by poisoned honeybees, contaminated sport fish, and closed swimming beaches; and higher insurance premiums stoked by antibiotic-resistant infections and increased cancers caused by a thinning ozone layer. (See Chapter 3, strawberries and tomatoes.)

These factors are known as *economic externalities*—the costs of a privately profitable activity that are passed on to others. And, in many cases, those receiving the bill for the passed-along costs are children and members of future generations. Yes, as I discovered in the library basement, someone with a Ph.D. has attempted to quantify the externalized costs of pesticide use. His estimation is that, in the United States, only about half of the total cost of using pesticides to grow conventional food is included in the price of the food itself. By that logic, buying organic food is a good deal.

"If the public could only see the real price tag of the food we buy, purchasing decisions would be easy," writes Andrew Kimbrell, the director of the Center for Food Safety, commenting on these findings. "Compared to industrial food, organic alternatives are the bargains of a lifetime."

The conundrum for us organic-buying parents on a budget is that we are shouldering both the full costs of the food that we are feeding to our own families—produced on that 0.7 percent of agricultural acreage that is managed organically—as well as the externalized costs cast off by the other 99.3 percent. And all of our children, whether their bodies are constructed out of Ding Dongs or heirloom apples, will eventually be paying for damages not incorporated into the bar codes that beep their way through the convenience store checkout lanes. An agricultural system that is costly to society even though privately profitable is not a problem that can be solved by individual consumer choice or educational campaigns about the glories of farmers' markets. This is a structural problem that requires a structural—political— solution. Which is why agricultural policy and commodity pricing rules are as much issues of parenting as car seat recalls.

Of all the ingredients in a pizza, the one that has been with us the longest is wheat flour. Bread first appeared in the human diet at about 8,000 BC, which makes wheat about as old as goats. What distinguishes wheat from other domesticated grains is an abundance of gluten. By trapping carbon dioxide bubbles released from live yeast, the protein gluten allows bread dough to double in volume by rising. The milled flour of no other grain can do this.

Gluten provides both farmers and bakers a language to describe the various varieties of wheat, but each profession brings a different vocabulary to the table. Farmers talk about hard wheat, which has high gluten content, and soft wheat, which has less. Kansas became a wheat producer in the mid-nineteenth century when Mennonites brought to its prairie a hard wheat variety from Crimea. In general, hard wheats grow in dry areas, soft wheats where it is humid. Farmers also talk about spring wheat and winter wheat. Planted in the fall, winter wheat overwinters in the field and is harvested the following spring. Spring wheat is sown in the spring and harvested in the fall. Durum is an exceptionally hard spring wheat.

Bakers, on the other hand, refer to bread flours and cake flours. Bread flour is milled from gluten-rich hard wheat. Cake flour is milled from soft wheat, which allows for crumbliness. All-purpose flour is a mixture of the two. Durum wheat is for pasta. (Try making a loaf of bread from durum flour, and you end up with something akin to a paving stone.)

Everything about wheat is big. Wheat flour is the single most consumed food in the United States. On average, it makes up 7 percent of the daily diet—twice that for young children. In

2009, the amber waves of U.S. wheat fields filled 59 million acres—exceeding the combined acreage of all the National Parks. The fields themselves also trend toward the gargantuan, and 20,000 acre farms are not unheard of. (My mother's father raised six children on 160 acres.) That kind of bigness is made possible by pesticides—and, once achieved, guarantees that pesticides will continue to be used. With the right, very big machine, an acre of wheat can now be harvested and threshed in six minutes flat. Nevertheless, with crops located a hundred miles from the machine shed, moving equipment into far-flung fields costs money and takes time. Therefore, big wheat operations, by necessity, plant only wheat rather than many different crops in rotation. Vast tracts of land growing only one variety of one kind of crop allow pest populations to expand to equally vast proportions—at which point only chemical poisons can keep them in check.

There is no national pesticide registry in the United States. Farmers are not required—as are manufacturers—to report their chemical releases. So we don't know exactly how many tons of which pesticides are used on these 59 million acres of U.S. wheat. However, the USDA does periodically survey farmers in various wheat-growing regions about their pesticide habits. The compiled results are the closest thing we have to a portrait of chemical usage.

If you are the family breadmaker—or pizza or pasta maker—I invite you to visit the National Agricultural Statistics Service Web site. Take a look at its chemical use spreadsheet for wheat, which runs many, many pages. In 2009, eighty different pesticides were used on winter wheat. On spring wheat, sixty-eight. On durum wheat, forty-seven.

One of the names appearing there is 2,4-D, an herbicide that has been linked with birth defects. In studies conducted in Minnesota, Montana, North Dakota, and South Dakota, children who lived in counties that grew a lot of spring and durum wheat suffered significantly higher rates of birth defects than children who lived in counties where less wheat was grown. And within these wheat-growing counties, children who were conceived in the spring, during the time of planting and herbicide application, showed higher rates of malformation than children whose conceptions fell during other months of the calendar year.

Another of the names that appears on the wheat spreadsheet is the organophosphate insecticide chlorpyrifos, which has been linked to cognitive deficits in children. Emerging evidence also links it to autism. All children are exposed to these chemicals in the food they eat. Those living in the grain belt face the additional burden of drinking herbicides in their water. And those living in communities where crops are produced also must contend with exposure to pesticide drift in the air, which sometimes dwarfs exposures via the diet and water routes.

Large farms are leaky farms. Like pesticides, synthetic fertilizers also drift away from the fields they are sprayed on, and their nitrogen invariably washes into creeks and streams. It eventually ends up in the ocean where it contributes to algal blooms and dead zones. And this is how our flour-buying choices affect the health of the fish at sea.

In 2009, nitrogen fertilizer was used on nearly all conventionally grown durum wheat, 94 percent of other spring wheat, and 83 percent of winter wheat. Using two different USDA databases, I crunched the numbers . . . total acres planted . . . percent of acres treated . . . average rate of application per acre. . . . And I arrived at the following statistic: 2,968,000,000 pounds of nitrogen fertilizer were used to grow America's wheat in 2009. That's about ten pounds for every man, woman, and child in the United States. Almost all of these nearly 3 billion pounds were created from the fossil fuel called natural gas. Which was drilled out of the ground somewhere—often in somebody's backyard. (Whose backyard is a question we'll take up in Chapter 10.)

Organic farms that produce wheat, by contrast, tend to be smaller, and they usually grow other crops as well. This simple combination of modest size plus diversity provides the farmer an entire armory of potential tricks with which to outwit pests. By rotating crops, organic wheat farmers create a constantly shifting vegetative landscape that keeps disease and pest populations from exploding out of control. Smaller fields also have shorter distances from perimeter to center. Thus, along the edges of their fields, organic farmers sometimes plant insectories—botanical beds that serve as habitats for predatory insects that are the natural enemies of wheat pests. For fertilizer, organic wheat farmers can use either cow manure or green manure—which is not really manure at all but a plowed-under cover crop like hairy vetch or rye. It tends to stay put.

- Cost of 3 cups conventional flour (1½ cups whole wheat; 1½ cups all-purpose): $0.60
- Cost of 3 cups organic flour from Farmer Ground Flour of Newfield, New York (1½ cups hard red spring wheat; 1½ cups all-purpose): $1.26

Olive oil is about 3,000 years younger than wheat. Most scholars locate the botanical birthplace of the olive tree in what is now the border region of Iraq and Iran, where archeological evidence suggests that olive oil has been manufactured since at least 5,000 BC. References to olives abound in the sacred texts of Judaism, Christianity, and Islam. And its branches, leaves, flowers, and fruit appear in paintings throughout the ages. It's easy to see why. Olive trunks writhe from rocky ground and yet offer serene and restful silhouettes. Olive branches grow in full sun and yet their silvery leaves seem eternally bathed in moonlight.

Ripe olives are 15 to 40 percent oil. At least a dozen pounds of them are required to make a quart of virgin oil in a process of simple squeezing that has not really changed much over the years. But more is concentrated than just the juice of the fruit. When organophosphate insecticides are used to control the olive fruit's nemesis, the olive fly, trace residues can remain on the olives. Because insecticides are fat-soluble, they often find their way into the oil within. When olives are pressed, the concentrations of these residues can increase in the finished product by a factor of three to seven. A 2005 study tested for and found a dozen different pesticide residues at trace levels in olive oil. Washing olives prior to pressing them can sometimes lower residues in the finished oil, but the wastewater from the mill is then contaminated with pesticides and poses an environmental threat.

For my original pizza study, I interviewed Paco Núñez de Prado, the Spanish kingpin of organic olive oil. For an hour, we shouted at each other over a bad phone connection. *Olive grower* is a profession that men in his family have held, he told me proudly, for seven generations. In 2003, he oversaw 100,000 olive trees on four different farms as well as an olive oil mill and bottling plant. And he did it all organically. To control olive flies, he used bait infused with sexual attractants. With the males trapped, the females, he supposed, died of loneliness. And because they are not contaminated with pesticides, he could mix together the leftover olive residue with pruned leaves and branches to make organic fertilizer that is recycled back into his orchards.

A few years later, a Spanish study found that this type of olive production is significantly more energy efficient than conventional production. That is to say, organic olive oil has both demonstrably lower pesticide residues as well as a smaller carbon footprint. Given that double benefit, you might imagine a pricing structure that rewards consumers for choosing the organic alternative. That is not the case.

- *Cost of ¼ cup of conventional olive oil: $.52*
- *Cost of ¼ cup of bottled organic olive oil: $1.26 ($1.05 with Jeff's discount)*
- *Cost of ¼ cup of organic olive oil purchased in bulk: $.98 ($.83 with Jeff's discount)*

The tomato is the only ingredient of pizza native to the Americas. When Cortez conquered Mexico City in 1519, he sent its seeds to Europe where it was initially grown for the beauty of its fruit but not widely eaten except by a few bold Spaniards and Italians. In contrast to the reverence bestowed upon the olive, tomato fruits were viewed with suspicion. The tomato was reintroduced to America in the 18th century.

Vulnerable to fourteen different fungal diseases, the tomato plant is the delicate Victorian heroine of the horticultural world. Blight, rot, wilt, and canker are all words that appear in guides for the commercial tomato grower. Tomatoes are vulnerable to an equally impressive array of insect pests, some of which inject disease-causing viruses or deposit secretions that attract mold.

The tomato also requires the assistance of insect pollinators for its creation, and that fact, too, makes it vulnerable. Bringing pollen to the tomato's starry flowers is a task happily performed by bumblebees. A darkened tip around the flower's stigma indicates fertilization. Commercial varieties can self-pollinate but will not do so if the air is too cool or too still. Thus, one can find in any good agricultural library, step-by-step, full-color manuals on the art of performing tomato sex. This involves handheld vibrators, which are applied to open flowers. As too much mechanical stimulation is counterproductive, a light touch is recommended.

Commercial field tomatoes come in two basic types, neither of which has much in common with the backyard garden tomato. The first kind are fresh-market tomatoes, which we closely examined in Chapter 3. Fresh-market tomatoes are mostly grown in Florida. They are picked green, harvested by hand, and sold on the open market. The second kind are processed tomatoes, which are those destined for sauce and paste. Those who produce them are usually under contract to a processor. In contrast to garden tomatoes that putter along all summer, processed tomatoes all set fruit and ripen at the same time. They are thus harvested by machine and picked ripe. Most come from California. And, because California is the sole state in the union with a comprehensive pesticide registry, it's possible to investigate pesticide use in tomatoes for processing and find out quite a lot. In 2008, for example, 11,585,200 pounds of pesticides were used on field tomatoes in California, at an average application rate of 2.96 pounds per acre. More than half these 11.5 million pounds were sprayed on tomato vines in just one county—Fresno.

Now that I've become my own tomato cannery, I've become skilled at negotiating with farmers for bushels of organic field tomatoes. My technique is to walk up and down the stalls at the farmers' market and let it be known, in a voice slightly louder than normal volume, that I'm thinking about canning. Suddenly, I have men interested in engaging me in conversation about the virtues of certain varieties, offering free samples, slipping me phone numbers, promising discounts.

For me, haggling about price is not the point. I try to strike deals over timing. *Can you get me a bushel by Friday? No? How about a week from Friday?* With kids too young to help with a project that involves lots of boiling water, I put up tomatoes at night when everyone is asleep. It's a joyful task, but, unlike other after-hours activities—like, say, answering email or separating crayons from jigsaw puzzle pieces—it's not one to take on when exhausted, so I always quit after processing a single bushel. Converting fifty-three pounds of muddy field tomatoes into eighteen sealed and sterilized glass jars full of red, tomato-y hearts—all cheerfully cooling on my kitchen table—takes four hours. Then, a week or so later, I'll put another bushel behind glass. And so on. But baskets of tomatoes won't inertly wait in the hall closet until I am ready to tackle them, and they can't be safely canned once soft or bruised. The annual trick is to find farmers with tomatoes when I have time to can them . . . and find the time to can when the farmers have tomatoes.

- ♦ *Cost of one small jar of conventional tomato paste plus two fresh tomatoes: $3.70*
- ♦ *Cost of one small jar of organic tomato paste plus fresh organic tomatoes: $5.70 (or $4.70 with Jeff's discount)*
- ♦ *Cost of one quart of home-canned organic tomatoes standing in for above: $2.14*

Garlic is a lily with origins in Asia. Unlike the tomato, garlic has no sex life. No flowers. No bees. No fruits. No seeds. Domesticated garlic reproduces solely by cloning itself. Thus, garlic growers plant whole garlic cloves, which then sprout leaves and roots and bud off more cloves, each of which results, nine months later, in an underground bulb of twelve to fifteen cloves that is about the size and shape of a doorknob. This is not, admittedly, a lot of bang for the buck. Garlic farmers who save their own planting stock must hold back between 10 and 12 percent of their harvest. The miserly asexuality of garlic is one reason for its high price.

Gluttony, not lust, characterizes garlic. Insisting on lots of nutrients and water, garlic is what farmers call a heavy feeder. Such crops pose real problems for conventional growers because the soluble nitrogen the plants require can easily be swept away in the irrigation water. Garlic also dislikes growing alongside weeds and is prone to a number of fungal diseases. Conventional farmers address these problems with a battery of herbicides and fungicides whose trade names—Stomp, Prowl, Repulse, Squadron, Rout, Sabre, Torch—sound like designations for Special Operations forces. One of them is a suspected developmental toxicant. At least two can contaminate groundwater. Two are suspected carcinogens. Three are suspected endocrine disruptors. One is a brain poison. And one is methyl bromide.

These are the chemicals we spray into our environment in order to bring to market a food that many people buy for its health benefits. Which are real. Garlic has been shown to lower blood pressure and cholesterol, stimulate the human immune system, and slow tumor growth. Frequent garlic consumption lowers the risk for colon cancer and may well do the same for breast cancer.

Organic garlic growers also rely on a battery of weapons to defend their prized crop, but they go by other names: fire, flowers, mulch, and crop rotation. Some growers pull flame weeders behind their tractors to get a jump on the weeds. Some growers plant sweet alyssum and cilantro in their garlic fields to combat insect pests. (The flowers of both species support populations of serphid flies, which, in turn, eat aphid larvae.) To protect her many heirloom varieties from weeds, Margaret, Curator of Garlic, swears by thick slabs of oat straw mulch. Out at the CSA farm, garlic is protected by the sheer force of agricultural diversity, which allows for an incredible thirty-year crop rotation. Garlic is never planted in the same plot two years in a row, and, in fact, may not see the same patch of ground again for three decades. Eternal relocation shakes off pestilence. And mulch takes care of the weeds.

- *Cost of three cloves of conventional garlic: $.16*
- *Cost of three cloves of organic garlic from food co-op: $.23*
- *Cost of three cloves of Shang Tung Purple garlic from Margaret: $.32*

According to historian Joan Thirsk, cheese-making received a big boost from the Black Plague. During two remarkable years, 1348 and 1349, one third of the population of Europe died, and demand for food plummeted. As labor for cultivation became scarce, many farmers let their land go fallow, and cows grazed where crops had once been sown. Consequently, milk production increased, and techniques for cheese-making improved. An unexpected benefit emerged: Farm fields that had spent some time as meadows became more fertile. By the 16th century, there was no longer any reason to rotate fields into cow pastures, but peasant farmers lobbied the nobility for the right to do it, arguing that this practice enabled the land to "regain heart." In this way, cheese-making became enshrined in western agriculture.

During two remarkable years, 2008 and 2009, the U.S. economy collapsed and milk prices plummeted. As the wholesale price for 100 pounds of milk fell from $18 to $11—far below the cost of production—many farmers folded, and the rest barely hung on. Almost no U.S. dairy farmer made money in 2009. Organic milk prices declined less steeply, but the resulting price gap that opened between conventional and organic dairy products became too great for many shoppers to leap. Organic sales contracted.

No one was hurt worse by the economic downturn than dairy farmers who were in the process of transitioning from conventional to organic farming. The three-year transition period required to become certifiably organic is the eye of the needle for any farmer. It requires taking on the higher costs of organic practices without the ability to command premium organic prices. For dairy farmers, it means giving up antibiotics to prevent illness in the herd and forgoing artificial hormones to increase milk production and control reproductive cycling—and finding reliable veterinary care that doesn't involve chemical and hormonal solutions. It means finding a source of organic feed and access to pasture. To accomplish this, according to a recent analysis in *Review of Agricultural Economics*, most organic dairy farmers "utilize primarily unpaid operator and family labor" and a seven-day workweek.

Once finally certified, none of the organic dairy farmers I have met regretted the decision. Their cows live longer, are sick less often, and seem less stressed. And, as more than one organic farmer told me, the same improvements befall the cows' human caretakers. Happy cows make happy farmers. Less time is spent ministering to sick animals. Income goes up. Stress goes down. A sense of wellbeing fills both farmhouse and barnyard.

But when the price of conventional milk bottomed out, farmers just getting into organic production—because mothers like me had created what looked like solid demand—were wiped out. Most of them were young and just starting up. Many had small children of their own. All should have had years of chemical-free farming ahead of them. But, in a star-crossed moment, the economic collapse hit at a time when they were bearing all the start-up costs of organic farming and receiving none of its benefits. And nothing in the system reached out to catch them as they fell. Nobody bailed them out. Nobody pulled them through. General Motors and various Wall Street investment banks had softer landings than the transitioning organic dairy farmers of 2009.

As a parent who has continued to purchase organic milk throughout the economic crisis, even as my own income also took a hit, I would appreciate some institutional support for the farmers who build the bones of my children. Moms pushing grocery carts trying to decide between the $3 gallon of conventional milk and the $6 organic gallon should not be the sole safety net for the people who keep America's cows on pasture and off hormones.

Much of the economic vulnerability of dairy farmers—conventional or organic—arises from the fact that they have no say in the price of milk. They are what's known in the agricultural world as price-takers. But in a back-to-Middle-Ages trend, some entrepreneurial dairy farmers have responded to the milk crisis by . . . making cheese. And so became price-makers. An unexpected benefit: There are now a dozen artisanal cheese makers near Ithaca, New York. While the number of dairy farms in New York State keeps falling—further eroding the rural tax base and making upstate dairy communities susceptible to the schemes of landmen in the employ of the oil and gas industry (coming up in Chapter 10)—my home community is now on the New York Finger Lakes Cheese Trail.

- *Cost of 1 cup (grated) conventional mozzarella: $1.28.*
- *Cost of 1 cup (grated) organic mozzarella bought at the food co-op: $1.53*
- *Cost of 1 cup (grated) Red Meck cheese made and bought at Finger Lakes Farmstead Cheese, located nine miles from my house: $3.68*
- *Cost of 1 cup (crumbled) goat milk cheese made at the Lively Run Goat Dairy, located four miles from my house and bought at the farmers' market located three blocks from my house: $4.50*

Can organic farming feed the world? Depending on whom you ask, this is either the right question or the wrong question.

For the defenders of industrial, chemically intensive agriculture, it's the right question, and the answer is no. Political scientist Robert Paarlberg, for example, argues that scaling up organic agriculture until it is close to 100 percent of total acreage would have disastrous consequences for at least two reasons. First, because organic crops have lower yields, the world's forests would be cut down and conscripted into agricultural service in order to meet everyone's food needs. Second, by disallowing petrochemical fertilizers, universal organic farming would compel a population boom of cattle to provide all the needed manure. And because organic cows require pastures to deposit their manure in, even more forests would disappear to support all the four-legged fertilizer-makers. Paarlberg reminds us that food crises are common across the continent of Africa where chemical fertilizer is scarce and encourages us to stop romanticizing preindustrial food systems. (More on his claims in just a minute.)

Likewise, in a 2010 analysis, a research team at Stanford demonstrates that the trend toward increased use of pesticides, fertilizers, and high-yielding strains of crops has, since 1961, helped food production keep pace with rising population growth. Along the way, say the authors, chemically intensive agriculture has also helped to tamp down greenhouse gas emissions. Were the world's growing population fed instead by extensification—ongoing conversions of native landscapes to farm fields—the carbon footprint of farming would be far higher.

Can organic farming feed the world? For critics of industrial, chemically intensive agriculture, the question is the wrong one . . . but the answer is yes anyway.

It's the wrong question because world hunger is created not by food shortages but by failure to get food to the people who need it. And it's the wrong question because, however spectacular its past successes, chemically intensive agriculture is dependent on nonrenewable petrochemicals. Like a Hollywood star who requires an entourage of bodyguards, stylists, and personal trainers, intensive agriculture can't perform in the absence of fossil fuel-derived fertilizers, fungicides, and weed killers. It's not sustainable. As it is now, 5 percent of global natural gas reserves is turned into nitrogen fertilizer. (All by itself, the United States consumes 22 billion pounds of nitrogen fertilizer a year.) Do we really want the whole world's agricultural system to ride a tandem bicycle with the oil and gas industry?

And even if the fossil fuel party could somehow flow on forever, high-yielding varieties may be unable to keep the high going. Already there are signs of diminishing returns, argues author and agricultural analyst Anna Lappé. These take the form of pesticide-resistant bugs, herbicide-resistant weeds, antibiotic-resistant pathogens, declining soil fertility, and the need for ever more petrochemical inputs to keep the whole enterprise pedaling forward.

Author and family farm advocate Terra Brockman offers another kind of critique: High yields of chemically produced corn and soybeans are not feeding the world. They are not even feeding the farmer who grows them. Central Illinois, for example, is carpeted in chemical-intensive, high-yielding corn and soybeans. But many rural towns in central Illinois—where soybeans grow right up to the back door and cornfields bristle near the football field—are food deserts. With taxpayer-subsidized crops in the field destined for feedlots, ethanol plants, processing plants, and export markets, the peculiar truth is that, too often, there is no food in farm country. Even in agricultural regions with spectacularly high yields, food insecurity abounds.

A 2010 health ranking study undertaken by the Robert Wood Johnson Foundation showed that some of the least healthy counties in the United States are located in bumper crop regions. And yet, for many of these counties, the list of their underlying problems includes the phrase "lacks access to healthy, affordable foods."

In Illinois, Brockman points out, only one rural county received a high health ranking. That was little Woodford—the third healthiest county in the state. On the map, this small agricultural county stands out as an island of health in an otherwise not-so-healthy zone. Because of an accident of geology—the last retreating glacier dumped a pile of rocks there—Woodford does not contribute much to the impressive tonnage of corn and soy harvested from the state. Its rolly terrain and steep ravines make it unfit for the giant machines of industrial farming. Instead, it's gone its own agricultural way. Woodford County is home to a cluster of organic farms that produce, with no chemical inputs, fruit, vegetables, meat, eggs, and grain—healthy food for people's tables at reasonable prices through farm stands and CSA subscriptions. Undoubtedly, other factors also contribute to the unusual healthfulness of Woodford's 35,000 residents—it has a county hospital and a tightly knit Apostolic Christian community—but the number of bushels of soybeans per acre isn't one of them.

Can organic agriculture feed the world? Beside the point or not, the answer is yes, say critics of industrial agriculture, as well as peer-reviewed, long-term studies. First, they refute Paarlberg's assertion that, under an organic regime, the world would have to choose between insufficient quantities of fertilizer and ripping down rainforests for manure-making cows. Organic farming does not require manure, and most organic farmers do not use it. Compost also works. And many organic farmers enrich their soil solely through rotating crops and planting legumes. (They're called *green manure* for a reason.)

It's worth pausing a moment to think about this fact: Legumes are plants that can pull naked nitrogen out of the air— which is totally unusable by living organisms—and turn it into ammonia, which is highly useable. It does so by studding the stubbornly unreactive nitrogen molecules with hydrogen atoms. Voilà: ammonia. That's the hard part. From ammonia, it's but two easy chemical steps to nitrate, the essential material for plant growth.

Legumes can perform this miracle because their root nodules serve as housing for nitrogen-fixing bacteria. The bacteria, with their special enzymes, actually do all the labor of chemical synthesis. When the legume dies—or the farmer plows it under—its carefully hoarded nitrogen is released into the soil and can be used by other crops. Legumes and their microbial workforce accomplish all this ammonia manufacturing while sequestering carbon dioxide and exhaling oxygen. (Cue song of thanksgiving to the humble clover.)

Carried out in massive chemical facilities under conditions of high heat and intense pressure, the so-called Haber process can also accomplish this feat—synthesizing ammonia for use as fertilizer—but consumes natural gas reserves to get the same result. The gas donates its hydrogen atoms; the nitrogen comes from the atmosphere. Voilà: ammonia. A German chemist, Fritz Haber, invented this method in 1913. It's still what we use to make synthetic fertilizer.

In a 2007 study, a team of biologists at the University of Michigan concluded that legumaceous cover crops could fix enough nitrogen to replace all the fossil fuel–derived fertilizer now in use. They thus dispute the idea that organic agriculture is constrained by lack of nitrogen.

More centrally, this same research team disputes the evidence that organic farming suffers from lower yields. In a review of 293 studies that compared yields of organic and conventional farms in both developed and developing nations, researchers found parity. In the United States, yields on organic farms were about 92 percent of the yields produced by conventional agriculture, whereas in developing countries, yields were actually higher on organic farms. The authors then used data from the United Nations Food and Agriculture Organization to ask the question: *What would happen if all farming became organic?* They concluded from their analysis that organic methods could produce enough calories to sustain the current population of the world without resorting to extensification.

Meanwhile, in Wisconsin, results of a twelve-year comparative study also found roughly equivalent yields in organic and conventional systems. Organic hay allowed cows to make just as much milk. Organic corn, soybeans, and winter wheat also performed equally well—except during years with wet springs when weeds took over. Under these conditions, the organic harvests were fully one-third less than conventional. From this, the authors concluded not that conventional agriculture was therefore superior but that research and development should focus on improving weed control for organic systems.

These results confirm the observations of many farmers who have made the switch from conventional to organic practices: Organic grain fields are inherently more resilient and outperform conventional fields, especially during conditions of drought and unpredictable weather. Organic soils hold more moisture, and crop rotations provide a diversified portfolio or crops with different ripening times. Bets are hedged; economic eggs are carried in many baskets.

Thus, two mutually exclusive narratives about the transformative potential of organic agriculture compete for our allegiance—and our grocery budgets.

Why this debate remains unsettled has much to do with the affiliations of those in the debating arena—Robert Paarlberg is an advisor to Monsanto—but it has also to do with the complex nature of ecology itself. A *Consumer Reports*-style rating of organic and conventional fields is of limited use. In the real world, organic farmers rarely plant just one crop and may, in fact, rely on the interactions between crops to control pests and boost yields. So you don't learn much by comparing the performance of two identical monocultures, one sprayed with chemicals and one not. Second, the highest yielding varieties in conventional systems are not always the highest yielding varieties in organic systems. (This is especially true for wheat.) So you don't learn much by planting identical seed in each of two fields—one conventional, one organic—and comparing yields under two different methods of pest control.

Also, an unpoisoned field is a dynamic field. Over the years, the ecosystem of an organic field diversifies—species richness increases, along with the relative abundance of each species. The soil itself evolves. Typically, an organic field that began as a conventional field shows a dip in yield during the first few years after its conversion and then rebounds as its ecosystem rebuilds. Thus, the results of a comparative study will change as the seasons go by and as living creatures take over from petrochemicals the job of dispatching pests and pulling nitrogen out of thin air.

A paper published in July 2010 in the prestigious journal *Nature* elegantly documents, in the potato fields of Washington State, just how murderously effective natural systems can be in the targeted assassination of pests. Compared to fields treated to the usual barrage of chemical pesticides, organic fields had fewer problems with potato beetles and produced bigger potato plants if they were managed organically. They also supported a greater diversity of species in its potato-y food web. And the relative abundance among the members of that food web was more evenly distributed. Indeed, the demographics in the organic fields revealed truly integrated communities. Nobody sat in the back of the ecological bus. By comparison, the food webs in the conventional fields looked like gerrymandered voting districts with a few common species dominating and minority groups achieving only token representation.

Through a series of experiments, the researchers were able to demonstrate that the key to bigger plants and superior pest control was precisely this even-handed abundance of species. Within an organic system, the potato beetle—the bane of potato growers big and small—apparently finds itself surrounded by enemies at every turn. While dining on the foliage of the potato leaves, it is soon beset by ladybird beetles and damsel bugs. And should it manage to shake off these girly predators, it sooner or later comes face-to-face with the common black ground beetle. (That's the shiny, jet-black scuttle-y insect with a pair of pliers for a face that's seen crossing sidewalks on summer evenings.) There are quite a lot of ground beetles in organic potato fields, as it turns out, patrolling around like U.N. peacekeeping forces.

Neither does the subterranean world of an organic potato field provide safe haven for pests. While seeking a private under-ground spot for pupating, the potato beetle is easily parasitized by soil-dwelling roundworms and fungi, including the particu-larly lethal *Beauveria*. A potato beetle infected by *Beauveria* fun-gus is a sight to behold. It resembles a small Volkswagen sprayed with artificial Christmas tree flocking.

Meanwhile, out in the chemically managed fields, everybody gets gassed—the beetle-y good guys along with the targeted pest. The structure of the food web skews toward a few dominant species, and the natural enemies of the potato beetle become rare. The potato plants are thereby vulnerable to catastrophic pest outbreaks . . . which only the application of *more* pesticides will knock back.

Ecosystems services is the name given to the helpful activities of other species that accrue economic value to us. They represent acts of unpaid labor. The bee that pollinates our tomato flowers. The earthworm that tills our soil. The bird that eats the weed seeds before they sprout. The *Beauveria* fungus that foams up the potato beetle.

In this, ecosystem services are the flip side of externalized costs. And yes, someone with a Ph.D. has attempted to quantify their global economic value: about $33 trillion a year with a range of $16 to 54 trillion. (At the time this estimate was calculated—1997—$33 trillion was twice the global gross domestic product.) In other words, other species on the planet are providing major inputs to the global economy. It is worth thinking hard about what this means. If we had to create machines and pay people to carry out the services that other species provide us for free, the total cost would exceed twice the world's GDP. Needless to say, replacing nature with machines and labor would have a devastating impact on us and the planet. And yet, as one recent review put it:

> [B]ecause most of these services are not traded in economic markets, they carry no price tags. There is no exchange value in spite of their high use value that could alert society to changes in their supply or deterioration of underlying ecological systems that generate them.

Admittedly, attaching values to pollination, nitrogen fixation, and soil improvement is an odd exercise. (Soil turnover by earthworms, by the way, priced out at $25 billion/year.) In another time or context, ecosystem services might just be described as *blessings*—which we would then be called upon to praise with prayer rather than monetize. Nevertheless, the concept of ecosystem services offers a way of compelling us to pay attention (as does prayer) to the myriad ways in which the natural world supports human existence. It also explains why, for example, yields in organic fields gradually increase over time.

In essence, organic agriculture is a form of farming that replaces synthetic chemicals with ecosystem services. At the same time, by eliminating chemicals—many of which are toxic to the organisms providing those very same services—organic farmers help shore up ecosystem services for future generations. And herein lies a lucky paradox: by recruiting more ecosystem services and relying more heavily upon them, organic agriculture sets the stage for maintaining them. It sows the seeds of its own preservation.

Ecosystem services provide capital goods. And that is why an organic food system accrues value in the long run. Or, to use language from another time and context: *It abides.*

Here in the moist Northeast, the wheat is mostly soft. It's low in gluten and not valued for baking. With few varieties suited to these growing conditions and no access to markets, wheat production has waned. But the Northeast Organic Wheat Project, a consortium of bakers and farmers, seeks to change all that. In an attempt to bring wheat back into farmers' rotations, it's retrieving seeds from heritage strains on the brink of extinction—including from seedbanks—and field-testing them. The local Farmer Ground Flour I used in my most recent pizza test is part of this ongoing project. While researching their big experiment in preparation for conducting my little one, I ran across an observation about the particular flour in my heritage pizza dough that made me pause: *The taste will change as the soil evolves.*

So, flour was more than just an inert matrix whose job it was to trap air bubbles. It was living thing. Of course, I knew that already, just as I knew that the distinctive flavor of certain wine grapes comes from, say, chalky soil. Yet, the idea that soil imparts the taste of bread was a new thought for me. Something to ponder while kneading. The *terroir* of flour.

In 2009, the wheat yield in New York State averaged 65 bushels per acre. It mostly came from small farms. I checked the U.S. Department of Agriculture stats. The national average was 44.4.

ELLIS HOLLOW PIZZA

Mix 1 cup warm water, a tablespoon sugar, a tablespoon dry yeast, and a tablespoon of flour in a large bowl. Let sit for ten minutes. When bubbles appear, add 3 cups flour (half whole wheat; half all-purpose), ¼ cup olive oil, and a teaspoon of salt. Mix. When the dough becomes stiff, turn it out onto a floured surface. Knead until shiny and smooth. Place back in the bowl. Cover and let rise until doubled in bulk—about an hour. Meanwhile, grate 1 cup of cheese, dice 3 cloves of garlic, chop 2 tomatoes, grease a cookie sheet, and preheat the oven to 475°. Ask a small child to punch the dough down. Roll it out and stretch it to fit in the cookie sheet. (Freeze any extra dough.) Brush with tomato paste. Layer on chopped tomatoes. Sprinkle with garlic and cheese. Bake for 10 to 15 minutes, until cheese bubbles and begins to brown. Serve with green and orange vegetables. Announce that the pizza is too hot to cut, so we might as well eat our vegetables first.

CHAPTER FIVE

The Kitchen Floor
(and National Security)

THAT MOST inconsequential of holidays, Groundhog Day, has decidedly consequential origins. Called *Imbolc* by pagans and *Candlemas* by early Christians, the date that hangs at the midpoint between winter solstice and spring equinox was set aside not just for prognostication but for general inventory. It was a multitasking sort of affair: You watched for seasonal omens while also taking stock of the family's remaining food reserves. *A farmer should on Candlemas Day have half his corn and half his hay.*

On the morning of February 2, 2004, I was trying to figure out how the next twenty-four hours were going to play out, while keeping an eye on various supplies that were running low.

Like sleep. I was running very low on sleep. Jeff had been away from home for most of the past two months, working as an artist in residence at an esteemed studio called Sculpture Space—a much needed and much anticipated opportunity for him—and I was in charge of the household. At Sculpture Space, Jeff had forklifts, cranes, sand blasters, and drill presses at his creative disposal. With only two days of his residency remaining, he was undoubtedly wielding many impressive tools as he began the work of installing his new piece in the exhibition area. Meanwhile, I had my usual research and writing deadlines, which I mostly chipped away on at night while the kids slept, no cement mixer required. I had been managing it all admirably well, so I thought, but the cumulative effect of weeks of single parenting was starting to mount. That morning I was up early after having stayed up late. The forecast was for heavy snow all day. That meant shoveling. It was also Faith's snack week, which meant that the responsibility for five consecutive days of snacks-for-twenty-five-kindergarteners fell to me. But all I had to do was hang on for two more days, and then Jeff would be home again.

And I had a menu plan. Today's snack was trail mix. That was easy enough. I found sacks of raisins, nuts, and sunflower seeds in the cupboard and poured them into a mixing bowl and gave them a stir with a spatula. Done. The snow was coming down harder now, but I still had enough time to take out the dog and shovel out the car before the kids woke up. Done.

By 8 am, both kids were up, dressed, fed, and the breakfast dishes in the sink. I'd located all of our snowsuits, mittens, boots. The car was de-iced. The driveway drivable. And somehow I still had 15 minutes to spare. I glanced over at the kids. Faith was reading on the couch and Elijah was contentedly playing his toy piano. It was too good to be true.

I could take a shower.

It was the excited thought of a starving person who, standing at a bakery counter, realizes the proprietors are otherwise occupied. *I could steal a doughnut.* I considered the possibilities. If I announced my showering intentions, there would likely be an insurrection of some kind. But if I just snuck away, it was possible no one would notice.

Five minutes later I was back in the kitchen, wrapped in a towel. The backs of my knees were not even wet, but I felt, nevertheless, triumphant. It had been a glorious five minutes— although I had observed, while furiously shampooing, that the bathtub drain seemed not to be draining. I'd deal with that later. I peeked into the living room. Faith was still reading. And Elijah was nowhere in sight.

That's when I heard a voice from upstairs say, *Mama! I making trails!* And I noticed the mixing bowl of trail mix was no longer on the kitchen table. And as I walked toward the stairs, I noticed, yes, indeed, there were trails of trail mix that could lead me to my son.

I making trails! I making trails!

And that percussive sound I was hearing, like gravel splattering a windshield—that would be the sound of walnuts and sunflower seeds, catapulted by a spatula, hitting the bedroom window. I had to sit down in the middle of the staircase . . . not out of shock but because the trail mix was so thick on the top steps, I was afraid I would slip. Elijah came running toward me with a spatula in one hand and an empty mixing bowl in the other.

I grabbed his arm to prevent him from spinning out on the nuts under foot, and we sat together, hugging each other and rocking back and forth. He was laughing. I wasn't. But I wasn't crying either. All things considered, I thought I was handling this moment pretty well. Let's see. I now needed a snack for twenty-five kids and had no ingredients—but surely I had a box of graham crackers stockpiled somewhere—and Faith was going to be late, so I needed to call the school secretary, and I would have to take the dog in the car with us because, if home alone, she'd snuffle up all the flung-everywhere raisins and vomit. I would sweep and vacuum when I got back, thus losing an hour of work time but I could stay up again late tonight and still make my deadline.

While running through this mental inventory, I began turning the whole episode into narrative. In fact, even as the events were unfolding, I was shaping the story, revising it in my mind, imagining how I would regale Jeff with it when he came home tomorrow night. This would be my artfully told family tale. It had legendary potential. And in the telling of the story to Jeff, I would finally be able to laugh. But right now, I was tired and had multiple problems to solve, and I was not laughing.

As I carried Elijah through the living room, I paused at the couch and asked the child reading there, *Why didn't you come get me?*

She looked up at me and blinked. *I was reading.*

Perfect. That would be the last line of my story.

But this was not to be my story. In fact, I forgot all about the trails of trail mix until, many months later when I discovered dried cranberries inside my slippers and the memory of that morning came back—and seemed inconsequential. The real story of that day took place many hours later.

It began with a decision to deal with the clogged drain after the kids were asleep. Midnight plumbing is, in fact, a terrible idea. In a different version of this story, I would wisely live with a congested bathtub until Jeff returned. But the way the snow was falling seemed ominous—fast, thick, sideways, and in the form of small, granular flakes that almost always means lots more where that came from. The weather seemed to encourage vigilance. Its message was, *Do not ignore problems; there is enough stuff piling up around you already.* So, I poured two gallons of water into a stockpot and turned on the burner. The plan was to pour boiling water down the drain and see if that helped. While the water was heating on the stove, I initiated bedtime rituals. After two stories and three lullabies, everyone was sleeping.

Including me.

Some time later—how much time? minutes? hours?—I woke with a start. *The water.* I hurried downstairs. In the darkened kitchen, the red coil of the stovetop was still glowing, the pot's contents at a rolling boil. Everything was okay. I hadn't burned the house down. In that lucky, grateful moment, I could have turned off the stove, brushed my teeth, and gone to bed. But I didn't.

Instead, I looked around for potholders. Seeing none, I pulled the sleeves of my sweater over the palms of my hand, removed the lid, grabbed hold of the opposing handles, and hoisted a pot of boiling water.

Two gallons of water weigh 16 pounds. The stockpot proba-bly weighed another four. I felt my grip start to slip as I headed across the kitchen floor. Something was sliding under my socks—trail mix maybe—and I hurried toward the bathroom. More slip-ping. The pot lurched dangerously, the still-simmering water sloshing now, and I realized with calm astonishment (is there such a thing?) that I couldn't hold on long enough to get to the bathtub. So I heaved it—the whole pot—in the direction of the tub and backed away as fast as I could, averting my face.

Here's what could have happened next: The pot could have banged inside the tub, and a roiling, steaming tsunami could have been safely contained. The story I had for Jeff could have been a cautionary tale of my narrow escape from harm. *Can you believe it? I could have been really hurt. What was I thinking?*

But what really happened was this: The pot hit the side of the tub and, absurdly, boomeranged. And for the second time today, something in my house was catapulted through the air.

There are memory lapses from this point on, but I was later able to reconstruct the forensic facts of the situation from the location of my splash burns. Some were on the back of my hands: Water must have sloshed over the edges of the pot before I let go. (And maybe that's what prompted me to throw it.) Some were on my neck and face: I must have turned—like Euridyce, like Lot's doomed wife—to see what was flying toward me. I had only a few blisters on my legs. The truly large and serious burns were on the tops of my feet: The arc of flying water must have been low.

As for my memory of the events that followed, I recall sitting on the kitchen floor, trying to tear my steaming wool socks off. I remember the smell of wet sheep and an inability to make my fingers work. *Too hot. Too hot.* A sentence from a Red Cross lesson floated by: *The longer the exposure to heat, the more serious the burn.* I remember thinking, *This is going be bad.* I remember sitting on the kitchen counter with my feet in the sink, cold water pouring from the tap. And I had the phone—although how I retrieved it from across the room, I don't know. I remember hearing that the roads were closed. I remember discovering that Jeff's cell phone was off. I remember Caryl—the friend and nurse who lived across the swamp—saying she would come. *Just hang on.*

I must have pulled my socks off under the icy water. Something that looked like wet Kleenex was hanging from my left foot. I remember thinking, *Is that my skin?* And I remember pain of many sorts, like different colored ribbons unfurling one after the other. There was a scream balled up in the back of my throat, but I willed it to stay there. I willed my body to stay on the counter. I willed the cold water to keep running. I willed the deaf dog to not bark. I willed my blood plasma to remain in its vessels.

Upstairs, the sound of breathing. I matched it, breath for breath. The inhale. The exhale. I willed my children to stay sleeping. Out in the hiss and swirl of the night, I willed headlights. I willed the sound of a shovel. I willed time to move forward.

And then there were headlights and the sound of a shovel.

Over the next six weeks, I learned a lot about burns. The first lesson was that burn prognostication is a tricky business. Before they get better, burns often get worse. Or they get worse and don't get better.

Much depends on how deep into the dermal layer the damage goes, and even the best burn experts have a hard time assessing depth of injury. The visual appearance of burns can mislead. My plastic surgeon said, "I can't tell yet for sure, but I predict you'll need a skin graft." My hydrotherapy nurse said, "Let's prove him wrong."

And I did. More than two weeks after the accident, the day before my scheduled surgery, the sign that we all watched for, finally appeared: tiny, white islands of epithelial tissue erupted from the red lagoon of the wound bed on the top of my left foot. The emergence of these islands meant I had not entirely boiled away the dermal infrastructure—blood vessels, collagen, lymph ducts—needed to grow my own skin back. The surgery was joyously cancelled.

The second lesson was patience.

Burns are more serious than other wounds of the same size—mammalian skin has not evolved to recover from damage inflicted by flames, steam, or boiling water—and they take a long time to heal. My pearly islands of new skin would require another month to create a continuous land mass. Along the way, I suffered a serious infection—triggering new feelings of appreciation for the miracle of antibiotics—and endured twice-daily dressing changes. Many of these took place at the hospital wound care clinic. In an excruciating process called *debridement*, the burn's oozy slough of protein and dead cells was removed—along with the sulfa-based cream from the previous dressing change. This could happen in one of two ways: by scrubbing the burn with gauze sponges or by blasting it with jets of water in a hydrotherapy tub. Water might have seemed the less agonizing option were it not for the fact that churning room-temperature water looks identical to boiling water. The tub itself resembled a giant KitchenAid mixer, and I learned to close my eyes when the toggle switch on its motor was flipped to *ON*.

Thus, I had plenty of time to ponder the sequence of events that had led me to this place. Like my own father—who, in the 1960s, was one of the first dads in the neighborhood to install shoulder belts in the front seat of his car and enforce seat-belt compliance in the back seat—I placed a high premium on safety. Handling large volumes of boiling water late at night while groggy with accumulated sleep deficit violated my self-image, if nothing else. And yet, my afterhours home repair scheme gone terribly wrong was the direct result of my parental effort to protect my children from harm. Because I didn't want them anywhere around me while transporting a dangerous material—I don't even like having my kids in the kitchen while pasta is cooking or being drained—I had taken on a hazardous job at a hazard-prone time of day.

Children were ever-present as patients in the waiting rooms of the burn and trauma centers where I sought treatment and second opinions. At the local wound care clinic, there was often a child in the whirlpool tub on the other side of a partitioning curtain. From my hydrotherapy nurse, I learned that young children are more likely than adults to get burned and consequences for them are worse. Their skin is thinner. The surface area of a burn covers a bigger percentage of their total body area. They are more likely to contract life-threatening infections. They also need to grow and burn scars cannot. Even when they survive their injuries, their skin eventually becomes a too-tight covering for the lengthening bones and muscles under the surface. Burned children require more surgeries, more therapy.

Whenever I despaired at the latest setback in my own recovery or wondered how many more dressing changes I would have to endure, I made myself stop. *You didn't burn your children. And that's the only thing that matters.*

By March, I was wearing a pressure sock. By April, I was walking in sheepskin boots. By June, I was in street shoes again.

And so my skin regenerated, like a wallpapering job in slow motion. While recuperating, I joined a research project that was investigating an emerging health and safety threat to children for which no parents, no matter how wily or willing to serve as a human shield, can offer protection: the plastic building material called polyvinyl chloride.

It was an unsettling choice of subject matter. The burning of PVC creates dioxin—that was the problem the 9/11 mothers had asked me about two years earlier—and it also generates black, choking smoke. As such, PVC building materials and home furnishings—flooring, wallpaper, sewer pipe, shower curtains—complicate search and rescue missions during house fires. Those were topics I didn't want to think about right now, let alone research. I couldn't even deal with my teakettle. I cringed every time I heard the stove click on. And now I was going to look at data on building fires, incinerators, and explosive vapors.

But I needed paid work, and writing a briefing paper that summarized the environmental health effects of PVC building materials was a job that I could do with my feet elevated. The U.S. Green Building Council had issued a solicitation for data, and the Healthy Building Network, an organization that advocates for health-based green building standards, had responded by submitting many dozens of documents and reports. My assignment was to craft a reader's guide to the technical evidence.

The council's interest in the topic was fueled by a debate within architecture and design circles over the question of whether PVC belongs in buildings that are certified as "green" by the Leadership in Energy and Environmental Design (LEED) rating system. For all its apparent inertness, PVC has a number of menacing qualities, most of which can be attributed to the fact that it is the only common plastic to contain chlorine. PVC is 56 percent chlorine by weight, and it is this ingredient that makes vinyl an environmental wild card. And because the council was interested in the whole life cycle of the product, I needed to understand how it was manufactured, how it was used, and how it behaves after disposal, as in, say, a landfill fire.

The story of PVC turned out to be a familiar one. It is an inherently toxic material that does not keep its toxics to itself. Children are more highly exposed than adults to the chemicals shed from PVC, and they also suffer risks unique to their status as children. But the story was bigger than that. Because the manufacture of PVC *always* involves environmental releases of some of the nastiest chemicals ever synthesized, PVC also threatens child health in the communities where it is fabricated. Moreover, some of PVC's explosively flammable feedstocks have been identified as potential weapons of mass destruction. The Chemical Security Act, introduced by then-Senator Jon Corzine, had identified the untracked production and transport of such chemicals—including the industrial feedstocks for PVC—as threats to homeland security.

At the beginning of PVC's life history is chlorine gas, a wicked poison that turns into hydrochloric acid upon contact with moisture. It kills by burning the airways of those who inhale it. Victims suffocate in their own body fluids; those who survive are often incapacitated for life. This is why, after World War I, chlorine gas was outlawed by international agreement as a chemical weapon. But it's still prevalent as an industrial byproduct. Making PVC was one way to get rid of chlorine gas, which is generated, for example, during the manufacture of caustic soda. Now PVC manufacturing consumes more than 40 percent of the chlorine gas produced in the United States, with most of it going into building materials.

In this, the story of PVC reminded me of the story of pressured-treated (CCA) wood: a common building material containing large amounts of a potent toxicant, originally invented to solve a particular industrial problem, but subsequently promoted for all kinds of household purposes. In the case of PVC, its ascendency after World War II had routed other less toxic substances from the marketplace. One of them was linoleum, made of linseed oil and jute. The last U.S. linoleum floor maker closed its doors in 1975. Vinyl flooring took its place. Genuine linoleum is now an imported product.

Like CCA wood, PVC enjoys close contact with children. Kitchen floors. Raincoats. Lunch boxes. Shower curtains. Art supplies. Backpacks. Wallpaper. Toys. Shoes. And, like CCA wood, PVC sheds its ingredients onto and into the bodies of children. For CCA, the dislodgable transgressor is arsenic itself. For PVC plastic, it's DEHP.

DEHP is the oily phthalate plasticizer that's used to make vinyl flexible and prevent it from cracking. As we saw in Chapter 1, phthalates are a footloose group of synthetic chemicals that easily migrate into indoor air, house dust, and food. Virtually all Americans have phthalates in their blood, according to the results of Centers for Disease Control testing in a representative cross-section of the citizenry, but children have higher levels than adults. Airborne exposure is linked to earlier birth in pregnant women. Phthalates are hormone disruptors and, as such, pose unique risks to children. They are associated with impaired genital development in boys, for example, and earlier puberty in girls. Studies in children show associations between phthalate exposure in the home and the risk of asthma and allergies. (More on these effects in Chapters 6 and 9.) A vinyl floor contains an impressive load of phthalate plasticizers—between 4 and 20 percent by weight. PVC flooring—especially when it is damp and begins to degrade—represents a special respiratory hazard to children. It's linked in pediatric studies to respiratory distress and wheezing.

And, as with arsenic-treated wood, there is no good way to get rid of vinyl. CCA wood is lethal when burned, a threat to groundwater when landfilled, and poisonous to the recycling stream. Ditto PVC.

In the way that it consumes vast quantities of natural gas, however, PVC more closely parallels the story of nitrogen fertilizer. So let's go back. The first step in the making of PVC is to combine chlorine with ethylene to create an oily liquid called ethylene dichloride. Traditionally, in the United States, the ethylene in ethylene dichloride comes from natural gas. (China makes its PVC from coal.) A probable human carcinogen, ethylene dichloride has a nasty habit, when spilled, of heading straight for groundwater (as does synthetic fertilizer). According to the National Toxicology Program, at least four Americans out of every hundred drink water that contains traces of ethylene dichloride.

The next steps along the PVC assembly line took me into frightening territory—into the world of chemicals prone to detonation. Here is the place where risks for harm include not just cancer and asthma but a long convalescence in a burn and trauma center. Here is the point where environmental health meets the war on terror, where a tankerful of feedstock is a potential weapon of mass destruction. Within this landscape lies the shrouded province of chemical security, where databases formerly available to researchers like me—about certain kinds of chemical hazards, for example—have been quietly pulled from the Web in the name of homeland security.

Ethylene dichloride is the raw material for vinyl chloride, a potent human carcinogen linked to liver, blood, and brain cancers. Vinyl chloride is explosively flammable. A sweet-smelling vapor that liquefies under high pressure, vinyl chloride is transported via train car to PVC plants. Once there, the small vinyl chloride molecules are bonded together in big vats to form long chains of polyvinyl chloride. In so doing, the feedstock turns into a solid polymer called *resin*. All by itself, PVC resin is not very useful. It does, however, readily combine with other polymers, like vinyl acetate. Vinyl chloride and vinyl acetate together make a copolymer that works well for making floors. Vinyl acetate is also a carcinogen, and it is also explosive. Workers who handle it wear special nail-less shoes to avoid striking sparks.

Even absent explosions and upsets, PVC plants are leaky places. As part of my work, I needed to understand the public health threats created when chemicals from PVC plants drift to the far side of the fence line. Fortunately, I had available to me the research findings of Wilma Subra, a MacArthur Award–winning chemist, who had studied the air quality in the neighborhoods surrounding various PVC plants in Kentucky and Louisiana. Subra had documented consistent patterns of dioxin and vinyl chloride exposure to area residents—including children. In February 2004, Subra had presented her data at the first public meeting of the U.S. Green Building Council's vinyl task force in Washington, D.C. She had been warmly received, and the council's members asked thoughtful questions. After Subra's presentation, however, more than twenty different vinyl lobbyists spoke on the virtues of PVC.

According to the Toxics Release Inventory—the self-reported list of chemical emissions that manufacturers submit each year to the EPA—the nation's twenty-one PVC facilities were among its biggest polluters. While looking at this spreadsheet, a name jumped out at me: Formosa Plastics in Illiopolis, Illinois. I knew exactly where the funny-sounding community of Illiopolis was located. A village of about nine hundred souls, Illiopolis sits in the rural center of the state about an hour south of my hometown. You get to Illiopolis by driving down old Route 121, which, when I was a teenager, was the road favored by motorcyclists and the envied owners of Pontiac Firebirds. But I had no idea that Illiopolis was home to a PVC plant.

I dug a little deeper. Formosa Plastics Corporation, U.S.A., a subsidiary of Taiwan-based Formosa Plastics Group ("the Prince of PVC"), had purchased the Illiopolis-based Borden Chemical a few years earlier. Borden was a name I recognized. When I was a kid, Borden made Elmer's Glue. We were all very proud of that fact and had imagined railroad tankers full of Elmer's, emblazoned with its Elsie the Cow mascot, heading out from Illiopolis to schoolrooms across America. Now those tankers apparently carried vinyl chloride and vinyl acetate.

Surrounded by corn and soybean fields, Formosa's Illiopolis plant ranked within the ninetieth percentile in the nation for air releases of carcinogens in 2001. This distinction was entirely attributable to the 41,000 pounds of vinyl chloride that it released that year. On top of this, Formosa released 40,000 pounds of vinyl acetate. In 2002, Formosa's vinyl chloride emissions fell to 31,000 pounds, while its vinyl acetate emissions rose to 45,000 pounds. In other words, for two years running, this facility had released into the air *each day* about 200 pounds of cancer-causing chemicals.

Here's the spooky part of the story. Three weeks after I finished my briefing paper, at 10:40 p.m. on April 23, 2004, Formosa Plastics' PVC plant in Illiopolis, Illinois, blew up.

A photograph of the explosion appeared in *The Ithaca Journal.* It was the first time in my memory that photographs from that part of the world appeared in the local paper here. Indeed, the disaster made headlines around the globe, and it was spectacular by all accounts. The blast killed four workers outright—a fifth would die in a burn unit twenty days later—and sent into the night sky a hundred-foot fireball. This ball, once it faded from view, left behind a dark, hovering mass that drifted slowly over the landscape like some kind of evil UFO. Four towns were evacuated, several highways closed, a no-fly zone declared, and 300 firefighters from twenty-seven surrounding communities battled the flames for three days. Their efforts were complicated by the power outage—triggered by the shock wave—that disabled the water supply. Formosa Plastics ran the wells that provided the town's water.

The U.S. Chemical Safety and Hazard Investigation Board (CSB) was summoned to determine the cause of the explosion. Its chairwoman, Carolyn Merritt, estimated that it would take a year to figure out why the plant exploded. Terrorism was ruled out. But beyond that, only two facts were known for sure. One: A large amount of vinyl chloride, the vaporous feedstock of polyvinyl chloride, had been immediately released into the air before the whole facility blew skyward. Two: The explosion originated in the reactors where vinyl chloride and vinyl acetate were being mixed together. The resulting inferno destroyed the reactors and the adjacent warehouse and ignited the PVC resin stored in it.

In the end, it took the board three years to determine the cause. In their 2007 report, CSB investigators deduced that the incident occurred when a night-shift worker who was cleaning an empty reactor became confused and mistakenly opened the bottom valve on an operating reactor, releasing its explosive contents. The report noted that two similar mistakes had occurred in the recent past—albeit without catastrophic consequences—and yet Formosa had not implemented corrective actions. The report concluded that Formosa had not adequately addressed "the potential for human error." As someone who, with catastrophic consequences, once threw a pot of boiling water at my bathtub in the middle of the night, I found this explanation credible.

By early May 2004, Illiopolis had vanished from the newspapers. I talked to a pediatrician friend who practiced in the area to see what she knew. Gail had been vacationing when the explosion happened. "I read about it in the newspapers in Maui," she said. "And now I get back home, and it's like it never happened."

On the other hand, Illiopolis was still making headlines in the trade magazines. In May, the Web site ebuild.com disclosed the contents of a communiqué from Armstrong Industries to the U.S. Securities and Exchange Commission. In it, Armstrong warned that the explosion might disrupt the manufacture of its vinyl floors and thereby impact financial results. Meanwhile, as reported in *Floor Covering Weekly*, the three kingpins of flooring—Armstrong, Congoleum, and Mannington—all announced price increases in vinyl composition tile due to the explosion-induced PVC shortfall.

It turned out that, at 200 million pounds per year, the Illiopolis plant was a major provider to these three customers of a particular kind of PVC resin needed to make vinyl floors. Who knew? (I certainly had not been able to ferret out that kind of information for my briefing paper.) Chemical chains of custody are carefully guarded industry secrets. Who sells what chemical to whom is not a query public affairs officers are normally happy to answer. In order to expose the connections, it takes a disaster of some magnitude to break them.

I went to Illiopolis in June 2004. By then, I was in regular foot-wear, although I had to be careful about blisters. I drove a rental car from my mother's house down what is now called I-155 and then took a series of blacktops through thigh-high corn until I crossed the railroad track. I pulled off onto a gravel service road, got out of the car, and there it was. Twisted sheets of metal. Knotted piles of pipe. A blackened hull. A gaping hole. And rubble. Rubble like I had not seen since I had stood at Ground Zero. Which is exactly where I felt I was.

Everyone who made a pilgrimage to the World Trade Center site soon after 9/11 will tell you that you can't appreciate the sheer enormity of the devastation unless you see it in person. And I am saying that again now.

The Formosa Plastics Plant sits on a 20,000-acre site. Strangely the storage silos were still standing. The hopper cars were still on their tracks. A field of corn rustled in the wind. I checked to see which way its leaves were blowing to make sure I was upwind from the ruins.

The last worker to die from injuries sustained during the explosion was Randy Hancock, age fifty. Randy had clung to life in the Springfield Memorial burn unit for nearly three weeks until he heard the news that his wife, Linda, also a worker at the plant, was already dead. According to family members, he had, in his last days, repeatedly used his right hand to gesture toward his left, pointing to the place where a wedding ring would be. From this, they inferred he wanted news of Linda. When finally told that she was gone, Randy slipped into a coma, and then he was gone too.

On the Illinois prairie, the prevailing winds blow from west to east. Two miles east of Formosa sat the village of Illiopolis, where tiny bungalows shared residential streets with formerly grand Victorians. At the very west edge of town sprawled Illiopolis High School—"Home of the Pirates"—along with the middle school, the elementary school, and the prekindergarten. All that stood between the Pirates' playing fields and Formosa was corn.

I drove past the school on my way to talk with Beverley Scobell, who lived behind the spire of the Catholic church in a Victorian four-square. Beverley was an editor at *Illinois Issues*. We knew each other's writing. She and her husband, Hank, had served as volunteer firefighters in Illiopolis for more than two decades, although they were no longer active members the night Formosa blew up. They originally joined the firehouse, Bev said, for the social life. The firefighters' parties were the best in town.

Hank and Bev had gone to bed early the night of the accident. Hank was asleep and Bev was watching television when the power went out. Seconds later she heard the boom. Hank went right to the firehouse. Bev went to the home of a disabled elderly neighbor, Connie, and helped her get dressed by flashlight, presuming that an evacuation order was imminent. In fact, there was already such an order, but few residents of Illiopolis knew about it. Because of the blackout, the emergency siren failed to go off.

When Hank came back from the firehouse, he reported that the wind, miraculously, was blowing to the south-southwest—away from the town. Nevertheless, and against a current of evacuees driving east, Hank, Bev, and Connie all got in the car and headed west, toward the cloud. The plan was to deliver Connie to her nephew's house in Mechanicsburg, but by the time they got there, that community was also under evacuation. So they doubled back. In the end, Connie spent the night in a sealed room at a firehouse in a nearby village. Bev and Hank went on to Springfield.

"We never believed or bothered to know what went on out there," said Bev about Formosa Plastics. "It was just a fact of life. We've lived with it for so long."

Bev introduced me to Rayeanna Stacey, Illiopolis's coordinator for emergency services. A certified emergency medical technician, Rayeanna drove the van that runs with the fire trucks. She was also the village clerk and the school bus driver. Her husband, Dennis, was the superintendent of public works and first captain at the fire department.

Rayeanna was testy about the issue of the siren never going off. She had been complaining for years to Formosa about the need for an emergency siren that worked during a power outage. For that matter, Rayeanna added, she had also objected repeatedly—and to no avail—to the many vinyl chloride tanker cars that sometimes spent the night parked on the tracks right in the center of town. All it would take is a couple of teenage pranksters and nobody in Illiopolis would wake up in the morning.

The night Formosa blew up, Rayeanna and Dennis were at their cabin an hour south of Illiopolis. Alerted by pager, they made the trip home in much less time than that. Dennis joined the firefighters at the scene of the disaster. Rayeanna went to Christine's Diner, which had been turned into a command center for emergency responders. Before the weekend was over, she would transport four dead bodies in her van. Rayeanna emphasized what a miracle it was that the initial blast did not set off a chain reaction of even larger explosions. Because some of the first responders were employees of the plant, they were familiar with the layout and could direct others to critical locations. In her professional opinion, if all the chemicals in and around the plant had detonated, there would be nothing living within a five-mile radius.

Rayeanna said, "We never trained for something like this." And then she cried.

Before I entered the Springfield Memorial burn unit, I had to pass through a windowless room containing a sink with a foot-operated tap and a list of instructions. Thorough disinfecting of hands was a requirement for entry. Anyone with a communicable disease was ordered to leave. I was on familiar ground here. I knew the drill.

After washing up, I stepped inside. Directly in front of me was a glass wall with a bed on the other side. In the bed lay what appeared to be a gray, granite sculpture in the shape of a reclining man. A nurse approached and asked if she could help. I said I wished to see Bradford Bradshaw. The nurse said that decision would be up to Donna, his companion, who was in the cafeteria eating lunch. Brad had just come through surgery and was resting. Then the nurse looked through the same glass pane where I had looked, and I realized that the sculpture on the bed was in fact the man I had come to see: Bradford Bradshaw, age forty-seven, a longtime Formosa employee.

When Donna returned, she said Brad had just learned that he had permanently lost the vision in his right eye. He was burned over 60 percent of his body. He had not yet been told about Randy Hancock's death. But if I wanted to talk to Brad, she would ask him. I said that I could see that my timing was off. *Thanks, though. And give him my best.*

I sat out in the parking lot for a long time. Bradford Bradshaw was the last blast survivor still hospitalized, and he was my secret reason for coming to Illiopolis. I had nurtured the unlikely idea that we would meet and swap stories—inspired, I guess, by my memory of the wound care clinic where all of us burn patients bonded and felt ourselves a tribe apart from the others (the diabetics with their unhealable sores). But I was way out of my depth here. Bradford wouldn't view me as a member of his club—nor should he. My research on PVC, which I had imagined telling him about—why would he even care? And whatever he might know about how much PVC resin was on site the night of the explosion—which was the key to figuring out how much dioxin could have been released during the three-day inferno—well, I'd have to ask somebody else.

On June 16, the Illinois EPA hosted a "public availability session" in the high school cafeteria. The flyer I had picked up in the local library had indicated that the session would strictly focus on the health and environmental concerns arising from the explosion at the Formosa Plastics Plant and would offer no comment on the ongoing investigation regarding its cause.

I found this don't-even-ask clause chilling. More off-putting to the Illiopolians I spoke with was the "availability" concept. Shortly after the blast in April, the village had held a bona fide public meeting that drew five hundred people. It was a worthwhile event, said those who attended. The various investigators made speeches and answered questions from the audience. By contrast, officials at the upcoming venue would be stationed at tables, and individuals would have to approach them one on one. If you wanted to go and listen, you wouldn't learn much.

When I pulled into the high school parking lot, it was filled with overcoifed reporters standing in front of television cameras. Once inside, I realized why the interviews were taking place outside. Without air conditioning, the cafeteria was besieged with swarms of gnats. Nevertheless, I thought the press was missing a great photo opportunity. Formosa's table was located by the "Home of the Pirates" banner that displayed the school's mascot: a one-eyed pirate in skull-and-crossbones regalia.

I approached the Formosa table first. It was staffed by Roe Vadas, the local plant manager; Peter Gray, an environmental engineer; and Rob Thibault, the manager of corporate communications. I wanted to know what chemicals were on-site when the explosion happened. I asked the question a variety of ways— What about those storage silos? Are they full or empty right now?—and received different versions of the same response. *That's business information. We're not going to go into that. That's proprietary. Some are full. And some are empty.*

When I asked if anyone in Illiopolis knows what kind of chemicals are stored and used at Formosa—like, say, Rayeanna Stacey, emergency response coordinator—the answers got even murkier. I was told there was something called a "risk management plan" that had been written and submitted by Formosa, but it wasn't a document that I could have. Of course emergency response had a copy.

The Illinois EPA (IEPA) table was across the room, staffed by, among others, Joe Dombrowski, the designated project manager of the ongoing environmental investigation. Yes, a written list of chemicals stored on site at Formosa had been supplied to the

IEPA. No, I couldn't have a copy. I could file a Freedom of Information Act request . . . but it would probably be denied.

Behind the table was a map of the county over which was superimposed a series of concentric rings and arrows, with the site of the Formosa plant as the bull's eye.

This was truly interesting. It depicted wind direction and speed during the first seventy-two hours after the explosion. Essentially, during the three days the fire burned, the wind made a complete 360-degree turn. It also rained during this time. When I asked if copies of this map were available to the public or if it was posted on a Web site somewhere, the answers were no and no. When I asked if the smoke plume itself had been mapped, I was told it wasn't done that night and couldn't be done now.

The second half of that sentence was simply untrue. There are several good computer models for mapping wind-dispersion patterns of toxic materials. That's how radioactive fallout is mapped. Dioxin dispersal can be mapped the same way. I'd seen such maps myself.

On the far side of the room were the tables for the Illinois Department of Public Health. From its representatives, I learned that no one knew whether cancer rates in this community were higher than normal over the years since the PVC plant had been operating and that there were no immediate plans to conduct such a study. The department would be glad to run the numbers, I was told, if someone in the community asked for that analysis to be conducted. So far, no one had.

The public availability session was starting to feel like the Q and A after a college lecture—without the lecture. To learn something at this event you had to come up with exactly the right question, a task I was failing.

I looked over at the Formosa table and decided to give it one more try. *Let me make sure I have this straight. If someone in Illiopolis came up to you and asked what chemicals you store out at your plant, your answer would be what exactly?*

I received another harangue about proprietary information and corporate competitiveness. I was told I should learn something about how vinyl is made. Read a chemical engineering book. Go to vinyl.org. At some point, I quit listening because the glowering scowl on the Formosa plant manager's face came to resemble so closely the expression on the face of the pirate mascot above our heads that I got distracted. Formosa. Pirate. Formosan pirates. I dimly recalled that there was such a thing. Their ships had terrorized the coasts of Taiwan during the Dutch occupation.

Outside in the parking lot, I ran into Rayeanna Stacey. I asked if Formosa had ever provided her a chemical inventory. She said no.

The truly startling results were released weeks after the public availability session. Like this one: "A vessel containing radioactive material was located in a building adjacent to the explosion site. Following the explosion, the vessel was checked and secured by the plant safety officer." Meanwhile, the dioxin test results showed higher than background levels of dioxin in the soil collected from twelve of thirteen sites located on and around Formosa's property. The highest on-site sample had dioxin concentrations of 65.8 parts per million, which is ten to sixty times greater than average background levels—but below the actionable level of 100. A sample of soil collected below someone's downspout revealed a dioxin level of 125.8—but subsequent testing of a composite sample collected from various points in the yard revealed lower levels, and the homeowner was advised not to be concerned.

Formosa spokesperson Rob Thibault was quick to say that Formosa may not be the source of the dioxin "given the history of the site." And, about this, he was certainly correct. The Formosa Plastics Plant was originally built in 1942 as a federal munitions factory, and munitions manufacturing can also generate dioxin. Between 1942 and 1945, when Illinois was the nation's leading producer of ammunition, this plant turned out fuses, propellants, boosters, and ninety-millimeter rounds. But as the Sangamon Ordnance Plant, it had never blown up. For a hundred-foot fireball, you apparently needed to be making PVC.

On July 13, 2004, Formosa began loading its unexploded stock-piles of vinyl chloride and vinyl acetate into two dozen railroad cars for transfer to its other plants. The image of ninety-ton tanker cars full of liquid explosives rattling across the plains—to where? Texas? Louisiana? Delaware? New Jersey?—was deeply unsettling. It was a reminder that the low whistle of a freight train in the night is no longer just the sound of lumber, coal, and grain on the move. Although originally laid to carry such goods, the train tracks that roll through the backyards of count-less small towns and big cities are now used to transport all manner of dangerous chemicals.

Each year in the United States, 1.7 million carloads of hazardous materials travel the rails. These include the highly explosive feedstocks for America's floors. According to the Argonne National Laboratory, vinyl chloride ranked eighth among the top 150 hazardous materials most heavily shipped by rail. Vinyl acetate was twenty-fifth. (It's an eye-opening list; other items include molten sulfur, bombs, and "warheads, rocket.")

In June 2010, here in upstate New York, a train derailed in downtown Binghamton. The cars involved remained upright. Good thing, because they contained chlorine and anhydrous ammonia—two gases that are lethal on inhalation.

Clacking tanker cars. A nostalgic train whistle. Predawn darkness. Now insert into this picture a terrorist armed with materials of the kind once manufactured by the Illiopolis plant during its incarnation as a bomb factory.

It's hard to write these words—as if the very act of describing horrific possibilities has the power to make them come true. Faith and Elijah strongly believe that this is so. If they knew what I was saying here, they would cover my mouth with their hands and whisper, *Don't speak it, Mama.* But I have stood at Ground Zero, and I can imagine a hijacked tanker train and a major metropolitan area. I can imagine a truck, a suicide bomber, and a vat of vinyl chloride. I can imagine a PVC plant as a weapon of mass destruction.

Such scenarios have been the motivation behind various chemical security bills that would compel chemical plants to use safer alternatives to inherently dangerous technologies. In 2006, in a horror movie of a report entitled *Terrorism and the Chemical Infrastructure*, the National Academy of Sciences endorsed this approach, noting that chemical plants would be much less attractive to terrorists were they not reliant on large stockpiles of explosive, lethal chemicals. "The most desirable solution to pre-venting chemical releases is to reduce or eliminate the hazard when possible, not to control it." But, so far, after intense lobby-ing by the chemical industry, all such bills requiring a switch to safer chemicals have floundered. In July 2010, the opening para-graph of an article in *Homeland Security Newswire* began like this:

> The U.S. chemical industry breathes a sigh of relief: A Senate panel votes unanimously to extend current chemical facilities security law . . . . The industry worried about modifications . . . that would make the measure more stringent—for example, by requiring the chemical plants to replace the more toxic and volatile chemicals they used with inherently safer technologies.

So while the chemical companies are relievedly sighing, what are concerned parents doing? Mostly consulting Web sites. To avoid PVC toys, you can check the Ecology Center's toxic toy database. To find PVC-free school supplies, download the list at the Center for Health and Environmental Justice Web site. For PVC-free building materials, there's Healthy Building Network. For the sans-PVC baby nursery, there's . . . . And so on. Every purchase requires research. Every goody bag brought home from every birthday party requires vetting.

I have five complaints about this individualized approach to a public health menace. First of all, I don't have time for this. Nobody does. And parents who think they do use up their quota of environmental energy on vinyl avoidance and have no attention span left for other crucial matters of parenting, like the need our children have for clean water (Chapter 10) or a stable climate (coming up in the next chapter).

Second, playing the role of the PVC police makes me an enemy of my children. Example: For his birthday, Elijah received from a family friend a Curious George raincoat. It was vinyl. As he joyously pulled the coat from the plastic packaging, the smell of phthalates was overpowering. Even he noticed it was weirdly smelly and made him cough. But he *loved* Curious George, and he *loved* the raincoat, and he wanted to wear it in the bathtub and in his bed, but I took it away and never gave it back. I was distinctly unpopular for awhile after that.

I could have handled the confiscation more cleverly, but sometimes I am too tired to be clever. I would like my nation's chemical policy to recognize that I should not be the only one standing between my son and a toxicant with demonstrable links to testicular abnormalities (Chapter 9). Especially since, when I'm the one playing the role of regulatory agency, I'm forced to take actions that solve no real problems—like land-filling Curious George so he can leach his phthalates into some-one else's drinking water. (What? I was supposed to donate a toxic raincoat to Goodwill?) Needless to say, landfilling vinyl is also a complete waste of its nonrenewable starting point, natural gas. (Or in this case, because the coat was made in China, coal.)

Third, as a matter of principle, toxicity should not be a consumer choice. Believing that we can buy safety for our children with money and knowledge leaves those with neither in harm's way.

Fourth, as a renter, there was nothing I could do about my PVC water pipes and kitchen floor.

Fifth, as long as there are railroads and industrial dependencies on acutely toxic and violently flammable substances, we all remain, no matter how personally vigilant, a nation of Illiopolians, living just downwind of potential catastrophe.

Back in Ithaca, home from my mom's house, Elijah dragged his box of toy trains and wooden tracks into the kitchen and excitedly described a plan to construct a great railroad. His engines pulled little tankers labeled FLOUR, OIL, and MOLASSES. Derailments were frequent. Around him I swept the floor. Here was the spot where I had slipped on my way to the bathroom with a cauldron of steaming water. And here was the spot where, two years earlier, I had dropped my ash-covered book bag and floured the tiles with Ground Zero dust. This floor was developing its own biography.

Now I wondered if Bradford Bradshaw—or any of his five dead co-workers—might have had a hand in stringing together the molecules that make up this floor. It was more than a remote possibility. Prior to April 23, 2004, the Formosa plant in Illiopolis made fully *half* of all the flooring-grade vinyl in the United States. Now when I looked at these floral-patterned tiles I saw tanker cars full of explosive vapors rattling through towns while people slept. I heard the silence of emergency sirens that failed to go off. I imagined the hushed urgency of evacuees taking to the roads. I thought of Connie, elderly, disabled, in a sealed room at a firehouse shelter while a mushroom cloud of toxicants passed overhead.

In 1961, when I was exactly Elijah's age, diplomatic relations between the United States and the Soviet Union suddenly soured, and the nation was gripped in a frenzy of fallout shelter construction. A basement or backyard bomb shelter, stocked with canned goods and a first-aid kit, was promoted—by both government and media—as a sensible response to the nuclear endgame that was possibly coming. *Protect Your Family. Build a Home Shelter*, the Federal Civil Defense Administration exhorted. My father did not build one, but our neighbors, the Mattheessens, did, and although I was too young to remember its construction, I do remember that Karen Mattheessen and I liked to play in it. We hosted tea parties there—running quickly through the basement to escape the rays of pretend radiation.

In hindsight, the strange, temporary faith in the family fallout shelter, as a pod of protection against unimaginable devastation, was clearly delusional. As we now know, in the event of all-out atomic war, a backyard shelter is useless. But, more than that, as sociologist Andrew Szasz has observed, the shelters' false promise of individual security may actually have increased the possibility of nuclear hostilities—both because their soothing presence made war seem less unthinkable and more inevitable and because their construction distracted people from the collective task of preventive action. Peace activists, anti-nuclear scientists, and religious leaders argued as much at the time. *Stop digging. Defuse the crisis.* Eventually, their antishelter message won the day.

In the proliferating Web sites that offer tips for how to keep our children safe in an increasingly toxic world, I see signs of fallout shelter folly. At the very least, it seems to me, the 1961 shelter-building fad offers 2011 parents some timely questions. Is the goal to craft for our own children a personal refuge from dangerous chemicals like PVC? Or is this both an illusion and a distraction from some larger engagement? Why are vinyl-free lunch boxes worthy of articles in parenting magazines but the ongoing attempts of the petrochemical industry to undermine chemical security measures not?

Are we desperately seeking blueprints for a shelter against environmental catastrophe? Or are we attempting disarmament?

CHAPTER SIX
Asthma
(and Intergenerational Equity)

Wε MOVED. From a log cabin by a beaver lodge to a Victorian house by a fire station. From a frog-filled hollow to a kid-filled village. From cords of wood and a wellhead to a water main and a sewer bill. From a sleeping loft and a back deck to three little bedrooms and a front porch. From a lease to a mortgage.

The only thing that was staying the same was floor space. The new house, like the cabin, was just over a thousand square feet. A cottage, really. And a not entirely charming one. There were broken windows, Edgar Allen Poe–like shrubbery, and dry rot. But Jeff had construction skills. Top on his to-do list was to build, in the branches of the sturdiest of the backyard walnuts, a tree house commodious enough to serve as the rec room that didn't come with the main house—and private enough to be a hide-out for those with sibling complaints.

Of the two of us, I was the one less filled with elation. Indeed, I'd walked into the real estate closing ceremony preceded by the same sense of teeth-gritting doom that I'd felt while walking to my Ph.D. oral exams. Giving up my life-long identity as a rakish renter felt like a final concession to parent-hood. Family life was no longer going to be one big, extended camping trip. But even I could see that, sooner or later, bedrooms would be useful. And, frankly, the middle-of-the-blizzard burn accident had made physical isolation less alluring.

There was one other insight that kept me turning the pages of all the closing documents and signing my name on each one: Life in the woods with two children—one going on six and the other turning three—had increasingly meant life in a car. The ditched, shoulderless road that connected us to the rest of the world not only murdered newts and toads, it carried trucks and a 50 mph speed limit. Nobody could tricycle along it, and trips to obtain groceries, Band-Aids, and cash involved car-seat buckles, books on tape, and drive-through windows. I could see where this was going. Every library visit, music lesson, play date, soccer game . . . would involve driving. Living with the foxes and the beavers would increasingly mean operating a machine whose emissions undermined the conditions that foxes and beavers needed to live. The contradiction was hard to justify. So we chose a small home in a walkable village with a bus line that connected us to Ithaca where Jeff would be teaching art (at the school our kids would attend).

The advantages of our new habitat were soon apparent. A sidewalk stretched from our door out to a craggy maple tree and then connected with another sidewalk that headed down the block toward Main Street. Here was a track, upon which the wheels of a stroller could roll, that linked me to coffee, library books, postage stamps, hardware displays, bank tellers, and a bus stop. The fields of our new CSA farm lay within biking distance. So did a state park with a swimming beach and hiking trails. Three blocks away, a music conservatory operated out of an old church; it offered piano lessons and dance classes. The built environment and open spaces surrounding our new house seemed to provide the best elements of both rural and urban life.

I sat on the front porch (spongy floorboards) and grinned. To be sure, the village sidewalks—century-old slabs of stone—were neither plumb nor true, but this was evidence that they had outlasted a generation of street trees whose roots must have lifted them and then, in dying, set them down uncrumbled but askew. Looking at the misalignments, I tried to guess where trees had stood in 1840. From a geologist neighbor, I learned that our sidewalks are a form of shale—the mother of slate—created from marine sediments. That's when I noticed the marks of a vanished ocean on the walks' rippled surfaces.

The kids were not interested in geology lessons. They seemed bewildered. Faith especially was lonesome for the cabin and for a particular tree behind it—a cherry—that she considered a special friend. Elijah, too, seemed subdued. He was also coughing. In fact, he had been coughing for a couple weeks now—even before the move. This was not unusual. Whenever respiratory infections blew through our household, it was Elijah who suffered the worst. Long after the rest of us had recovered, he would slog along—colds morphing into bronchitis—and he was sometimes still coughing when the next round of pathogens swept in and the cycle started anew. But this time, he wasn't sick and seemed to have no other presenting symptoms. Just the cough.

I touched his forehead, and he looked up at me. *My chest hurts.*

Lungs are two vineyards separated by a heart. Inside their lobes, hanging like clusters of grapes from the ends of delicately branching airways, are the lung's alveoli. Here is where the Earth's gaseous layer and the inside of our body meet.

It is not a small space. (The combined surface area of all 300 million air-filled alveoli—the estimated average for a pair of healthy, adult lungs—exceeds the total area of our skin by a factor of twenty-five. That's about the size of a tennis court.) But it is an intimate space. The diameter of a human hair is about 70 microns. The width of the alveolar membrane is . . . one micron. On one side of the micron, atmosphere. On the other, blood. The vanishing thinness of the alveolar boundary is what makes breathing our most ecological act.

The exchange between organism and environment that goes on across that border—within the ribs' bony cage, behind layers of recoiling elastic fibers, inside the spongy lobules of the lungs—is at once highly protected and terribly exposed. As my old human anatomy textbook notes, respiration provides us with "methods of expressing emotions, such as laughing, sighing, and sobbing." At the same time, chemical pollutants have greater access to us at the respiratory junction than anywhere else. The substances we inhale into our permeable alveoli—that subsequently enter us—are a function of our energy policy, our systems of transportation, the Clean Air Act, tobacco regulation, trends in home decorating, dry cleaning practices, pest control methods, and laws governing workplace exposures. Big policy issues swirl around the wellspring of laughter, sighs, and sobs.

Because the boundary between environment and organism is an air-water interface, keeping the alveoli inflated is a tricky business. Wet surfaces want to cling together—like pages of a book left out in the rain. To counteract this tendency, special cells coat the alveoli with surfactant. This slippery fluid prevents lungs from sticking together and collapsing during exhalation and allows carbon dioxide to exit the blood plasma through the alveolar membrane and flow into the atmosphere. Moving in the opposite direction, during inhalation, is oxygen, which jumps the alveolar micron, enters the bloodstream, and is quickly embraced by red blood cells. (Less soluble in water than carbon dioxide, oxygen relies on hemoglobin to ferry it around.) Meanwhile, specialized lung cells with janitorial duties wander through the alveoli and sweep up any inhaled debris. They are called, satisfyingly enough, dust cells.

Lungs enjoy a leisurely development. Unlike the heart, they are not needed at all during prenatal life, when the placenta serves as the site of gas exchange, so they take an unhurried approach to getting their anatomical affairs in order. Lung buds sprout during the fourth week of pregnancy. As the tubular airways begin to grow out in all directions, the lungs resemble two jacks in a child's game. The alveoli do not form until late pregnancy, and the ability to make surfactant comes last of all. At birth, the alveoli must suddenly make the transition from water to air and, like a sky full of parachutes, undergo rapid inflation.

The just-in-time production of surfactant allows for this. This is not to say the fetal lungs are collapsed during the aquatic life of pregnancy. They are, as we now know, fully expanded—but full of fetal lung fluid rather than air.

Compared to those who arrive via Caesarean section, babies born vaginally have an easier time of it. The stress of the birth process itself stimulates their adrenal glands to produce gluco-corticoids, which, in turn, enhances surfactant production. Stress is not always bad. (As mentioned in Chapter 2, glucocorticoids also play a role in glucose metabolism and cancer prevention. They are a steroidal family of multitaskers.)

Premature babies are the ones caught in the roughest spot. They have far fewer alveoli at the ready, and, if born before 32 weeks, they lack surfactant to inflate the ones they do have. Injections of glucocorticoids can hasten the maturation of lung tissue in preemies, but the speed comes at a cost: Their lungs will ultimately contain fewer alveoli. Thus, with or without steroid treatments, preterm birth is a severe disruptor of lung development. If born too soon, you grow a different pair of lungs than you would if you weren't. These differences persist into adulthood and increase the risk for chronic lung diseases at all stages of life. And one of these diseases is asthma.

Elijah's delicacy went beyond his lungs. He was prone to eczema, and he'd had a couple of mild but noticeable reactions to routine vaccinations. I'd convinced our family medical prac-tice to spread out the inoculation schedule and administer his boosters one at a time rather than in clusters.

And then, out of the blue, while accompanying his dad on an afternoon of errands, he suffered an inexplicable bout with hives and anaphylactic shock. Thanks to a quick-thinking medical receptionist, who sent Jeff straight to the emergency room after hearing his report ("he's covered in spots and com-plaining that his hands and feet hurt"), we'd averted disaster. By the time I joined Jeff in the ER, I didn't even recognize my own son. His eyes were puffed shut, and his hands belonged to Mickey Mouse.

And so we entered the world of allergists and EpiPens—spring-loaded syringes full of epinephrine that work to counteract anaphylaxis. These would, from here on out, accompany Elijah wherever he went. Hypervigilance comes with this territory, especially since, in our case, the triggering agent was a phantom one. The allergy test results came back inconclusive. As did the pulmonary tests. Although Elijah didn't exhibit the classic asthmatic wheeze, he certainly had hyperreactive airways. We were told that he might outgrow this. Conversely, it could worsen and develop into full-blown asthma. Or perhaps he had a cough-variant asthma already. Either way, watchful waiting was the byword.

To be on the safe side, we followed the allergist's advice and eliminated dairy and nuts from his diet. We instituted higher standards of housekeeping. To avoid respiratory infections, we initiated obsessive handwashing rituals and swift quarantines of the sick. I pursued conventional medical approaches. Jeff pursued the alternatives. Thus, in addition to the allergist, Elijah was under the care of a homeopathic healer and a Russian acupuncturist. Gradually, over the span of a year, he seemed to improve. But it was a long time before a case of the sniffles did not always lead, inexorably, to sleepless nights of coughing, croup, vaporizers, and steamy showers. And often enough, it still does.

To parent a child with allergies and asthma is to enter a paranoid and all-consuming world. In this, we were not alone. Many other parents in Faith's school were living here, too. Jeff, the lunchroom supervisor and after-school coordinator as well as the art teacher, became aware of just how many kids were coping with these conditions. He knew exactly whose names were written on the EpiPens and asthma nebulizers stashed in the nurse's office. On the front door of the school, a poster of a peanut shell with a slash through it told all who entered, even the preliterate, that children with peanut allergies were members of this community. For a few years, the school not only disallowed peanuts but banned all tree nuts from the premises to accommodate two students with life-threatening nut allergies.

Ergo, no lip balm (palm oil). We reflexively read the fine print on the back of hand lotion bottles and boxes of crackers. And breathed secret sighs of relief when those boys graduated.

Asthma has its own taxonomy, and many different forms exist. One type is simply triggered by inhalation of respiratory irritants. (Adult-onset asthma is often this type.) By contrast, atopic asthma, like eczema, is a kind of allergy. Which is to say, it's an inflammatory disorder.

Inflammation—from the Latin, *to set on fire*—is the least discriminatory part of our immune system. While other elements of immunity bring sophisticated approaches to the surveillance of foreign enemies and devise elaborate sting operations to apprehend them, the motto of inflammation is *burn 'em out and shoot 'em dead*. When it comes to homeland security, inflammatory immunity is a vigilante mob wholly uninterested in the presumption of innocence until proven guilty. That's not a bad approach if, for example, you have no skin on your feet. With open wounds, there is no time to sort out the innocent bystanders from the truly wicked. A kill-anything-that-moves credo is useful—up to a point. The problem with go-go inflammatory agents is that they can easily overdo it or head out after the wrong enemy—sometimes mistaking benign things like cat dander for dangerous pathogens. The result is an allergic reaction.

This is exactly what happens with atopic asthma. Following sensitization to some kind of allergen or irritant, the lung's airways become hyperreactive. Further exposure provokes inflammation, which serves no good purpose at all. Chronically inflamed airways become obstructed and narrow, and breathing becomes difficult. During an asthma attack, the smooth muscles around the airways contract sporadically, constricting airways further. Different asthma sufferers have different triggers. The common ones include dust, dust mites, mold, pollen, pets, gasoline vapors, perfume, and cold air.

What sensitizes the lungs to the triggering agent in the first place? Asthma triggers are not the same as asthma causes. But neither are they independent of each other. Here is where the terrain of asthma gets complicated, and sorting out the causal chain of events is like following a trail through the woods that loops back around on itself.

The weight of the evidence suggests that experiences in prenatal life or infancy can alter the development of both the immune system and the lungs in ways that heighten inflammatory responses. The result, in many people, is asthma. But by what molecular pathway? No one knows, but some evidence suggests that both stress and chemical exposures can be part of the story. Stress can disrupt the developing adrenal gland. Adrenal hormones communicate with that part of the immune system involved in inflammation. They also, as we have seen, play a role in guiding lung development (glucocorticoids). Reprogramming the adrenal gland, which affects both how the developing lung gets constructed and how it responds to allergens, is one of a number of potential influences on lung and immune system development. At the same time, exposure to environmental chemicals can also influence programming of the immune system as well as alter the branching pattern of respiratory airways and the architecture of the alveoli in ways that undermine pulmonary functioning throughout life.

In other cases, a pollutant may aid and abet inflammatory responses. This appears to be the case with phthalates, the plasticizer used to soften PVC. In their ability to incite the infiltration of white blood cells, phthalates fan the flames of inflammation. Lab studies reveal how the phthalate DEHP may act as the agent provocateur of wheezing: When metabolized, DEHP induces the production of an enzyme whose job it is to make inflammatory weaponry out of subcellular materials. In so doing, phthalates issue a treasonous call to arms.

Certainly phthalates are linked to child asthma. In a groundbreaking 2007 study that examined chemicals in indoor air, researchers identified six household activities consistently associated with child asthma: recent painting, renovation, cleaning, and the acquisition of new furniture, new carpets, or new wallpaper. Digging deeper, the team discovered that two chemical suspects were repeatedly found at the crime scene: formaldehyde (an airway sensitizer that is found in textiles and in the glues holding particle board together) and phthalate plasticizers in house dust. These results corroborated an earlier Swedish study that had uncovered an association between asthma and phthalates in children's bedrooms: The higher the phthalate levels, the higher the risk for asthma.

Because phthalates are hormone disruptors as well as respiratory irritants, the Product Safety Improvement Act, passed by Congress in 2008, now bans six types of phthalates from children's toys. But this law does not protect children from phthalates leaching from flooring, carpet backing, or wallpaper, for example, nor does it offer them protection during prenatal life, which would require limiting exposures of their mothers.

The upside of renting is that is allows for blithe inattention to large parts of one's surroundings. Like bathtub tiles. It doesn't matter if they are too fussy, too retro, too pink, or exactly reflective of one's taste in home décor. Because you can't change them anyway. And there is no danger that anyone *else* is going to read into the tilework a statement about your stylistic sensibilities because . . . you are a renter! When I held a lease, I could safely ignore the furnishings and go back to whatever paragraph I was working on. Or head outside for a run. Or plan a camping trip.

Inattention had been my lifelong approach to many objects around me: flowerbeds, shrubbery, mailboxes, crown molding, newel posts. (*Newel post.* That was a new term for me.) But suddenly, as a homeowner, I had to have opinions about everything. Was the light fixture above the dining room table to my liking or not? More to the point, why did it flicker when the kids jumped on the beds?

Worse than figuring out stylistic preferences, I was now responsible for the maintenance of every damn thing. I had never before in my life examined a fuse box with a flashlight. Or peered into soffits. Or cared about the pitch and capacity of gutters and downspouts. Or wondered if the venting system for a sump pump was up to code or not. (*Not* was the unfortunate answer to that query.) Our semiruined carriage house, which, in my renting days, I might have appreciated as a Romantic object of decay, was stressing me out. And its surly inhabitant, a raccoon with an attitude problem, I viewed as an agent of entropy. In some hidden way, it was undoubtedly hastening the decomposition of the foundation. Surely, the homeowner thing to do was evict the raccoon. But I didn't.

Then I realized the obvious: Having responsibility for our surroundings allowed Jeff and me to align them with ecological principles. Heat. Light. Food. Waste. I now possessed much more control over their cycles and flowcharts. In fact, I could redesign the household—not with the fantasy of creating an organic pod within an otherwise toxic world but with the aim of running a laboratory, an incubator for sustainable ideas.

In this, Jeff was way ahead of me. He had prior experience gutting rooms and repurposing them—albeit with different ends in mind—because he had once worked as a contractor for a high-end decorator. His clients wanted wine cellars and wainscoting and had decided opinions about wine cellars and wainscoting. So Jeff didn't believe me when I announced that I had no opinions *whatsoever* about doorknobs. (The one in the bathroom had come off in my hand.) He was right. As he described to me the many ways by which doors could open and close—and in which heat, light, and children flowed through doorways—I realized I did have opinions. I just never knew I had them.

Our approach to home renovation was to pay cash. That meant making changes gradually. It quickly became apparent that there were different categories of work competing for our money and Jeff's time. There were things that would make the house more energy efficient—and contribute to future environmental goals—and there were things that would contribute to the current health, safety, and happiness of the occupants. Devinylizing the house—which would improve indoor air quality for Elijah and protect us all during (God forbid) a house fire—fit into the latter category. Insulation and window replacement, the former.

Stripping the walls of vinyl wallpaper was a cheap, fast, and easy decision, as was ripping out the vinyl-backed carpeting. But replacing the vinyl kitchen floor was deferred for a while out of a need to capitalize a project from category one: The replacement of a fickle and wildly inefficient boiler, which the plumber said dated back to the Kennedy administration. (When I relayed this news to my tax accountant, he responded blandly, "Well, at least it's a *Democratic* boiler.")

Time passed. Finally, during a summer vacation, we got to the kitchen floor. Having sent the kids to the tree house and me to my office, Jeff began the work of ripping out the PVC tiles. He soon called me to take a look. Under the floor he was removing there was another floor and under that one, still another. The middle layer was a handsome, black and speckled tile floor that, Jeff guessed, dated back to the 1930s. I conjectured that it might be real linoleum. We imagined for a while what the kitchen might look like if we simply restored it. But the more Jeff excavated, the more we could see how damaged it was. By the basement stairs, a big swath of it lay in shards. Those broken pieces allowed us a glimpse into what appeared to be the original floor—a green-painted, wide-planked wood floor. Hardwood maybe. We looked at each other. That could be nice. Jeff said that he would crowbar up the top two layers of flooring and call me when he was finished.

Back in my office, listening to the sound of floor demolition, I began to wonder if the street were being repaved. I smelled asphalt but couldn't see any dump trucks up or down the block. Then something clicked in my mind. *Asphalt tile.* It was a type of early synthetic flooring manufactured in the first half of the twentieth century. I had read about it while researching the history of PVC. So I typed "asphalt tile" into an online encyclopedia for building inspectors, and up popped a photograph of what looked exactly like the speckled black floor in our kitchen. And under the picture was a warning: If uncovered during a renovation, this flooring should not be disturbed as it may contain up to 70 percent by weight asbestos fiber.

Asbestos. The mineral with jagged, microscopic fibers that lodge deep into the lungs when inhaled. First identified as a death trap in 1898. Easily released into air when disturbed. Known cause of lung cancer and mesothelioma.

Just at that moment, Jeff called me. *I think you should see this, Sandra. It's backed with some kind of fiber.*

It turns out that dialing 911 is an appropriate thing to do in situations like these.

With the help of the fire chief, who arrived at our house within minutes after I called, Jeff sealed off the kitchen from the rest of the house, created a negative airspace using fans, and located an EPA-certified lab to help with remediation. At night, he slept in a tent in the backyard. The kids and I moved into a nearby hotel. I signed them up for day camp and worked at the library. In the evenings, we hiked to the waterfall and picnicked at the beach. Forthcoming yet nonchalant about the problem that made us flee the house, I tried to make our exile seem like an adventure. *We are just being extra careful.* Still, it was hard to explain why we couldn't go home. Faith asked if daddy and I were getting divorced. Elijah cried.

Ten days later, the kitchen was covered with a layer of new (formaldehyde-free) plywood. Air sampling conducted throughout the house showed no traces of asbestos. The kids and I checked out of the hotel.

It was a sober homecoming. Jeff and I both felt like nominees for Worst Parent of the Year. Somehow all of our respective experience in home renovation and environmental health had not helped us avoid a serious environmental health hazard while conducting home renovation. In attempting to remove one respiratory threat from our son's life, we very nearly exposed him to a much worse one—and may indeed have exposed ourselves. We created a mess that required our entire year's renovation budget to solve. What was wrong with us? We had so carefully tested the house for lead and radon—and indeed asbestos—before we signed the purchase offer. Why had we not done due diligence on the black floor before ripping it out? If only I had entered "asphalt tile" into the search engine ten minutes sooner.

By the way, Jeff said. *The green floor was covered with lead paint.*

There was nothing to do but laugh. Essentially, our many-floored kitchen was a toxic archive of every terrible building material of the twentieth century: lead (destroyer of brains); asbestos (destroyer of lungs); and PVC (see Chapter 5). As a partial explanation for how such a thing could even happen, I found this sentence in the online encyclopedia for building inspectors:

> One reason that so much asbestos was used in flooring tiles was simply the wish to find an application for asbestos waste product from asbestos mining operations.

A long time ago, somebody made a decision to turn kitchen floors into a burial ground for mining waste. Not so long ago, a father unfamiliar with this decision—who was not even *born* when the decision was made—inadvertently released it into his kitchen. And had to empty our bank account to clean it up.

This is how one generation with unresolved environment problems leaves unexploded ordnance for future generations to trip over.

To talk about asthma's statistics requires facility with millions and billions. Asthma accounts for 14.4 million lost days of school every school year. (It's a leading cause of school absenteeism.) It affects 7.1 million children—and each year, it kills 600 of them. (It's the number one chronic childhood disease.) Asthma carries an impressive price tag: $20.7 billion in annual costs. (It's the number one cause of child hospitalization and visits to emergency rooms.)

What role environmental pollutants play in the story of child asthma is a hotly debated question. The most reliable guess is that it accounts for 30 percent of cases, but because asthma, like cancer, is a disease with multiple causes, that's a soft number. Nevertheless, evidence for a contributing role of pollutants comes from multiple sources. The first is time trends. Since 1980, childhood asthma has doubled in incidence, and severity has worsened. A doubling in the rate of a disease over a twenty-five-year period means that it's unlikely that dust, mold, or smoking is driving the trend. As a recent analysis of the issue points out, our homes have not become twice as dusty or twice as moldy. Moreover, asthma rates do not differ between humid and dry regions. Cigarette smoking is down, yet asthma rates are up. What about exposure to airborne contaminants?

At first glance, air pollution, like house dust, seems to have a reasonable alibi. Outdoor air quality has also mostly improved during the time span in which child asthma has worsened, as gauged by the falling levels of those air pollutants regulated by the EPA. Nitrogen oxides, sulfur dioxide, and vaporous compounds are down. Ozone (smog) is also down—although not by a lot. Particle pollution yields a more complicated picture. Particles were down, but then rose again after 2002, mostly because of coal-fired power plants.

But the air pollutants we monitor and regulate under the Clean Air Act are only a tiny fraction of the total. Diesel exhaust, for example, contributes to the air we breathe high levels of ultrafine particles (less than 2.5 microns), which are sticky and act as miniature taxicabs for other noxious chemicals like formaldehyde and sulfuric acid. Ultrafine particles and chemical hybrids are not routinely measured in outdoor air. So perhaps asthma is related to them—or something else in our air not monitored and regulated.

Or perhaps children today show greater immune sensitivity to the air pollutants we do monitor and regulate.

Or perhaps the sensitizing exposure is not a classic air pollutant. Evidence from both epidemiology and lab animal research has uncovered links between pesticide exposure and asthma. Organophosphate pesticides in particular can induce spasms in bronchial tubes and contribute to airway hyperreactivity by altering the functioning of nerves that supply the muscles of the airways. As we have seen (Chapter 3), children are exposed to organophosphates through their food and possibly also when they are sprayed into their environment. One such pesticide, chlorpyrifos, was banned for household pest control in 2001, precisely because of its ability to interfere with children's neural pathways (more on this in Chapter 8), but it is still used in agriculture and in urban pest control.

Other clues can be found in the demographics of the disease. Asthma disproportionately affects poor children, black and Hispanic children, and urban children. Children born prematurely are at increased risk for asthma, as are obese children. The concurrent rise of pediatric asthma with increased rates of preterm birth (now 12.5 percent of all births) and obesity (now affecting 16.9 percent of children) certainly suggests possible connections.

Children living in polluted areas also have higher incidence of asthma. But children living in polluted areas are more likely to be non-white and poor—making difficult the task of teasing apart socioeconomic factors from environmental exposures. Nevertheless, geographic patterns in Europe also show connections between asthma and air pollution. A recent study shows that Dutch children who live near busy roads are more likely to have asthma. (They also have more ear, nose, and throat infections, more colds, and more flu.) These patterns mimic those found in California: Asthma rates double among children with the closest residential proximity to busy traffic lanes. And, independent of home location, California children who attend schools located on busy roads also have elevated rates. (Poverty also matters: Given the same exposure to traffic-related air pollution, kids of lower socioeconomic status are more likely to develop asthma. Poorer children have higher inflammatory markers, suggesting they are living closer to the threshold of disease.)

So there are two things we can say with a high degree of certainty. One: Outdoor air pollution is geographically associated with high rates of child asthma. Two: Outdoor air pollution exacerbates asthmatic symptoms. When air pollution goes up, the lung functioning of asthmatic children goes down, and hospital admission rates go up. We even have molecular evidence for this trend: On bad air days, asthmatic children have increased markers of inflammation in the condensation of their exhaled breath. And their lung functioning gets worse.

But can outdoor air pollution *cause* asthma? Answering this question requires painstaking studies that involve following asthma-free children over time and noting who gets asthma and under what conditions. These sorts of prospective studies have been carried out by teams of researchers in both New York and California. All together, they show that, indeed, early-life exposure to fine particles, ozone, diesel exhaust, and a group of combustion byproducts called polycyclic aromatic hydrocarbons is associated with the onset of asthma. In particular, prenatal exposure is a risk factor for its development.

Teams of investigators at Columbia University's Center for Children's Environmental Health have been monitoring cohorts of children since before their births. (These are the scientists from Chapter 1 whom we saw equipping pregnant women with personal air monitors.) In a study published in 2009, they report that exposure to traffic exhaust during pregnancy reprograms a gene in ways that increase a child's chances of developing asthma.

Similarly, researchers in California followed a group of children from early life onward and discovered that traffic-related air pollution was associated with the onset of asthma. This 2008 study was particularly convincing because an air pollution monitor was placed outside the home of each child in the study. (Earlier studies had relied on centrally placed air monitors.) In a 2010 study, a team of California researchers followed a cohort of asthma-free kindergarteners. Those who attended a school with high levels of air pollution—independent of the levels of air pollution at their homes—were 45 percent more likely to develop asthma. Furthermore, California children who participated in at least three outdoor activities in communities with high smog (ozone) levels were 30 percent more likely to develop asthma.

Air pollutants impede the breathing of children in a variety of ways. Particulate matter buried deep in the lungs can excite the release of cytokines. These are proteins that can call a state of emergency, deputize immune cells, and send them off after the bad guys. Cytokines are the sheriffs of inflammation. More cytokines means a twitchier respiratory system. In addition, prenatal exposure to air pollutants can alter the development of immune cells. Polycyclic aromatic hydrocarbons (emitted by cars and coal-burning power plants) can change the ratio of white blood cells in the umbilical cord, skewing the nascent immune system in a more inflammatory direction.

Air pollution can also interfere with the treatment of asthma. Asthma drugs work by relaxing muscles that wrap airways, compelling them to dilate. But air pollution can continue to provoke inflammation, constricting airways.

Even among asthma-free children, air pollution stunts lung development. Up to 80 percent of the alveoli in an adult lung develop after birth, with alveolation continuing until at least age eight. Some pulmonologists believe air sacs are still forming during adolescence. But, as of age 20—at the outside— you have all the respiratory surface that you will ever have. Air pollution causes permanent changes in developing pulmonary structures. The result is less surface area for respiration and a smaller lung volume.

Studies show that teenagers in southern California who grow up in more polluted places have smaller lungs and diminished lung functioning. These alterations raise the risk of chronic obstructive pulmonary disease in later life and, in childhood, raise the risk for bronchitis. And with rising levels of air pollution also come rising rates of middle ear infections.

Did you ever dream that you had to go back to school and take a math exam? This actually happened to Jeff in his waking life. All certified art teachers in the state of New York must have on their transcripts passing grades for college-level coursework in math and science. Such requirements were not in place for art majors in the free-wheeling academic world of our youths—not even, apparently, for the art majors of Amherst, his very serious alma mater. Thus, some number of decades later, he had to make up the deficits. He was a good sport about it all—tackling college algebra first and then astronomy.

I was surprised at his choice of astronomy. Of all the sciences, it's the one I know least about, so I was not in a position to be helpful. But he believed it would be the most *visual* field of study and thus most compatible with the way his own mind worked. And soon he was interpreting Hubble telescope imagery and had no need of my assistance anyway.

There was an unexpected benefit to all this required knowledge. Through direct observation and some kind of calculation involving math, Jeff figured out the angle of the moon as it passed over our house and cut a hole in the roof over our bedroom so that moonlight would trace a path directly across our bed. (And installed a skylight in the hole.) This was a home renovation firmly in category two—improves happiness of occupants—although a case could be made for number one—aligns household ecosystem with environmental goodness. An open skylight, Jeff noted, would serve as our summertime air conditioning, drawing cool air up the stairwell at night.

And so all four of us took to the bed to engage in astronomy, looking straight up through the ceiling, through miles of air and atmosphere, into the firmament above. Moon. Stars. Clouds.

Sometimes Jeff gave lessons on the properties of light.

Sometimes we all just lay together in silence.

Sometimes, if we didn't know where one child or another had disappeared, Jeff or I would find him or her supine on our bed, staring up. Night or day, it became the preferred place to have a private talk, check the weather, or recoup during a time out. The moon-aligned skylight brought a measure of infinity into a small house.

On the first night that I awoke to a waterfall of silver in my eyes—my husband breathing next to me, my children breathing in the next room—I realized I was home. This was my house. And I matched my breathing to theirs.

By all the pathways we have seen, air pollution wrecks the health of the whole family. First, air pollution during pregnancy is linked to lower birth weight and preterm birth. Both of these conditions, all by themselves, raise the risk for childhood asthma. Second, air pollution—both the indoor kind and the outdoor kind—makes children's asthma worse, and growing evidence suggests it serves as a direct cause of the disease. Third, air pollution reprograms pulmonary development in ways that stunt lung growth, a condition with lifelong consequences. Fourth, air pollution alters immune functioning. And fifth, it increases the frequency of bronchitis and ear infections. In short, early-life exposure to air pollution sows suffering for children and misery for parents. It diminishes respiratory health, fills up emergency rooms, incurs medical costs, and steals time, money, energy, and sleep.

The air pollution-asthma link is also part of a much larger story. Among adults, air pollution contributes to cancer (with tobacco smoke, radon, and diesel exhaust among the substances conferring the greatest risk). Lung, breast, and bladder cancers all have demonstrable links to air pollution. Air pollution also contributes to diabetes. More specifically, in both the United States and in Europe, traffic-related air pollution has been identified as a risk factor for type-2 diabetes in women. Again, its ability to induce inflammation, which contributes to insulin resistance, appears to be the link.

In addition to all this, according to the EPA, about 20,000 Americans die prematurely each year from cardiovascular problems attributable to exposure to air pollutants, most notably, the byproducts of fossil fuel combustion. Small sooty particles (those less than 2.5 microns)—which can easily cross the alveolar threshold and enter the bloodstream—are thought to be largely responsible. By multiple mechanisms, their inhalation contributes to heart attack, stroke, and high blood pressure. Fine and ultrafine particles, for example, make clotting factors in the blood stickier and more prone to coagulation. Stoked by air pollutants, chronic inflammation also irritates nerves, altering their electrical signals, which in turn, contributes to abnormal cardiac rhythms.

In other words, the combustion of fossil fuels reprograms not just the lungs and immune systems of our children but the pace and tempo of our own heartbeats.

These results raise some questions. Given the breadth and depth of the evidence published in the scientific literature, why are we not reading about the dangers of ozone and diesel exhaust—and traffic and coal-burning power plants and phthalate-laden flooring—in parenting magazines? The ones I pick up in the doctor's office waiting room (while waiting to see if Elijah has bronchitis yet again) or in the pharmacy (while waiting to refill the EpiPen prescription) have lots of information about *managing* asthma with medication but none about preventing it through the provision of cleaner air.

Why is the only person interested in talking with me about our local coal-burning power plant a childless college student out canvassing for an environmental group?

Why is there no analysis of the proposed amendments to the Clean Air Act in the Web sites of online support groups for mothers of children with asthma and allergies? I entered "clean air act" into the internal search engines of many such sites and came up with nothing. (On the other hand, the testimony of the American Lung Association's director before the U.S. Senate Committee on Environment and Public Works provided an excellent overview of the issues.)

So why can't parents become conversant with the Clean Air Act and its National Ambient Air Quality Standards, whose various rules affect our children so intimately? The American Petroleum Institute certainly is.

Originally passed in 1970, the Clean Air Act is the federal law that requires air pollution limits to be based on the latest available science. As the science shifts, so too must the law. And therein lies a mighty battle, with arguments over what constitutes "maximum available control technologies" and who should receive exemptions, and so on. Yes, it's complicated, but so is figuring out a vaccination schedule for an allergic child or staying *au courant* with ever-evolving car seat/booster seat regulations. (These vary by state, by type of vehicle, and by weight, height, and age of the individual child.) And yet I've seen plenty of articles in parenting magazines, online and off, that help shepherd families through the byzantine intricacies of car seat rules. I've seen none about what cars are doing to our air.

In 2009, more than half of Americans lived in counties that received failing marks for either ozone or particulate matter. Given the direct consequences for children, that news, when it broke, should have been a big topic among parents: What needs to be done to bring half our nation into compliance with the Clean Air Act? Instead, what children with asthma, and their parents, heard about air pollution was how to "limit their exposure as much as possible by decreasing their time outside when particulate levels are high." (A government Web site, www.airnow.gov, allows you to monitor your local particulate levels.) But physical activity is exactly what we and our children are supposed to be doing *more* of in order to avoid obesity, which in turn is linked back to increased asthma risk in children.

Surely, growing up indoors and sedentary is not the answer to air pollution. There is no substitute for clean air. And we can't go shopping for some. And since the only way to ensure that our children have access to healthful air is by not polluting it— why aren't mothers of children with hyperreactive airways across the land marching on Washington, demanding investments in green energy and public transportation?

Happily, there are plenty of evidence-based reasons to believe that political action would bring quick benefits. When air quality improves, children's respiratory health rebounds. In Switzerland, an 11-year decline in airborne particulate matter was followed closely by immediate improvements in children's lung functioning. Likewise in China, after the government closed a polluting coal-fired power plant in 2004. In Atlanta, while traffic was reduced for the 1996 summer Olympic games, ozone levels fell by 25 percent, and child hospital admission for asthma fell by 19 percent. When the steel mill in Provo, Utah, closed down in the mid-1980s, hospital admissions for children with asthma dropped by half.

And when it reopened, they went back up.

Just when I felt truly settled in to our skylit Victorian cottage—the piano teacher's phone number committed to memory, the origin of the basement leak identified, peace with the carriage house raccoon brokered—along came a job offer in a different time zone.

At first I viewed the invitation in the way a sober alcoholic might view an offer of a drink. *You don't know who you're dealing with, pal.* As though the offer of a steady salary, an academic post, and retirement benefits were wicked temptation. I worried that I was still too much in love with sleeping bags and on-the-road-again adventure. The job description seemed perfect. My first interview was a laughter-filled conversation with colleagues whom I could easily imagine at a potluck dinner. And Jeff could be considered for a post in the art department at the same university.

Soon we were on a sleeper train heading for the Midwest—the kids each promised a turn in the top bunk in exchange for open-mindedness. Jeff and I both had campus interviews scheduled. Mostly, we hoped, as a family, to get the lay of the land, and look at the area schools, neighborhoods, food co-op, public library.

During the train ride, I pored over local air quality data. There were a couple of things I had noticed during my first interview that were bothering me. One was the coal-burning power plant on campus, and the other was an old-style trash incinerator downtown that served as the municipal utility. Among the nation's incinerators, this was a famously polluting one. Indeed, the dioxin generated by it had been traced all the way to northern Canada (using a computer model that could also have been used to model dioxin emissions from the Illiopolis PVC inferno). What I learned was troubling, but I set it aside on the grounds that all communities have their environmental issues.

Here were the highlights of our visit: Jeff was offered a position. He was also invited by the city council of a nearby town to consult on a public art project. We were all treated to a dinner of local, organic foods. Faith made a friend, attended a day of fourth grade, and impressed a geography teacher by correctly identifying the location of the Snake River (Idaho). I received a standing ovation for my public lecture. The provost apologetically rescheduled our meeting because an asthma attack landed her in the emergency room. And Elijah started coughing. He coughed during the whole week of our visit, and when we came back home, he stopped.

Sometimes when I'm faced with weighty decisions, scraps of Scripture enter my mind—indelible remnants of a childhood spent in a hymn-filled Methodist church. Within my youth group, I was the resident expert on the fifth book of the New Testament, The Acts of the Apostles, which is where, said I, the plot got truly complicated. But as the train rattled east again, the words that came to me were from the second book, the gospel of Mark:

> *For what shall it profit a man, if he shall gain the whole world,*
> *and lose his own soul?*

Soul is a Germanic word, perhaps referring to the sea. In Latin, *spiritus*, which means to breathe, as in *inspire*. In Greek, *psyche*, to blow, as in air, referring to the breath of life. In Hebrew, *nephesh*: life, the soul, which is to say, *that which breathes*. My son, who had once floated in a sea within me, needed to breathe. And no salary or retirement benefit was profit enough to compensate.

I turned the job down, fully aware that for many parents—and for myself at an earlier stage in my own life—the choice between a job and a child's well-being too often comes without the option of favoring the latter. But as long as I could squeak out a living, I couldn't choose to relocate an asthmatic child near a trash incinerator. For clean air, I was willing to forego retirement benefits.

Lungs exist at a place where two environmental crises meet.

Their ability to respire—to exchange oxygen for carbon dioxide across a membrane as fine as gold leaf—is compromised by the toxicity of emissions from fossil fuel combustion. About half of these emissions come from power plants (coal and natural gas) and another third from transportation (petroleum). The chemical adulteration of the planet with toxic pollutants derived from coal, gas, and petroleum is one crisis.

At the same time, the combustion of fossil fuels contributes heat-trapping gases to the atmosphere—most notably, carbon dioxide. These are destabilizing the planet's climate. Global climate change is, thus, the second crisis.

But the problems of toxicity and temperature are not independent of each other. Higher global temperatures accelerate the creation of toxic lung pollutants, such as ozone, nitrogen dioxide, particles, and carcinogens. And they accelerate the evaporation of liquid pollutants, like gasoline. By raising the heat, you raise the air's toxicity. Higher temperatures also increase levels of pollen, dust mites, and fungal spores. In all these ways, climate change is an asthma trigger.

I have now walked us right into the blockbuster topic of global warming, and there are many dozens of things I could say at this point. Global warming is a Biblical epic with a cast of thousands, from polar bears to floods, and many complicated subplots. It is easy to feel that you've arrived late to the theater and find the storyline confusing. It is easy to wish you could skip this show altogether. (As if.)

For the legions of busy, multitasking parents who walked into the middle of the climate movie, here's what's already happened: In 2007, the Supreme Court ruled that heat-trapping gases fall under the Clean Air Act. In 2009, the EPA declared that six of these gases constitute a threat to human health but has not yet regulated them because it's waiting for Congress to pass climate legislation.

For the sake of this discussion, I'm going to focus on just two other scenes of the movie that each relate to the air we breathe—with others to come in the next chapter.

Scene One: In a 2008 study, Stanford University engineer Mark Jacobson demonstrated that upticks in the average temperature of the planet lead to significant increases in human deaths due to air pollution. Specifically, an increase of one degree Celsius (1.8 degrees Fahrenheit) kills, each year, an additional 1,000 Americans from exposures to ozone and particulate matter—and leads to many more cases of asthma.

This is not a possible threat; this is an outcome already measurable. The combustion of coal and oil has already increased the global temperature by nearly one degree Celsius, with much of that increase occurring in the last few decades. Global climate change is, thus, already contributing to the burden of child asthma and, unless mitigated, will add even more rocks to the pockets of asthmatic children in the years to come.

In 2009, a year after the Jacobson report, members of the European Respiratory Society—physicians—called for immediate political action on climate change on the grounds that the deterioration of air quality wrought by higher temperatures is disproportionally killing their patients—people with asthma and other chronic respiratory infections. Whereas a one-degree increase in average ambient temperature results in a 1 to 3 percent increase in the premature deaths of the general public, it leads to a 6-percent increase in deaths among people with preexisting respiratory conditions.

In other words, if you have a child with asthma, he or she is two-to-six times more likely than everyone else to perish from global warming-induced air pollution.

Those are the stakes.

In the same year, on this side of the Atlantic, researchers from the Columbia Center for Children's Environmental Health convened a remarkable conference that explored the ways in which both environmental crises—the crisis of toxic chemical exposure and the crisis of climate change—interact to erode the health of children.

Of particular interest was how climate change magnifies the problem of toxic exposures. Overseeing this event was molecular epidemiologist Frederica Perera, who has been studying the effects of prenatal chemical exposure for thirty years. Perera oversees the long-standing studies of mother–newborn pairs— including the 9/11 cohort and mothers and children in China and Poland—the results of which we've already explored. It is Perera's team that installs individual air monitors in the backpacks of pregnant women and traces the flow of air pollutants from mother to child, recording the damage to fetal blood cells along the way.

If anyone deserves to feel overwhelmed by the power of climate change to magnify the toxic effects of air pollutants, it would be Perera. But that was not the note she struck in her remarks at the conference. First, she meticulously documented the myriad impacts of fossil fuel dependency on child health and development—lower birth weight, preterm birth, stunted lung growth, developmental delays, cancer, asthma, allergies, heat stroke, drowning, malnutrition, diarrhea, malaria, encephalitis, and psychological trauma. Then she meticulously documented how the economic costs of doing nothing exceed the cost of mitigation—through investments in green technology and greater efficiencies in power generation, building design, and transportation. In other words, quitting fossil fuels would both save us money and avert disaster. Win–win. Her conclusion was urgent but hopeful: "Our addiction can be cured. We do not have to leave our children a double legacy of ill health and ecological disaster."

Another conference speaker, climate strategist Michel Gelobter, put a finer point on the addiction analogy: Fossil fuels are like cigarettes for the planet. "They take money out of our communities, they pollute our environment, they kill children, and they waste our lives." A smoking cessation program for the planet necessitates nothing less than the remaking of the economy and the will to leave the remaining oil and coal in the ground and do something different. Devoted to realizing this vision, Gelobter said that his work was inspired by a deeply felt appreciation for the injustice of climate change. "Some people have used more of the atmosphere than they had a right to, than was sustainable, and are precluding the use of that atmosphere and those resources for others, have precluded it, and will continue to do so unless we take action."

Essentially, the unrestricted burning of fossil fuels is turning air into the atmospheric equivalent of an asbestos-filled kitchen floor. While I listened to Gelobter talk about intergenerational inequity to an audience full of Ph.D.s and M.D.s, it occurred to me that my children had simpler words for it: *That's not fair.*

It also occurred to me that Perera and Gelobter are the voices that parents of children with asthma and allergies need to hear while waiting for prescription refills in the pharmacy. All 12.4 million of us.

Here is the second scene from the climate movie: It's set in the top layer of the ocean, which has absorbed 93 percent of the excess heat trapped by global warming gases.

Floating around on the surface of the high seas are plankton. By definition, they are drifters, blowing where the currents send them. Some of them are animals. Some of them are plants. Both groups are struggling.

The animals—zooplankton—have two problems. First, they depend on plant plankton (algae) for their food source, and when the foundation of the food chain is in trouble, everybody north of the foundation also suffers. But zooplankton also have a second, separate problem. Many members of the zooplankton community are the larval forms of various shell-bearing species. They spend their youth at sea as directionless wanderers. Then they settle down and calcify. But now they are having trouble doing that.

About a third of the carbon dioxide contributed to the atmosphere from cars and power plants is absorbed by seawater. Here, it turns into carbonic acid. This transformation is changing the pH of the ocean's surface—indeed, has already changed it, from 8.2 in 1750 to 8.1 today. (The pH scale is logarithmic. Each whole pH value is ten times more or less acidic than the preceding one.) And an increasingly acid ocean makes it difficult for marine animals to secrete shells. (They dissolve.) An acidifying ocean thus endangers corals, oysters, barnacles—along with their seafaring offspring, the zooplankton.

The zooplankton's green counterparts, the phytoplankton, also have a problem. Because they are restricted to the sunlit surface of the ocean—indeed the upper layer is the only place where plants grow in the open ocean—phytoplankton depend on upwelling of water currents from the deep to bring them nutrients—like nitrates—otherwise lost to gravity. This plankton-filled surface layer is warming up, and its higher temperature inhibits its ability to mix with the cooler, nutrient-dense layers below. As a result, in eight of ten ocean regions, the abundance of plant plankton is declining.

Given that phytoplankton are microscopic, how do we know this? Satellites keep track of the changing color of the ocean's surface, but the better data come from a decidedly low-tech oceanographic instrument called a Secchi disk. And, thanks to 100 years of its continuous use by oceanographers, researchers can construct a timeline of plankton density. This shows that the world's plankton stocks are waning.

A Secchi disk is a black-and-white plate the size of a Frisbee that is mounted on a tape measure and lowered into the water from the side of a ship or boat. The point at which the disk disappears from view is called the Secchi depth. It's an indirect measure of how much chlorophyll is in the water. The shallower the Secchi depth, the murkier the water, and the greater the density of chlorophyll (phytoplankton). The deeper the Secchi depth, the clearer the water, and the lower the density of plankton. Secchi disks, standardized tools of oceanographic observation, are used on lakes as well, which is where I'm familiar with them. You gently slip the disk into the water on the shady side of the boat and watch as it disappears into the green depth. The point at which the black-and-white pattern disappears into the greenness is where you stop and take a reading.

Out in the open ocean, Secchi depths are increasing. According to a 2010 analysis of a more than a half million oceanic readings going back to 1899, global phytoplankton stocks are down 40 percent. The 1950s marked the beginning of their decline, which has continued at a rate of 1 percent a year. The investigators found a strong correspondence between this long-term trend and warming temperatures on the surface of the ocean.

Phytoplankton make half the oxygen we breathe.

Back at home, lying under the skylight, Elijah and I watched a windstorm. Leaves blew by. Birds that looked like leaves blew by, as though not flying under their own power. Twigs and spruce cones pelted the shingles.

Where does the wind come from?

It wasn't the first time a child asked me this question, but it's one I've always loved to consider. And I knew where it would lead. As soon as I explained the rotation of the earth and the origin of air currents, the follow-up question would be, right on cue—

But where does the air come from?

There are many ways to go with a query like this, but I like to bring it around to the oxygen cycle because it's an invitation to converse about photosynthesis, my very favorite biological reaction. It happens also to be the world's most complex biological reaction. No fewer than 100 proteins are required to spin sunlight, water, and carbon dioxide into molecules of sugar and oxygen.

At 3 billion years old, photosynthesis is an ancient process, and it explains many mysteries. The ripple-etched sidewalk in front of our house, for example, that was once the floor of a shallow sea: Its shale slabs are made up of the compressed bodies of photosynthesizing plankton and those who fed on them. And all those fossil fuels: The reason we call them *nonrenewable resources* is because the creatures whose bodies comprise them died before the world's plants had filled the atmosphere with enough oxygen to decompose their own corpses. The undecayed dead were squashed into petroleum, gas, and coal. That doesn't happen on the oxygen-rich planet we inhabit now.

And yet, in spite of the explanatory power of photosynthesis, the essence of the whole operation—the splitting of a water molecule—is an enigma. Biophysicists still don't know exactly how chloroplasts manage it.

Elijah, the oxygen you are breathing in right now is a gift from green plants. Plants make their own food out of the sun. They breathe oxygen out. And then you breathe it in.

And then what?

And then you breathe carbon dioxide out. And the plants breathe your carbon dioxide in. That helps them make food . . . and make more oxygen for you.

So we're nice to each other.

CHAPTER SEVEN

The Big Talk
(and Systems Theory)

ON AN UNUSUALLY warm September night—Jeff was in class and I was working late at my desk—I sent the kids upstairs to put on pajamas. When they reappeared, claiming an animal was under the bed, I pointed the way back to their room. *Pajamas. Now.*

Within minutes they were back downstairs. *Mama, we think you should come.* An electronic device of some kind was ringing. It occurred to me that we didn't have any toys that sounded like that. The kids solemnly followed me upstairs. Piercing bleeps came from everywhere and nowhere all at once. Like a tape loop used to disorient the enemy. Like the ring tone from hell.

I didn't think to ask Faith and Elijah to leave the room. I just started tearing it apart—the beds, the bookshelf, the toy box. Finally, I dismantled the registers, and there it was inside the guts of the baseboard heater. But *what* was it? Bigger than a mouse. Coppery fur. A greasy face. Ugly. Finally, I noticed the folded umbrellas of its wings, and a Latin name scrolled into my mind—*Eptesicus fuscus*. The Big Brown Bat. It scrabbled along a pipe. Without the metal cover to create a resonating chamber, the volume of the bat's chirps dropped to a series of pitiful twangs, like the plucked string of an unplugged electric guitar.

I ordered the kids out of the room, shut the door, and stuffed a towel under it.

In upstate New York, 2 percent of big brown bats are infected with rabies. I was aware that if I failed to capture this one, my kids would be compelled to undergo rabies vaccinations, in accordance with protocols set forth by Centers for Disease Control guidelines. A friend of mine had endured the whole series of shots—and her kids, too—because her husband, after a chase scene involving a tennis racket, had opened a window and released the bat that had awoken them all by swooping through the bedrooms. Bats have razors for teeth; their bites can be undetectable. And, although the odds are 98 percent that any given bat is rabies-free, rabies is a disease with a 100 percent fatality rate. All this, she had told me as a cautionary tale whose moral was, *Don't free the bat.*

I squatted in front of the disassembled heating register and devised a plan. What I needed was a long-sleeved shirt, a large yogurt container, and leather gloves. But these were all located in different parts of the house, and I didn't want to leave my intruder while I gathered the tools for its arrest. Through the door, I asked Faith to bring me the phone book. On the inside cover, alongside the numbers for the sheriff, fire department, and suicide counseling, was the after-hours number for the Rabies Prevention Hotline. I asked Elijah to bring me the phone.

(This is the last story I'm going to tell about emergency phone calls. I promise.)

After two rings, a live person from the county health department answered. And with that phone call, a well-oiled public health apparatus swung into motion.

Within fifteen minutes, a wildlife removal specialist was standing next to me. Within another fifteen minutes the bat was inside a bucket in my freezer. By morning, its frozen corpse was in the hands of a pathologist. Twenty-four hours later, the head of the county's rabies prevention program called me.

The bat was rabid.

He said that we needed to come to county health for an interview. During the conversation that ensued, he asked if anybody had been sleeping in the room. No, but we had all slept there the night prior to the night of the bat discovery. He asked if the children had been alone in the room with the bat. I had to admit that they had. He asked about the kids' encounter with the bat. He asked for the birth dates of each child. He wrote down the answers and was quiet for a while.

Then, with the kids out of the room, he reviewed with me the CDC guidelines: An encounter with a bat qualifies as a potential rabies exposure if a sleeping person awakens to find a bat in a room or if an adult witnesses a bat in a room with a previously unattended child, a mentally disabled person, or a drunk person. Elijah, as an unattended child, was on the cusp of what's considered old enough to be a reliable narrator about whether direct contact with a rabid animal had occurred.

I nodded silently. Earlier in the week, my son had walked into the kitchen and announced that he had learned to fly a helicopter. *And, Mom, I'm fully licensed.* I didn't repeat that claim to the head of the rabies prevention program.

He said that the decision to undergo the vaccination series was up to me. Ours was not a clear-cut case. Only I knew how trustworthy my children's stories were. I was encouraged to decide swiftly. The initial shots needed to be given within seventy-two hours of exposure. The remaining ones were given over twenty-eight days, according to a strict schedule. If I decided to go forward, county health would make all the arrangements.

For some reason, I felt the need to argue. I was giving an upcoming lecture at the University of Montana. I had to travel. How was that going to work? No problem. County health would arrange to send serum there by refrigerated courier. It would be waiting for me when I got off the plane. Oh, and by the way, he added, if your insurance does not reimburse, the county will pick up the cost of the vaccination series—which would be several *thousand* dollars—because *we don't want anyone making this decision on the basis of money. We want to err on the side of caution here.*

Those words were so amazing to me that I asked him to say that last part again. I had enough outstanding medical bills that

Jeff and I were nearly prevented from obtaining a mortgage for our house. In my whole life, no one has ever said to me: Look, we don't want you to forgo a cancer screening because of financial worries, and so the government will guarantee payment. Or even: Look, we don't know if you have been exposed, but we are removing carcinogens from your neighborhood because we want to err on the side of caution.

Back at home, I conducted some interviews of my own, calling each child over to the couch for a private chat. Faith claimed that she never saw the bat when she was alone in the room with it. She only heard it, and it sounded like it was under her bed. Elijah gave me the same basic story but with a twist: *And then, the bat flew around and landed on my hand.*

I blinked at him. *Real? Or pretend?*

Real.

I suggested we get a snack. How about some apples? After we cored, sliced, and ate four apples and drank two cups of tea, I casually asked him to tell me again about the bat. So he gamely repeated the whole story, word for word, until he came to the end. *The bat flew around the room and landed on my hand. It was carrying a tiny . . . gun.*

We had the shots. County health really did make all the arrangements, our insurance company paid without protest (amount billed: $5,798.04), and a nurse at the university clinic in Missoula, Montana, was standing by when I got off the plane on the day that my third round of shots was due. And so my children and I were afforded 100 percent protection against an environmental disease to which we may have been exposed. Or not.

Within the United States, bats are the most common source of rabies, which is an incurable and almost-always fatal form of encephalitis. And yet, *common* in this context still means something incredibly rare. Between 1997 and 2006, nineteen people in the United States died of rabies, and encounters with bats were involved with fourteen of these deaths. Basically, the odds of contracting bat-related rabies are far less than the odds of winning the state lottery. (The Organization for Bat Conservation likes to point this out.) Rabies kills, on average, one American a year. Sometimes two.

A century ago, the death rate from rabies was a hundred times higher, and most cases were caused by bites from domesticated animals. The virtual elimination of rabies deaths in the United States—which, by necessity, involves prevention and prophylaxis—is a triumph of public health. And accomplishing it required an all-hands-on-deck approach. Pets were vaccinated, legislation passed, strays captured, surveillance and reporting systems established, virology labs funded, and public awareness raised. Rabies prevention was—and is—a concerted, multidisciplinary effort involving federal agencies, local agencies, state laws, emergency room doctors, pathology labs, veterinarians, animal shelters, and wildlife biologists.

The precautionary, tightly coordinated world of rabies control into which my children and I were temporarily swept was impressively efficient and comprehensive. Even after the crisis was over, the search for primary prevention continued: The wildlife removal guy crawled along my porch roof with me and showed me how to identify the tiny entry points that bats use as doorways. (Their wings leave a telltale oily stain.) We tented the house with bird netting, allowing roosting bats to leave—they freefall before opening their wings—but not re-enter, and so humanely banished the source of our problem. Jeff caulked the holes after they left in a final gesture of good riddance.

And I could go back to work investigating environmental problems to which all children are exposed but for which no emergency hotline numbers appear in our phone books and no animals bleat SOS signals from the walls of our homes.

In doing so, I began to wonder why we don't bring a rabies approach—with its urgent, multitiered, take-no-chances, can-do lines of attack—to climate change. The central lesson of rabies— that an ounce of prevention is worth a pound of cure (okay, there is no cure)—seemed to apply in spades.

Climate change is now the biggest health threat to children. These are not children of future generations. These are children living today. (Like the two who are making a mess in my kitchen right now.)

Unless significant actions are taken very soon to dramatically curtail fossil fuel emissions, the human toll will almost certainly become catastrophic, according to the *British Medical Journal*. A report jointly compiled by the University of London and the prestigious medical journal, *The Lancet*, reached the same conclusion.

The American Academy of Pediatrics has issued its own policy statement about the menace of climate change for children, as has the World Health Organization, which notes that, of the deaths presently attributable to climate disruption around the world, 85 percent of those who perished were not yet adults.

Climate change and bat bites are dissimilar in at least three ways, and these differences help illuminate the obstacles blocking an aggressive approach to mitigating the dangers of the former. First, rabies is a disease that possesses what physicians call high specificity. If I had foregone a rabies vaccination, and, months later, fell into a convulsive coma and died, we would all know what killed me. By contrast, if I go on to die of chronic pulmonary obstructive disorder or cardiac arrhythmia, we won't know whether the higher ozone levels created by higher temperatures caused my death . . . or not. The summer heat wave that I suffered through in the final weeks of my pregnancy—and the crummy air I breathed as a result—could be responsible for Elijah's reactive airways. Or not. The damage caused by climate change may be dramatic and global, but it's not specific. It can hide in plain sight. Climate change manifests as shifting weather patterns, but it's in the nature of weather to be changeable. By contrast, to look into the eyes of a rabid animal is to confront a concentrated, undeniable danger.

Second, to remove bat roosts from one's attic is met with silence on the part of the bats. To remove fossil fuels from the economy is met with the opposite of silence by their manufacturers. If bats controlled as much of the global economy as Exxon, we would undoubtedly tolerate a much higher rabies death rate. (And aspersions would be openly cast on the posited link between bat attacks and rabies. Debate would swirl around the claim that bat saliva is a medium of transmission. In spite of government reports flatly asserting that "rabies transmission by bats is unequivocal," the public would remain ambivalent and confused.)

Third—and this is the difference I want to focus on—climate change is a topic that is surrounded by considerable social silence. This is hardly the case for rabies. The death of Houston teenager, Zach Jones—the only death from rabies in all of the United States in 2006—made national news. ("Texas Rabies Death Spurs New Concern About Bats" was the headline in the *Los Angeles Times*). In its aftermath, hundreds of people captured and turned in bats, overwhelming Houston's rabies control lab.

At about the same time, I discovered that my own bat story was welcomed at dinner parties. Even people who were terrified of bats—truly phobic—would listen in rapt horror as I described how a rabid bat the size of a chipmunk had used its wings to paddle across the baseboard heater behind Faith's bed. My kids held court on the playground with their own versions of our bat story, and the vaccination Band-Aids on their shoulders were admired as badges of courage.

By contrast, climate change tends to be a conversation ender among friends. Not withstanding the few legions of valiant activists who are indeed at the barricades, talk of climate change does not incite riots in the hearts of men. Or even curiosity. Instead, a miasma of quiet anxiety hangs over the whole topic. Even among individuals who admit feeling truly concerned, news about climate change seems to disperse attention rather than gather it. Certainly, the drumbeat of evidence for its accelerating destructive power has not prompted alarmed citizens to inundate government labs with biological samples.

And nowhere is the silence thicker than around the human health consequences of climate change. Although several august medical authorities have clearly identified climate change as the number one threat to children, leading news organizations have failed to report on the public health implications of climate change. So discovered a 2010 survey, which also reports that the

American public largely considers climate change a future environmental problem, not a here-and-now health problem—and certainly not a direct menace to their own children on par with, say, bullies or pedophiles.

Until recently, my own discourse about climate change followed a schizophrenic pattern. In public, I spoke boldly. In a commencement address I gave at a rural college in coal-is-king Pennsylvania, I made climate instability the centerpiece of my remarks. (Not all of the trustees were in a congratulatory frame of mind after I sat down.) Meanwhile, inside my own household, I followed a strict code of silence, underlain by my personal credo that childhood should be a time of wonderment and make believe, not for conversation about catastrophe. I'm particularly opposed to the idea that children should be pressed into public service as atmospheric junior rangers, believing instead that frightening problems need to be solved by adults who should just shut up and get to work.

But then things started to happen.

First, Elijah asked to be a lightning bear for Halloween. *It's my totem animal*, he informed me firmly. *Lightning bear is Elijah's word for polar bear,* Faith hollered up the stairs, which made me guess that she was the one who had put him up to the totem thing. But I didn't need convincing. I agreed that a lightning bear was a great thing to be for Halloween and suggested we make a costume from scraps of flannel and a chenille bedspread. (A background in animal vivisection does offer some practical life skills.)

To get a better idea of what the ears should look like, Elijah and I searched together through his picture books for drawings of polar bears, but there were none. I was afraid to go online because I didn't want images of starving, drowning animals popping up on the computer screen. Finally, I came across a file box in my office about polar bears. But this, too, contained, among other things, papers about the shrinking ice floes on which the bears hunt. I rifled through them quickly and wondered if Elijah could read well enough to figure out what these reports were about. They contained projections that melting sea ice would bring about the extinction of polar bears within his lifetime. I ended up hiding the whole box from him and set out to create a pattern by amending one for a lion costume.

As I pinned the fabric together, I wondered if his costume would outlast the species. It was more than possible. It was likely. And I wondered if any other mother of any other generation before mine had entertained such thoughts. It was unlikely.

Within the same year, three other things happened.

In November, Elijah came down the stairs for breakfast one morning and asked his sister for a weather report. Faith walked out onto the porch, spread out her arms in the manner of Saint Francis, and came back inside. *It's global warmingish.* He nodded, and they both dug into to their cereal. This could have been an opening for a conversation. I didn't take it.

During a weird balmy spell in January, when daffodils bloomed along the south wall, the front yard filled with mud, and a mosquito flew through the upstairs bathroom, I overheard a conversation on the playground. One child said, *I know why it's hot. Do you?* Another said, *It's because the earth is sick.* Other nearby children, hearing this, gathered around. They formed a circle, and they all nodded silently. I said nothing.

In early April, Elijah and I walked home from the library— no leaves to offer shade, the community bank's time and temperature sign reading eighty-four degrees—and he turned his ingenuous face to mine to ask, *Mama, is it supposed to be so hot?* I smiled as reassuringly as I could.

And changed the subject.

With that, I realized that the pedagogical wisdom of even a few years ago is already quaint. When the environmental catastrophes were occurring in some far-off biosphere—like tropical rainforests—it was sensible to defer lessons about the planet's problems for a more age-appropriate stage of development and send the kids out to the woods, to the park, to the backyard. But when the local ecosystem itself starts to shift—when the seasons around you stop making sense—maintaining silence in front of children starts to feel like a lie.

Here is one of my early memories: I was playing in the family room after dinner while a TV anchor reported the war news. When I looked up at the screen, I saw a Vietnamese child on fire. My father then stood up, walked over to the television set, clicked it off, and left the room. And my mother said that it was time to get ready for bed.

Was that the parent I was becoming?

Something else was bothering me, too. In spite of the popular truism that having children invests you in the future of the planet, the lived reality was that the parents around me seemed to be cocooned within the same hear-no-evil zone as their kids, inhabiting a world where talk about mass extinctions and rising sea levels never penetrated. When mothers gathered for playgroups or band concert rehearsals or swimming lessons or book clubs, we might comment on how tough it was to keep kids entertained during five solid days of rain, but we didn't speculate about why record-setting rains keep arriving or the latest report on the vanishing biodiversity. (*One in four mammal species is headed for extinction. Discuss.*) It's as though we needed to protect ourselves from terrible knowledge along with our kids.

For all these reasons, I decided that it was time to sit down with my own kids and have the Global Warming Talk. I had carried off the Sex Talk—and its many sequels—with grace and good biology. Surely, I told myself, I could rise to this new, albeit awful, occasion. As a parent, I am tasked with the story about the birds and bees. Therefore, am I not also tasked with the story of where all the birds and bees have gone?

On the surface, procreation and climate change seemed opposite narratives. Sex knits molecules of air, food, and water into living organisms. Climate change unravels all that. The ending of the sex story is the creation of a family. Climate change is the story of de-creation, ending with what biologist E. O. Wilson calls the *Eremozoic Era*—the Era of Loneliness.

I did discover that the two tales shared a common epistemological challenge: They are both counterintuitive. In the former case, you have to accept that your ordinary existence began with an extraordinary, unthinkable act. In the latter case, you have to accept that the collective acts of ordinary objects—cars, planes, dishwashers—are ushering in things extraordinary and unthinkable (dissolving coral reefs, dying plankton, mosquitoes in January). So, I reasoned, perhaps the same pedagogical lessons apply. During the Big Talk, keep it simple, leave the door open for further conversation, offer reading material as follow-up.

Of which there was no shortage. In fact, a veritable cottage industry of children's books on climate change has sprung up. There are fairy tales featuring carbon-breathing dragons and futuristic sci-fi novels in which child protagonists provide hope and leadership in the face of rising seas. The nonfiction selections range from the primer, *Why Are the Ice Caps Melting?* (Let's Read and Find Out!), in which lessons on the ravaging of ecosystems also offer plenty of opportunities to practice silent *e*, to the ultra-sophisticated *How We Know What We Know About Our Changing Climate: Scientists and Kids Explore Global Warming*, by foremost environmental author Lynne Cherry, in which middle school readers are cast as co-principal investigators.

This literary subgenre is impressive. Reading its various offerings, I found myself admiring the respectful tones and clear explanations. These books describe global warming as a reality that no longer lingers in the realm of debate. The children's books profile heroic individuals fighting to save the planet—in ways that kids can get involved. To read the children's literature is to see the world's people called to a greater purpose, working ardently and in concert with each other to solve a big problem— and enjoying a grand adventure while they're at it.

Is this the fiction under which we all should be laboring? I don't know. I do know that fatalism, which afflicts many adults but almost no children, is a big part of what's preventing us from derailing the global warming train that has now left the station. I do know that we grown-ups also need visions of effective challenges and radical actions that can turn into self-fulfilling prophecies.

I would like to tell you that, from these books, I pieced together the ideal Big Talk on Climate Change for young children. I did not. Talking with my kids about global warming was—and continues to be—a complete bummer. But through conversation, I did accomplish a couple of things. I managed to affirm the authority of my children's own observations—*you are right; apple trees are not supposed to bloom in December*—and I opened a new line of communication. *If you want to talk more, I'm available. I'm paying attention. I'm not pretending nothing's wrong.*

After I published an essay on the perils of talking to children about the climate crisis, one of my readers wrote to tell me a story:

During the Cold War, when the specter of nuclear annihilation hung in the air, a teacher asked her third graders how many of them thought that nuclear war would happen. Only one child did not raise her hand. And in response to the question, *Why not?* the lone dissenter answered, *Because my parents are working to stop it.*

Her point was this: The way we protect our kids from terrible knowledge is not to hide the terrible knowledge, or change the subject, or even create an age-appropriate story about the terrible knowledge, but to let them watch us rise up in the face of terrible knowledge and do something. The immediate lesson for me was: Stop acting like a Good German around your kids and let them see that you are a member of the French Resistance.

Happily, for parents, transforming oneself into a climate partisan in full view of one's offspring is not that difficult. It turns out that the work of achieving deep cuts in carbon emissions is carried out in two very different arenas, one of which is visible to children. But let's talk first about the one that is not: the political arena.

Although climatologists, economists, and ecologists may disagree about which specific road map should be used to direct us away from petroleum and coal dependency, no one argues with the fact that fossil fuels will continue to be dug out of the ground and burned as long as they are the cheapest form of energy and nobody has to pay for using the atmosphere as a dump site. We need a strategy that makes them uncheap in comparison to the renewable alternatives. This means ending subsidies for, levying taxes on, and collecting carbon fees from fossil fuels. These would then be used to accelerate the development of clean energy.

There is also broad agreement that what's preventing the enactment of such solutions is feeble public engagement. Says NASA climatologist James Hansen:

> Actions needed for the world to move on to clean energies are feasible. The actions could restore clean air and water globally. But the actions are not happening. . . . Concerted action will only happen if the public, somehow, becomes forcefully involved.

What might a forceful public involvement in the climate crisis look like? Possibly a lot like the civil rights movement. There would be marches, teach-ins, sit-ins, direct actions, speeches, music, art, and appeals by the faith community. Instead of lunch counters, think coal plants. As climate writer and activist Bill McKibben points out, this kind of political action has multiplicative effects. The civil rights movement didn't desegregate the South one lunch counter at a time. Instead, its leaders dramatized the events of one lunch counter to force a national change.

But swapping out fossil fuels for carbon-free, renewable energy is only half the battle. And it's the eye-rolling, dream-on, pie-in-the-sky half. And that's because, at the scale we now require, replacing all the energy contained in the supercharged carbon bonds of fossil fuels with alternatives like solar, geothermal, and wind power is nigh impossible. Thus, the unsexy, unpleasant, other half of the battle: changing the scale. Ending our dependency on fossil fuel is going to require dramatic reductions in energy consumption. With the willingness to make deep cuts, the whole project becomes doable.

This second arena of change—the child-visible one—is partly located in our own homes, the stage on which we parents play the role of lead actor. It's the place where a thousand molehills really do a mountain make. Individual residences are responsible for 21.1 percent of total U.S. carbon dioxide emissions. Add driving and that number rises to 38 percent. This is not a trivial figure and, in one respect, is good news because it means we don't have to wait around for political change before making immediate and radical transformations within our own spheres of influence. We can break the spell. We can prepare the way.

Indeed, many climate change analysts view the significant household contribution to climate change as the silver lining in a black cloud of otherwise ominous data. Because it represents the collective actions of individual people, rather than institutions, households are seen as a leverage point for swift change. As expressed in the title of one recent paper, "Household Actions Can Provide a Behavioral Wedge to Rapidly Reduce U.S. Carbon Emissions." According to another, we householders could quickly shrink our collective carbon emissions by nearly a third (equals 11 percent of the U.S. total) just by changing our "selection and use of household and motor vehicle technologies." By contrast, any lowering of emissions from other sectors—industrial, commercial, agricultural—is perceived, rightly or wrongly, to be thwarted by expensive infrastructure and the time required to implement complex new policy.

Maybe wrongly. I'm of two minds here. As someone who, not so far in the distant past, drove a car with more than 250,000 miles on the odometer and a coat hanger holding up the muffler while chanting *you can't break now*, I am not convinced that my "selection of motor vehicle technologies" is more amenable to rapid change than those of well-capitalized corporations. Yes, a Prius currently sits in my driveway. My bus-riding husband and I share it. We paid cash. But the time lag between the decision to select this motor vehicle technology and its execution was not brief. And the next motor vehicle technology I'd like to select—no car in the driveway—is held hostage by the fact that the village bus stops running at 6:30 p.m. and doesn't run at all in the middle of the day.

The presumption that householders are more malleable than business institutions sounds to me a lot like a cultural willingness to tell individuals how to behave coupled with a reluctance to hold the energy and manufacturing sectors to the same standards. And if the plan for blazing a path to a carbon-free future truly runs through households, shouldn't part of the strategy include rapid transformation of building codes, appliance standards, and public transportation?

And how about some bike lanes?

Some version of the same dreary list for how to green your house—the one that begins with adequate attic insulation—has been around since the Carter Administration, and it has not inspired a revolution yet. As recent investigations demonstrate, there remain significant psychological, economic, and institutional barriers to improved household energy efficiency. These include poverty; inconvenience; unclear metrics for estimating energy use; lack of actionable, coherent advice; and rental housing. As a homeowner with renter loyalties, I concede that the rental arrangement is a disincentive for action: The tenant receives bills for electricity and heat but can't choose the equipment or the source; the landlord controls the energy flow but offloads the bills. It's not a relationship to spur awareness. (For years on end, I wrote checks for heat from furnaces and boilers that I never even saw. It didn't seem strange at the time.) So, who is taking on that disconnect?

According to the most reliable estimates, we require a very rapid, 80-percent cut in fossil fuel emissions in order to avoid the kinds of temperature changes that would tip us, within the lifetimes of our children, into total calamity. Without policy solutions and a global strategy, volunteer efforts on the part of individuals—what children can see—are trivial in the grand scheme of things.

On the other hand, waiting around and doing nothing, as though bewitched, on the grounds that nothing one can imagine doing is sufficient to keep the ice caps frozen, is also a form of avoidance. Paralyzing despair is its own refuge from responsibility. So is cynicism. So is denial. Whatever the odds, we have to shake off the stupor, appreciate the severity of our situation, and get to work, heroically, ardently, and in concert, just like the characters in children's books.

And after reading a lot of them, I began to wonder: What if the heroes of the children's books on the climate crisis came to life? What if they were *us*? How would we live? What would we do? Judging by these stories, we would throw ourselves joyfully and wholeheartedly into work of public engagement (storming barricades, throwing out bad guys) without fear of looking foolish. We would develop skills that allowed us to do some big practical things (growing our own food, designing carbon-free cities). And because we are heroes—and why shouldn't we be; it's our own children's lives we are trying save after all—we would make whatever sacrifices were necessary. In this, the new crop of stories reminded me of some really old ones. In the Greek myths, sacrifices were serious undertakings and, when heroes headed out on their heroic journeys, obligatory. Done right, they could actually *change the course of nature*. Like calm the seas.

Bill McKibben again: "We have to do all that we can, whatever the cost."

So I have three suggestions for individual actions that can serve as symbolic starting points for heroism and have some practical, long-term value from a systems point of view. (More on systems thinking momentarily.) They are not intended to turn sleep-deprived parents of young children into the Martin Luther King Jr. and Rosa Parks of the climate crisis—although we sorely need a few of those, too. Yes, they involve sacrifice but not irritating inconvenience. (There's a difference.) They save rather than cost money; they can be implemented immediately; they promote health; and they are intended to inspire further action. But their central qualification for mention here is they are intended to serve as daily, visible reminders to all children of the family that our job as their parents is now, quite literally, *to change the course of nature, which has been placed, by human actions, on a terrible path.* (*Sacrifice:* from the Latin, *to perform sacred rites.*)

And I'm going to ask you to refrain from laughter until you get to the end of the chapter (and I've explained the systems thinking thing). Here they are:

1. Plant a garden.
2. Mow grass without the assistance of fossil fuels.
3. And replace the clothes dryer with evaporation.

Okay, so you're all laughing anyway—half of you because, given how far down the road of climate change we are already, you find these gestures pathetic and inconsequential, and the other half because, given how far down the road of frantic exhaustion working parents are, you find these gestures unrealistic and excessive.

But it's in the nature of heroes to shrug off snickering. Keep reading.

I am quite possibly the world's worst gardener. I would like to blame my yard for this. It's full of shade, and the shade comes from walnut trees, which, as any gardener worth her salt can tell you, exude from their roots a substance called *juglone*, whose job it is to take out the competition by wilting the neighbors. So, my garden now has raised beds (and, thanks to Jeff, brick walkways between them). These are supposed to help keep the tender roots of the garden plants away from the death-peddling juglone below. Of course, the problem of the shade remains. But I am a new gardener, and I am determined to keep at it. So far, I have grown impressive crops of mint, parsley, and lettuce. I once harvested enough peas to feed all four of us. One meal. As a small side dish.

I am upstaged in the gardening department by my own compost pile. All kinds of things take root in it. Two years ago, some member of the squash family—possibly a cross-pollinated wild hybrid of some kind—planted its seedling flag atop the steamy, south-facing slope and began to grow. Soon, its stickle-backed leaves were the size of placemats, and they were attached to a stem as thick as a child's wrist. When it started to climb the sides of the bin and spill over the top, blooming wildly with crepe-y orange flowers and shivering with bees, Elijah became so terrified that he bribed Faith into taking over his every-other-night compost-toting duty.

Elijah, you are scared of a squash vine?

It tried to grab me. I am NOT kidding. Please! You can have all my allowance.

When we came home from a camping trip, Elijah ran into the backyard to check on the compost predator.

Mom, come and look! It turned into a balloon tree!

And so it had. Grabbing hold of a tree trunk next to the bin, the vine had climbed a small maple and entirely covered it. Its flowers had become fruits—deep green globes, each the size of a party balloon. And so we had squashes—or something— dangling above our heads. They tasted good. I sautéed them with garlic and basil, grated them up for soup stock, added them to muffins and pasta sauce, and, in one form or another, we ate them all winter.

If only my garden were half as prolific. To be sure, I'm mostly slumming when I'm working in it. Because our CSA farm is only a half-mile away—and its hoop houses, hen houses, and root cellars provide us eggs and produce year round—I have the luxury of leaving the serious crop production to the experts and can still carry into my kitchen (via bicycle or sled) locally grown food, burning no carbon to do it.

It's likely that your garden will produce a much bigger fraction of your groceries than mine and thus save fossil fuels twice over—first, in the form of long-distance, refrigerated produce transportation and, second, in local trips to bring the food from market to kitchen. Regardless, the real climate value of the garden lies in three other places. The first is in building skills. If our children are going to grow up in a world of increasing environmental instability and declining oil reserves— and they are—it seems useful for them to know a few things about potatoes.

The second is in preserving the genetic diversity of seeds. In an unpredictable climate, we need as many varieties of as many fruits, grains, and vegetables as possible—the drought-resistant ones, the mildew resistant ones, the early blooming and late blooming kinds. Big commercial seed companies are, for the most part, not interested in the rare, the peculiar, and the untruckable. As a result, many traditional varieties are now on the edge of extinction. But gardens can function as living archives of genetic diversity, especially when backyard gardeners become seed savers. With almost no extra effort on your part, your garden can feed your family while also doubling as a gene bank.

As I think you can see, this is sounding pretty heroic already: *Son, I need your help in the garden today. Our job is to preserve 10,000 years of agrarian heritage.*

The real carbon savings of gardens, however, comes from the compost, and this is the reason I am issuing a universal recommendation for gardening. Together, food waste and yard trimmings make up 26 percent of the municipal solid waste stream in the United States. And, when buried in landfills, they become fuel for the production of methane, a long-lived gas that is twenty-three times more powerful than carbon dioxide at trapping heat. (See Chapter 4.) According to the U.S. Department of Energy, methane is now fully 10 percent of all U.S. greenhouse gas emissions—and its slice of the emissions pie is growing. Furthermore, of the thirteen major sources of methane, the largest single one is landfills. (At 184.3 million metric tons per year, landfills add more methane to the atmosphere than "enteric fermentation," which refers to, well, how the nation's 100 million cattle add methane to the atmosphere.)

Rerouting one quarter of the waste stream to compost bins would seriously tamp down the number one source of methane gas emissions. And, because it serves as fertilizer for the garden, compost obviates the need to purchase synthetic fertilizers, which are manufactured from fossil fuels, and which, when used as directed, add a second noxious greenhouse gas to the air: nitrous oxide (whose heat-trapping powers are 300 times more potent than carbon dioxide; it is also a precursor of smog).

So, to summarize: By planting a garden, you create the need for compost, and by starting a compost pile, three problems are solved:

1. You direct food scraps away from the methane factories called landfills.
2. You make homemade fertilizer with no fossil fuels.
3. You prevent the formation of a smog-making, heat-retaining gas.

And the best part of all: The compost pile, nature's Crock-Pot, requires almost no work whatsoever. Within months, dead leaves and old food will, all by themselves, transform into rich, black, loamy (non-smelly) humus. Depending on your perspective, this is either a holy blessing or the result of unpaid labor on the part of earthworms, fungi, and other members of the ecosystem service industry.

Depending on how you run your household, you can either send your eight- and five-year-old out to the compost pile on alternate nights, armed with a flashlight and a bucket of food scraps, because *come on now, we all have to pitch in around here*, or you can pay them.

And depending on your relationship to the various food-providing institutions that surround you—churches, temples, hospitals, schools, nursing homes, restaurants—you can go public with your newfound skills and make monumental changes. Bates College in Maine composts all of its dining hall food scraps and returns them to the farmers' fields that supply the (organic, local) food that is served there. The city of San Francisco has already instituted mandatory, curbside compost pick-up for urban residents. In short, the large-scale diversion of food scraps from methane-generating landfills is a doable project that awaits no technological breakthrough or venture capital investment. So, go do it.

Systems thinking is the idea that you can't know everything about an organization by just looking at its parts. You also have to look at the interactions among the parts because certain prop- erties *emerge* from the influence of one piece of the system on another. Systems thinking—and its heady conceptual underpin- ning—systems theory—has its roots in ecology. To understand the shape of a flower you need to understand the foraging behavior of the bee that pollinates the flower. And understanding the bee's foraging behavior requires knowledge about the distribu- tion of flowers across the landscape. In other words, the parts of the system (bees and flowers) are shaped by their reciprocal relationship (pollination) to each other.

Systems thinking has also been applied to human organiza- tions—like corporations—and to the complicated boundary between human organizations and the natural world. Systems thinking shows us that the potential of vegetable gardens to mitigate climate change goes beyond their ability to sequester carbon. We need to look at the whole system of which the garden is part, including the waste disposal system, the trans- portation system, the fertilizer industry, and the role of experien- tial learning in early childhood.

And since we're out here in the backyard already, let's apply systems thinking to the grass.

In my family, I am the designated mower. Thinking nar- rowly, I could mow my half-acre with a gasoline-powered mower, and it would consume an hour of my day and maybe— what?—a half gallon of gas? Or I could spend two hours pushing around a manual reel mower and burn no gas. A half-gallon is not a lot, and two hours is. In fact, few parents I know have two hours a week to devote to mowing. But let's widen the view and examine the whole lawn mowing system.

An hour of cutting the grass is not the same as burning a half-gallon of gasoline in a car. In fact, the mower emits in one hour the air pollution equivalent of driving an average car 200 miles. (Riding mowers are even worse.) Although the EPA required pollution reductions for new mowers starting in 1997— and may do so again by the time you read this—lawn mowers are still chimneys on wheels, and walking behind one is still more toxic than walking behind a car. Lawn mowers expose those who push them to benzene, polycyclic aromatic hydrocarbons, and fine particles. The exhaust from lawn mowers is especially rich in the chemical compounds that serve as raw materials for the creation of the lung destroyer, ozone. Five percent of smog is from lawn mowers.

Their hazards, of course, also include bodily harm. About 9,400 children are injured by lawn mowers each year, and their wounds are often complex and prone to infection. Many are disfiguring. There are fatalities. In fact, the epidemiology of lawn mower injuries is among the most gruesome subgenres within the medical literature. Consider "Stump Forming After Traumatic Foot Amputation of a Child—Description of a New Surgical Procedure and Literature Review of Lawnmower Accidents" and "Power Lawn Mower Decortication of Scalp and Skull." (Yes, I looked it up: *removal of the surface layer.*)

The risk of injury stops when the mower is turned off, but the toxic pollution goes on. Gasoline-related vapors, including carcinogenic benzene, continue emanating from the machine even when it is sitting mutely in its corner next to the snow shovel. Through evaporative emissions, lawn mowers—along with gas cans, chainsaws, leaf blowers, and, of course, vehicles—can significantly contaminate the air of the garages they are stored in—and, according to a 2007 study, the homes those garages may be attached to.

Each year, 800 million gallons of gasoline are required to mow the nation's grass, which covers 1.9 percent of the nation's land surface. Indeed, turf grass is the single largest irrigated crop in the United States. Photosynthesizing away in the sun, grass functions—as plankton and trees do—to take carbon dioxide out the air and exchange it for oxygen. Green lawns and city parks are thus potential allies in the fight against climate change. However, they don't always serve this role. The more that grass is mowed with combustion engines—and fed fossil-fuel-derived fertilizers, and blasted with gasoline-powered leaf blowers—the more that its carbon-sequestering powers are cancelled out. Indeed, many ornamental lawns are so intensively managed with fossil fuels, that they switch sides: like cars and cement plants—and unlike plankton and trees—they become net carbon emitters rather than net carbon collectors. And so breathes a strange new beast: grass that contributes to climate change.

Okay, you say. I'll switch to an electric mower.

You certainly can. The hydrocarbon emissions in your yard will drop to zero. But the mower draws energy from a power plant somewhere. And the cordless ones—which have batteries—can contribute toxic lead dust to the air.

Which is why I mow with reel mowers. (I have two of them.) My style is to mow the lawn a few rows at a time, about twenty to thirty minutes a day. Mowing thus becomes a daily ritual rather than a weekly task. As soon as my entire yard is mowed, I start over—although experientially, there is no start or end point. With a reel mower, cutting the grass becomes a Zen cycle of sickle blades that begins with the last rains of April and ends with October frost. Along the way, mowing satisfies some chunk of my hour-long daily exercise routine and so actually saves me time. Whatever minutes I don't spend mowing, I spend running.

Given its aerobic benefits, here's a marketing idea for the humble reel mower: Invite a team of design engineers to trick it out with a cup holder, a heart rate monitor, and an easy-to-read console that displays customizable workout programs. Now imagine entire suburban neighborhoods peopled with mower-cising athletes, their cardio rates rising into the maximum fat-burning zone as they pace determinedly back and forth across their yards. And, as an extra bonus, their grass gets cut.

I'm kidding. I wouldn't want to live there either. But my question is an earnest one: If we are willing to break a sweat in a gym by using muscle power to operate machines—costing us time and money, taking us away from our kids, and accomplishing no useful work—why not look at the carbon-free, hand-powered tools of lawn care as state-of-the-art exercise equipment?

From a systems point of view, a silent mower that needs no fossil fuel offers parents a number of advantages. Most notably, you can mow whenever you want—for example, during the early and late hours of the day when it's cooler. Or by the light of the moon. And because the machine is so twirlingly quiet, you can multitask. While pushing a reel mower, I've overseen play dates, supervised a trampoline party, drilled a child on the multiplication tables (her idea; not mine), and, on many summer afternoons, coaxed babies into napping. The weight of a reel mower—about 30 pounds—approximates a fully loaded stroller, and the shushing sound of the reel, like stroller wheels on pavement, is sleep-promoting. Babies and toddlers can slumber in the shade while you mow around them. (Of course, you should never leave a small child alone with any machine, reel mowers included.) By contrast, combining childcare with a gasoline mower is reckless endangerment. You can't mow during naptime, and, if you have children too young to leave in the house unattended, you can't mow at all because toddlers shouldn't be anywhere near an exhaust-spewing, rock-flinging, ninety-decibel machine as it advances through the yard. (Testifying to that are those 9,400 pediatric lawnmower injuries per year.)

I like to mow grass. It's meditative yet vigorous. It's good resistance training for triceps, deltoid, and trapezious muscles. I am therefore puzzled by men—and they are always men—who complain that reel mowers require too much sweaty drudgery, bog down in dense grass, cut poorly, and so on. How can this be? I weigh 125 pounds and cannot manage even a single push-up or chin-up. Boarding a plane, I look around for help to lift my bag into the overhead compartment. Yet I can manage a reel mower in dense grass. So also can the pair of thirteen-year-old boys who sometimes mow for me when I'm traveling or too busy. (They use my equipment. I'm not willing to ask someone else's child to walk in a cloud of noise and carcinogens while cutting my grass.)

I'm less baffled by the complaints lodged by those who believe the American lawn is an antiquated waste of time, energy, and biodiversity no matter how little carbon is expended in maintaining it. These critics fall into two camps—those who advocate for the conversion of turf into pollinator-friendly meadows of native plants and those who argue for edible landscapes of fruit and nut trees, berry bushes, herbs, and vines. Both ideas make sense and are worth exploring.

In the meantime, the use of fossil fuels for maintaining ornamental grass should be forsaken. Here's my idea for a universal philosophy of turf care: A lawn shall not exceed what its owner can comfortably cut with a reel mower. If it's so big that it requires gasoline to manage, it's too big. Plant something else. Like an orchard. In addition, exchange the synthetic fertilizer for some compost and replace the leaf blower and the weed wacker for a rake and a pair of shears. And as a reminder: In spite of all kinds of marketing imagery to the contrary, a strong machine doesn't make you strong. When engulfed in the blue haze of a two-stroke engine, it is fossil fuel, *not you, babe,* that's supplying the torque.

Men and women should have a higher power-to-weight ratio than their lawn equipment. It's sexier that way. And more heroic.

If what hangs us up about reel mowers is their inexplicable association with sissiness, the problem with clotheslines is their association with poverty.

Some communities forbid them outright, on the grounds that clotheslines lower property values. Which is a strange thing because dryers are not used in many places where people have money and enjoy their clothes. It would be hard, for example, to accuse the Italians of stylistic indifference. And yet, fewer than 4 percent of Italian households own clothes dryers. In fact, well-dressed people all around the world employ elegant, well-designed systems for evaporating water from their wardrobes in ways that don't draw from the power grid. Their houses are also smaller than ours with less space for stringing lines. Surely, we, too, can figure something out.

As it stands now, here is the choice: Loading a dryer with wet clothes takes one minute plus forty cents worth of electricity. Hanging a load is free but takes twenty minutes. (These are Jeff's estimates. He currently heads up the laundry service in our household.) A twenty-fold time difference is huge. It's a full order of magnitude greater than the time differential between mowing with fossil fuels and mowing without. And when you are washing eight or more loads a week, it matters. Added to this, the drying process itself is many times longer. On a breezy day, a cotton sheet might dry on a sunlit clothesline in thirty minutes, but a pair of jeans on a rainy Saturday will make you wait until late Monday.

A systems approach, however, can make the efficiency ratio look different. A tumble dryer is a clothes randomizer. What comes out is not a basketload of clean, dry, sorted, mated items. Another twenty minutes has to be devoted to folding clothes, hanging clothes, delivering clothes to closets and drawers, and chasing down the runaway spouses of all the jilted singleton socks.

In my own household, back in the days of the dryer, the baskets of clean but unfolded laundry formed the bottleneck between order and chaos. Everyone disliked the work of folding and sorting, and it represented yet another sit-down task, which is not healthy for anybody. The baskets just waited by the couch like so many guinea pigs whose litter needed changing. Sooner or later, somebody in search of a favorite outfit would start rooting around in the baskets, accelerating the rate of sock dispersal. Worse, dirty laundry would be tossed atop the clean. The bin labeled Home for Unwed Socks, which Faith tends, was overflowing.

By contrast, when you air dry, you can sort and mate as you go. If you use hangers instead of clothespins, it's possible to create a prêt-à-porter clothesline. (For baby clothes, a drying rack is faster.) Thus, the twenty minutes of hanging laundry incorporates the twenty minutes of sock mating, hanger work, and underwear segregation that awaits you after the buzzer goes off on the tumble dryer. As a result, our Home for Unwed Socks is now nearly unoccupied, and Jeff has to field fewer questions about where a particular beloved article of clothing might be located. And because our clothes aren't whirled around in tornadic heat, they last longer. (*Lint:* a mild-mannered word for disintegrated clothing.) As for evaporation, it just occurs on its own as one of the basic laws of physics. In this, a clothesline is more akin to a compost pile than a reel mower: The desired result, while slow, is autogenerated and can be achieved while everyone sleeps, which is not possible with a dryer. (A gentle reminder: "Never let your clothes dryer run while you are out of the house or asleep." So says none other than the U.S. Department of Homeland Security's U.S. Fire Administration.)

Clothes dryers pose hazards both direct and indirect. They might not seem as menacing as lawn mowers, but dryers are a leading cause of house fires, accounting for more than 15,000 blazes, 400 injuries, and 15 deaths a year. When not bursting into flames from lint buildup—and now I'm returning to the business at hand—a clothes dryer contributes, on average, 1,369 pounds of carbon dioxide a year to the atmosphere. More than half of all of the electricity that flows into our houses powers electrical appliances. After the refrigerator, the clothes dryer is the biggest consumer, using about \$1,500 of electricity over its 18-year lifespan. (I'm assuming you've already got the memo about the significant energy savings offered by washing clothes in cold water rather than hot.) Nationally, 3 percent of all household electricity goes to running clothes dryers. If we did nothing else but scrap them all, we could shut down a couple of coal plants.

Let me say that another way: In the midst of a global crisis, we are pumping carbon into the air in order to accomplish something—evaporation of water—that happens anyway. It you let the lawn sit overnight, the grass will not be shorter in the morning. Without inputs from somewhere, dinner will not get cooked nor houses heated. But wet clothes, happily, require zero assistance from the energy sector.

The drying system that Jeff invented for us is probably not nearly as sleek and unobtrusive as whatever the Italians do, but it has charms of its own. During the six weeks of the year that we call summer here in upstate New York, the clothes go out on a traditional flapping clothesline, as do, all year long, the sheets and towels. For most of the year, however, the laundry is hung on a series of retractable cords suspended over the stairwell, like the strings of a violin. During the winter, we hang laundry before bed. As the wet clothes dry in the draft flowing upward, they humidify the air in the bedrooms and prevent Elijah from coughing. Thus, I don't have to run an electric humidifier, which would further draw power from the grid. The moist air also allows me to keep the thermostat lower. In the morning, each kid is in charge of putting his/her own clothes away. Since we use hangers rather than clothespins, most things go right into their closet. Including pajamas.

Hanging laundry cannot stop global warming. The process that clotheslines—and reel mowers and compost piles—begin, however, is the denormalizing of fossil-fuel ways of living. They are daily reminders that we urgently need new choices within new systems. They are harbingers. They signal our eagerness to embrace much bigger changes. They bear witness to our children that we are willing to exert agency, that we are not cynical, that we respect their right to inherit a habitable planet. And they put the neighbors on notice.

The acquisition of new personal habits and new skills can change our thinking. It compels us to ask new questions. If all food scraps in the United States were composted, how much natural gas could we save? (Natural gas is the raw material for synthetic fertilizer.) What if homeowners associations encouraged, rather than forbade, clotheslines? (Project Laundry List is working on this.) What if all family homes and apartments had clothes-drying closets that doubled as humidifiers? What if landscaping services offered carbon-neutral lawn care? What if student athletes mowed their own playing fields with fleets of reel mowers as part of warm-ups?

Another world is possible. Creating it requires courage.

A few months after our final rabies shot, in February 2006, an observant hiker noticed something unusual about the bats hibernating in a cave a hundred miles east of Ithaca. Some of them had white muzzles—as though their faces had been painted with stage makeup. He photographed them, along with a pile of bat carcasses, also flecked with white.

A year later, New York State biologists found hundreds of dead bats inside several caves and also dying in the snow outside, which was even stranger because bats are torpid during the winter and do not normally rouse from hibernation until spring. They all had cottony noses. But why had they awakened and flown into the frozen, insect-less, sky? It was a complete mystery.

As the problem spread further, it was given a name: white-nose syndrome. Over the next four years, afflicted bats were found in 115 different caves and mine shafts, from Tennessee to Quebec. By 2010, a million bats had died of white-nose syndrome, many populations collapsed entirely, and wildlife biologists predicted regional extinction for the worst hit of the seven decimated species, the formerly common and widespread little brown bat. The headlines of four years earlier, describing the panic over rabid bats, were now entirely different: "Bats Perish, and No One Knows Why"; "NY State Bat Populations May Be Extinct Within 20 Years."

Each night of the summer, a little brown bat eats its weight in insects.

At this writing, whatever is killing the bats is still spreading and still a mystery. This much we know: The white fuzz is a fungus, and, when it grows on the exposed skin of a hibernating bat, it causes the animal to wake up, behave erratically, and burn up its fat reserves. Bats with white-nose syndrome starve to death. What we don't know is whether the fungus is the direct trigger for the die-off or a symptom of some other underlying problem. In May 2010, the U.S. Fish and Wildlife Service issued a statement saying, "The many people who enjoy watching the silent flight of bats through the trees or over wetlands in the night sky may no longer have that privilege." It also warned backyard gardeners that we may see, as a result of the bats' ongoing disappearance, increases in insect pests. The possible contribution of pesticides and climate change to the bats' malady is a topic of discussion among field biologists—as is the synchronous vanishings of fungal-afflicted honeybees and frogs.

A few months later, I accompanied Elijah on his nighttime compost run, and we saw a fluttery shadow overhead. *I like bats,* he announced firmly, claiming to remember nothing about the fully armed one that had once showed up in his bedroom. He did remember the shots, though.

I decided not to tell him about white-nose syndrome. We did decide to construct a bat house on the little hill above the magnolia.

And then we went inside and read *The Firebird*, a picture book adaptation of the Russian fairy tale and Stravinsky ballet. In it, our hero, Prince Ivan, finds himself in a lifeless world, whose streams are empty of fish and woods of deer. It is a barren land ruled by Kostchei the Deathless, who turns all living men in his realm to stone. Ivan is afraid. He accepts the full severity of his situation, which appears hopeless. How can he vanquish a foe who claims to be deathless? Who keeps demons as pets? He is counseled to leave. But, in the end, although afraid, he decides on a course of direct action and, with an assist from the firebird, ultimately prevails. Of course, he gets the girl. But this is the part that Elijah and I like best: Once the hero proves that the ruler of the kingdom is not omnipotent after all, the enchantment is broken. The statues rediscover their humanity. They become alive again.

Another world is possible. Creating it requires courage. Let's not be garden statuary. *Let's be Ivan.*

CHAPTER EIGHT

Homework (and Frontiers in Neurotoxicology)

THERE WAS a time in my own early childhood when I thought all children grew up to be teachers. My dad was a business teacher at the local high school and my mom a biology teacher who not only introduced microscopy and insect-collecting skills to the sons and daughters of Illinois farmers but also introduced them to the theory of evolution.

After she left teaching to raise children, mom became president of the PTA. My first job was setting up the folding chairs in the elementary school gym before she called the meetings to order. My Uncle Jack taught Spanish, my Aunt Annie home economics. Everybody my father invited over for poker games—which included beer, cigars, and *I'll see your ten and raise you five*—were, by day, teachers. Some of them, eventually, my own.

Surely, part of the unspoken commandment to teach came from the stories my parents told about their own schooling, all of which were instructive tales about the elevating power of education. All together, my mother and her five siblings were half the student body in her elementary school in Illinois farm country. But in that schoolhouse, a rolled-up map of the world awaited. Even the school's annual holiday party—with its standard-fare Christmas carols, handmade snowflakes, and cookie-encircled punch bowl—was a liberating departure from the cultural norms outside its walls. (The Bible-hewing church took a dim view of decorated, tinsel-covered trees.) By a stroke of good fortune, my mother, salutatorian of her high school class, received a full scholarship to the state teachers' college after the valedictorian turned it down. My father, likewise, became the first in his own family of nine children to go on to college, courtesy of the G.I. Bill. On their first date at the campus diner, my mother suggested they share a teabag to save money, and dad knew that she was the girl for him. So goes the legend. When they both landed teaching jobs in the same community, it became my hometown. Teacher by day and student by night, my father stacked one night-school class on top of another and, in due course, rose to the rank of community college professor. My second job was to help him grade his students' exams.

I made a pilgrimage to my old elementary school on the last day of my father's life, although I didn't mean to. The building itself—along with the baseball field, the playground, and the wooded path to the crosswalk—was razed years ago to make way for discount shopping. Out of protest, I had steadfastly refused to frequent that part of town whenever I came back to visit. But Elijah needed a haircut before the funeral that I knew was coming. ("In the *eventuality* of your father's death," was how the woman on the phone had referred to it, when I had called the community college to find out how to set up a memorial fund in advance of the obituary.) So, while my sister and my mom continued the vigil at my father's bedside—the final day, as it turned out, of a thirteen-day vigil—I retraced my old walking route to school in search of the one hair salon open on a Monday. The address in the phone book had put it somewhere nearby, and finally I found it, although nothing about the neighborhood was recognizable.

And then, while Elijah was being pumped up in his pneumatic chair, I spied, in the mirror above his head, a retaining wall whose stonework was as familiar to me as the cover of a favorite book. I turned and looked out the window.

I know those stones. I looked at them every day during math class.

I was standing in my fifth-grade classroom. Or, more accurately, in a hair-product-filled room built over the patch of ground that used to be my fifth-grade classroom before a low-rent strip mall replaced it. Here was the exact spot where the whole world had once opened to me in the form of *The Diary of Anne Frank, The Red Pony,* and *Twenty Thousand Leagues Under the Sea.* Here was the place where, in contrast to the inscrutable, we-don't-talk-about-it silence in my own family, I participated in discussions about Vietnam. It was the conversation I couldn't have with my father. And now, would never have.

A couple of storefronts down from the salon—probably where the military recruiting center now stood—would have been the lunchroom where, during seven years of lunch eating on long pink benches, I'd learned some lessons in socioeconomics. In the early years on the bench, bellied up to a long table, some of us ate neatly wrapped sandwiches packed by our mothers; some kids pulled scary-looking leftovers from greasy paper bags; and a few silent children had no lunch at all. Then came the hot lunch program. Everything changed. We all ate the same lunch, and no one knew who had paid for it and who was getting it for free. Our collective delight over identical fish sticks and fruit cocktail was some kind of lesson of its own. If my mother's school had offered respite from the constraints of religion, mine had provided a temporary suspension of the inequalities of poverty.

And somewhere out in the strip mall's parking lot was the ghost of the gymnasium—the place where I'd learned the rules of dodge ball and square dancing, where we'd all huddled during tornado warnings. Some chunk of that asphalt, striped with empty yellow parking slots, marked the place where my mother had run her PTA meetings, calling for someone to second the motion. All in favor say aye.

The Montessori school where my children attend—and my husband teaches—is a world apart from the grade school I remember—and almost unrecognizable from my mother's. The difference is more than pedagogical.

So what did you do today, Faith?

Same old. John played the sitar, and then we made African pancakes.

In December, our school celebrates the Festival of Lights, a parade of pageantry that manages to commemorate—along with Christmas—Hanukkah, Diwali, Kwanzaa, Eid al-Adha, Saint Lucia Day, and the winter solstice. Drawn heavily from Ithaca's international community, the student body itself is the map of the world. Faith and Elijah's school does not function so much as a countervailing force within a dominant culture but as a vibrant enactment of many cultures.

Maria Montessori developed her self-directed curriculum with street children in Rome. But in our school, a private institution with tuition bills, there is more ski-club membership than free-lunch eligibility, and I often wonder, as I listen to my own children talk about their friends, how their view of the world is shaped by that fact. As a third-grader, I had a much greater exposure than do my own children to economic disparity—the child of the banker and the child of the laid-off UAW worker sat side by side on the lunchroom's pink bench. But we were all white. We didn't know people who didn't celebrate Christmas. And whatever thoughts we might have had about them were probably shameful ones.

I know parents who take on second jobs and forego vacations to send their children to private schools, including ours. I know just as many who sign up for weighty mortgages in order to live within desirable public school districts. I know mothers with Ph.D.s who walked away from careers to homeschool their kids. In our household, Jeff's job as the Montessori art teacher— and consequent absence from the art studio—is what brings Latin instruction and Chinese New Year traditions into the lives of Faith and Elijah. With the nation's educational system in a state of perpetual, underfunded crisis, our parental choices are imperfect ones, fraught with anguish and sacrifice. And yet, no school, no matter how excellent, and no amount of parental sacrifice, no matter how costly, can protect our children from daily exposures to toxic chemicals that undermine their ability to learn.

Working at cross-purposes to the education of young children are a suite of chemicals known as developmental neurotoxicants—substances that impair the growth of the brain in ways that interfere with learning.

They take many forms, according to a major review of the evidence published in 2006 in the British medical journal *Lancet.* Some of them are heavy metals, such as lead, methylmercury, and arsenic. One is used to strip paint, turn crude oil into gasoline, extract natural gas from shale, and suspend pigment in some nail polishes (toluene). Some are long-outlawed compounds that still linger among us (PCBs). Another 200 chemicals are known to act as neurological poisons in human adults and are likely toxic to the developing brains of infants and children as well—animal studies strongly suggest that any neurotoxicant is likely also to be a developmental neurotoxicant—but scientific confirmation awaits.

Current laws do not require the screening of chemicals for their ability to damage or alter pathways of brain growth, and only about 20 percent of the 3,000 chemicals produced in high volume in the United States have been tested for developmental or pediatric effects. The evidence pulled together in the *Lancet* paper is one of the most comprehensive summaries available. (The EPA is using it as a starting point for its own database.) Parents struggling to pay tutors, tuition bills, and school taxes—who are, right now, clearing off a spot on the kitchen counter to sit down and offer help with homework—might consider taking a look at this compilation, particularly the review's central conclusion:

> The combined evidence suggests that neurodevelopmental disorders caused by industrial chemicals have created a silent pandemic in modern society.

In the basket of problems labeled *neurodevelopment disorders* are a variety of cognitive and psychomotor disabilities that have different names and changing diagnostic criteria. Mental retardation—increasingly referred to as *intellectual disability*—is one. Attention deficit disorder, with or without accompanying hyperactivity, is another. A third is learning disabilities, itself a basket of discrete disorders that are variously characterized by significant difficulties in listening, speaking, writing, memorizing, reading, or calculating. Dyslexia is a well-known type of reading disability. Its mathematical equivalent is called dyscalculia. Within the life-altering category of pervasive developmental disorders is autism—a continuum of problems that is now collectively referred to as *autism spectrum disorders*.

By *pandemic*, the authors of the *Lancet* study mean that learning and developmental disorders are unexceptional, cut across all walks of life and in all geographic regions, and are ballooning in prevalence. Changing diagnostic criteria, along with the absence of a nationwide registry, makes vexing the work of constructing precise time trends. The estimate most often cited by the medical literature is that developmental disabilities now affect about one in every six U.S. children, and most of these are disabilities of the nervous system. If accurate, the number of children with neurodevelopmental disorders now exceeds the number of children with asthma, which, as we have seen in Chapter 6, is also a problem of pandemic proportion.

By *silent*, the authors of the *Lancet* study mean that these disorders are subclinical. They don't announce themselves on an X-ray or in a pathology lab. There is no medical test to herald their increasingly familiar presence among us.

Child neurodevelopmental disorders do, however, leave economic tracks behind. At $77.3 billion per school year, special educational services, according to the most recent accounting, consume 22 percent of U.S. school spending. This is a cost we all pay. According to a 2006 Harvard study, the annual societal cost for mental retardation is $51 billion. For autism, it is $35 billion. The financial burden of autism is particularly cruel for families because affected children typically have special health needs and require more medical care on top of everything else. For a profoundly affected child, annual costs for care can exceed $70,000— or $3.2 million over a lifetime. Tellingly, even in the face of those kinds of financial costs, many of which are shouldered by the family, the demands of caring for an autistic child compel some parents to give up paid work altogether.

Of course, the child whose IQ is somewhat diminished by early-life exposures to neurotoxic chemicals—but not by enough points to trigger a diagnosis—also incurs economic costs to family, self, and community. These are not reflected in the estimates above.

Let's look more closely at the trends. Intellectual disability (mental retardation), which affects 2 percent of children (one in every 50), is one of the only disorders whose incidence is going down—in part as the result of the drop in children's lead levels over the past two decades.

This drop is dramatic—from an average of thirteen micrograms of lead per deciliter of blood when I was a first-grader to less than two today. Clearing lead from blood and brains of children was made possible by two public policy decisions: the 1977 decision to phase lead out of house paint, where it was used as a pigment, and the 1990 decision to phase lead out of gasoline, where it was used as an anti-knock additive. Those accomplishments, carried out over the objections, threats, denials, and obfuscations of the lead industry, are considered triumphs of public health. As well they should be. Lead exposure, at any level, is associated with distractibility, poor language skills, aggressive behavior, and lower overall intelligence. Removing lead from paint and gasoline lowered its levels in soil and house dust, and thereby slashed lead levels in children by a factor of six.

Lead is one of only two chemicals—the other is mercury—that we actually do regulate on the basis of its ability to sabotage brain development in children at background levels. At vanishingly small concentrations—at levels far lower than are required to swell the brain and produce the clinical, physical symptoms of poisoning, lead can paralyze the moving neurons within the growing brain of a child. As a result, the architecture of the brain is subtly altered in ways that compromise later learning. Many children labeled as mentally retarded in years past were, as we now know, actually lead poisoned.

In contrast to intellectual disability, other disorders are on the rise—including attention deficit/hyperactivity disorder (ADHD), learning disabilities, and autistic spectrum disorders. Four times more prevalent than mental retardation and twice as common among boys as girls, ADHD now affects nearly one in every ten children between the ages of four and seventeen. An estimated 5.4 million children are believed to suffer from ADHD, a condition that affects impulsivity and self-control as well as attention span. Of these, 2.7 million (4.8 percent of all children) are taking medication to control it. Its prevalence has been increasing an average of 3 percent a year. (Some, but not all, of the apparent rise in ADHD is fueled by greater cultural acceptance and relaxed diagnostic criteria.) In recent years, the rate of increase has accelerated. According to the U.S. Centers for Disease Control, the prevalence of parent-reported ADHD jumped 21 percent from 2003 to 2007.

Just ahead of attention problems are learning disabilities, which now affect 10 percent of U.S. children, according to the latest survey of a nationally representative sample. Of course, these are not mutually exclusive problems; some children are challenged both by attention deficits and by a learning disability.

Autism, first given a name in 1943, is a condition that is difficult to quantify, given the complex nature of the disorder and changes in how it's been defined over the decades. Children so affected may or may not have intellectual deficits. Indeed, some individuals with the variant called Asperger's syndrome have keenly developed intellectual skills. Along the spectrum of disorders that carry its name, impaired social interactions is the unifying trait. Other distinguishing features include repetitive behaviors, a narrow range of interests, limited symbolic thinking, an insistence on sameness, and—most notably—language deficits. About 40 percent of autistic children do not speak. Others do learn to talk but, often, far later than their peers. And a quarter of autistic children initially develop vocabularies and then, inexplicably, lose them, word by word, sometimes gradually, sometimes overnight.

Over the past two decades, the number of children identified with autism or autism spectrum disorders has increased tenfold. Not all of the upsurge represents a real increase in prevalence. Because the disorder was once blamed on poor parenting, shame and stigma almost surely obscured the true frequency of autism in decades past. In addition, as with ADHD, changing diagnostic criteria that now accept a broader array of signs and symptoms have brought the word *autism* to children who, in an earlier time, might have been otherwise labeled. Many doctors and teachers did not even know what autism was. Strong evidence suggests that some kids once categorized as retarded would today be identified as autistic.

And yet, the ever-longer shadow that autism casts cannot entirely be explained away by changing definitions and relaxed diagnostic criteria. An audit of carefully tended records in California suggests less than half of the apparent increase in autism is due to a more generous application of the label. But even after correcting for evolving diagnostic criteria, significant excesses remain. The Centers for Disease Control, which has established a monitoring network to track autism within selected urban areas and employs consistent criteria to do so, finds significant increases in the prevalence of autism among both boys and girls between 2002 and 2006. During this four-year span, the prevalence of autism spectrum disorders increased, on average, 60 percent among boys and 48 percent among girls.

Extrapolating from these data, CDC researchers estimate that autism presently affects 1 percent of U.S. children overall—about 730,000 individuals—and is four to five times more common in boys than girls. One in every 110 U.S. eight-year-olds is now on the autism spectrum. Among boys only, the rate rises to one in seventy.

Autism is the most swiftly rising developmental disorder, yet no one knows what causes it. It's likely that many pathways lead to the autism spectrum. Genetic factors clearly contribute—several variations in several genes have been identified as players, and identical twins are more likely than fraternal twins to be dually affected—but the genes involved are many and seem to create predispositions rather than destiny. A dramatic rise in incidence over a short time period certainly points to the potential role of environmental exposures. But what role? And what exposures?

Considerable early investigation focused on the question of whether the preservative, thimerosal, is linked to autism. Mercury-based thimerosal was used in children's vaccines until 1999 and is still found in some flu shots. For the majority of cases, the evidence does not point to a vaccine–autism link. As one thorough review of the topic concludes, "There are no reliable data indicating that the administration of vaccines containing thimerosal is a primary cause of autism." It's still possible that vaccinations play a role in a subset of autistic children—perhaps those with inborn genetic susceptibilities. But with autism rates still on the rise years after thimerosal was removed from vaccines, the search has shifted to include a much broader array of potential environmental contributors.

There are no answers yet, but there are some clues from the field of pharmacology. When taken by mothers during the first trimester of pregnancy, three drugs—thalidomide, misoprostol, and valproic acid—have been linked to increased risk of autism, as has infection by pregnant mothers with rubella (German measles). These results do not explain the increase in the prevalence of autism—these three compounds are rarely used medicines, and population-wide rubella vaccination has brought the infection rate of German measles far below what it was when I was a child—but they do point to early pregnancy as a window of vulnerability for autism. A 2010 review of the evidence thus recommended that future research efforts into the causes of autism seek to identify other environmental exposures—perhaps a common chemical agent—to which pregnant women are routinely exposed.

The very thick book, *Holt's Diseases of Infancy and Childhood*, copyright 1936, marches through an alphabetical litany of horror—rickets, scarlet fever, syphilis, tetanus, tuberculosis. Flipping through the copy that I found in the used bookstore here in our village, I tried to imagine what my life as a mother might have been like in 1936—the year that my own mother was six years old. I was about to conclude that my concerns for my children are entirely different ones than those that would have worried my grandmother until I reached the final pages of the book. Under "Miscellaneous Diseases, Lead Poisoning" is this complaint:

> The body seems to be able to repair all damage except that done to the brain. The prognosis of children who recover from the acute symptoms of lead encephalitis is bad. . . . Most of them are dull or obviously defective. . . . The treatment of lead poisoning is most unsatisfactory. It is obvious then that prevention of exposure is the main line of attack. Parents should be educated to recognize the possible harm of pica [eating things that are not food], paint chewing, and other common methods of acquiring lead. In spite of the rather high incidence of cases of lead poisoning, there are no laws in this country to prevent the use of lead paint in children's toys and furniture. In only three states is it necessary to label paint so that one may ascertain that lead is an ingredient.

Forty years before it was removed from paint, pediatricians had enough evidence for lead's ability to maim children's brains—catastrophically and irreversibly—to warrant discussion in a medical textbook. The only cure was understood to be prevention, but in the absence of government action, or even right-to-know legislation, parents were left to serve, as best they could, (ineffectively, as we now know) as their own poison control centers. This sounded familiar. And it reminded me that, even though the labels have changed, the evidence for an environmental connection to neurodevelopmental disorders is not exactly news.

Other than a keener appreciation for subtle damage at subclinical levels of exposure, what have we learned since 1936?

There seem to be four big lessons arising from the frontiers of pediatric neurotoxicology. The first is that the developing brain is more vulnerable than the adult brain, and the timing of exposure can determine whether and how severe the damage might be. PCBs, for example, are linked to decrements in remembering. Specifically, they interfere with recall ability and long-term retrieval of memories. They do so in part by disrupting the activity of thyroid hormone whose job it is during development to direct neurons to their proper places within the brain. The first cells to arrive then help direct the later ones. Thus, for PCB exposure, the earlier in development the chemically induced disruption, the more aberrant the final architecture in the memory centers of the brain.

The second lesson is that neurotoxicants can act in concert with each other. Prenatal exposure to lead contributes to the risk of ADHD, as does exposure to tobacco smoke. Both together, however, create a higher risk than either one alone. These findings indicate that neurotoxicants need to be regulated as a group rather than one by one.

The third is that elements of a child's social and nutritional environment can also be toxic to the development of cognition and can magnify the effects of exposure to chemical toxicants. Poverty and family stress are particularly detrimental. Attention deficit disorder is more prevalent among poor children, as are learning disabilities. Children living within dysfunctional families are also at increased risk for a learning disability.

(*Dysfunctional* is an overused term that tends to be thrown around with abandon. If you have ever wondered, while listening to two people yelling—and one of them is you—if *dysfunctional* is an applicable label for your own family, know that there are some established criteria. One of them is having a parent who "rarely discusses serious disagreements calmly." As long as the adverb *rarely* is part of the definition, I can honestly say that I am not that parent. But, when short on sleep and patience—as when simultaneously confronting quarterly estimated tax payment worksheets and a child who has made sibling annoyance a personal hobby—I have seen the road to that parent. And it's sobering to realize that more than emotional lives can be wrecked when tempers flare too often and too wildly. So can children's ability to learn.)

And fourth, maybe not surprisingly, is that the chemicals constructed by design to act as neurological poisons—the organophosphate pesticides—truly do so. And at levels common among children.

Organophosphates interfere with the recycling of the neurotransmitter acetylcholine, one of the messaging signals that flow between neurons. As described in Chapter 6, this disruption can affect the brain in ways that manifest in other parts of the body. By dysregulating the autonomic nervous system, organophosphate exposures can contribute to asthma, for example. Mounting evidence collected among various populations of children—from Harlem neighborhoods to the fields of California's Central Valley—all suggest that organophosphate exposure also affects cognition. For example, a small study of inner city minority children found connections between organophosphate levels in their umbilical cord blood collected at birth and attention problems at age three. These results were later corroborated by the results of a large study of children selected to serve as a cross-sectional representation of the U.S. population as a whole. As mentioned in Chapter 3, its main finding was this: Children with above-average levels of pesticides in their urine were twice as likely to have a diagnosis of ADHD.

The release of this study in spring 2010 triggered intense media coverage and lots of advice-giving to concerned parents. Wash fruits and vegetables well. Opt for organically grown food. Eschew pesticidal lawn chemicals. Avoid organophosphate pesticides when attempting to control insect pests within the home or on the family pet. With these admonitions, I felt myself back in 1936 with Dr. Holt: To prevent lead poisoning, tell parents to stop children from putting things in their mouths.

This sort of public health approach—surround kids with brain poisons and enlist mothers and fathers to serve as security detail—is surely as failure-prone with pesticides as it was with lead paint. Following all the popular advice, I do feed my children organic food, and my pesticide-free "freedom lawn" lets a thousand flowers bloom.

But my children do not live solely within the bubble of my kitchen and property lines. They occupy a much bigger ecological niche, and I cannot verify the agricultural origin of every food item served at every birthday party, summer camp, sleepover, recital, and library summer reading program event. I can't ensure that every backyard soccer field, every patch of lawn, and every pet in every neighborhood home they run in and out of is free of organophosphates.

Nor can I stop the wind from blowing. Even if I homeschooled my children and confined them to the treehouse, their bodies are connected to the rest of the world through the medium of air. Our house exists within a village community that is ringed by dairy farms and cornfields. The chemicals sprayed in those fields, on the neighbors' lawns, and on the driving range and putting greens of the village golf course can easily drift into our yard. And every study of pesticide drift indicates that they almost certainly do—just as every study of lead indicates that, sooner or later, lead molecules exit painted walls and window sills and seed themselves into soil and dust. Recall that the average four-year-old performs 9.5 hand-to-mouth interactions per hour. If lead—or pesticide—is in the dust or on the pet or inside the food, it's in the kid.

I am a conscientious parent. I am not a HEPA filter.

Moreover, consumer-based, parent-led enforcement against organophosphate exposure does little to protect the children of farmworkers, for whom exposures are even higher. Among farmworker mothers in the Salinas Valley of California, organophosphate pesticide exposure during pregnancy was associated with increased risk for pervasive development disorder among their children and with earlier birth, which itself impairs neurodevelopment.

If organophosphate pesticides are damaging children's brains at background levels of exposures and above, they should be abolished. After decades of dithering, abolition was the decision we ultimately took with lead paint. It worked. Educating parents to prevent the problem on their own did not work.

Those who argue that abolition for organophosphates is unrealistic need to explain how realistic it is to run a high-quality public school system when almost 9 percent of children can't pay attention and one dollar of every four must be directed to special educational services.

In fact, there is already some momentum for a lead-paint–inspired approach to organophosphate pesticides—although it is suffering from half measures and could use a big dose of step-it-up pressure by parents and teachers. The EPA is beginning to phase out a limited set of particular organophosphate pesticides. In response to the publication of the organophosphate–ADHD study, the agency released a statement that summarized its work in this area. In it, the agency pointed out that the cross-sectional data used in the 2010 pesticide–ADHD study had been gathered prior to 2004. (The wheels of science grind slowly.) Any effect from its recent steps to begin to restrict organophosphates is therefore not reflected in the results.

This is a valid point—although of limited relevance to those of us whose children, like the data in the study, originated before 2004. When brain-addling chemicals are phased out of production and use, no retroactive mechanism undoes the damage to children already born. Restitution is not possible. There is no back pay of IQ points. This is why further action, not just deliberative data-gathering, is required.

I doubt that I was the only parent less than comforted to learn from the EPA's 2010 announcement that the amount of organophosphates used on "kids' food" has declined by half—from 28 million pounds in the mid-1990s to 12 million pounds in 2004. My reaction to this news was along the lines of, *Twelve million pounds of pesticides with suspected links to ADHD are sprayed on foods intended for kids? Every year? That works out to a half pound of pure brain poison per U.S. child under the age of six per year. That's insane.*

As journalist Bill McKibben reminds us, we live on a climate-altered planet, one in which it's no longer possible to distinguish, even in the weather, the hand of nature from the hand of industrial pollution. A prolonged drought, a massive snowstorm, a powerful hurricane: These phenomena could be part of the natural order, or they could be a consequence of two centuries of intensive fossil-fuel combustion. From here on out, we'll never know.

Similarly, we live in chemically altered bodies now and think with chemically altered brains. It's no longer possible to discern, in our children, what part of their temperaments or cognitive quirks is innate and what part is derived from the cumulative impact of chemical exposures. Is a rush-ahead, chaotic, unfocused style the sign of a cheerfully unconcerned personality? Or a symptom of a subtle brain disorganization triggered by prenatal exposure to pesticides? The child who still can't decode a clock face by age ten: a normal variation? or the result of mildly elevated blood leads at age one?

And is the ongoing difficulty with telling time somehow related to the ongoing confusion about how to add fractions with unalike denominators?

Or is the source of the latter problem the math curriculum at school?

At the very least, trying to decipher all this is terribly time-consuming, expensive, and inconvenient for parents. Parent–teacher conferences. Parent–pediatrician conferences. Educational assessments. After-school appointments with therapy providers. Email exchanges with the math tutor. Tears shed at the kitchen table during homework time. All these activities require hours and hours of problem-solving, research, advocacy, and—most of all—continued acts of loving patience. Parenting a child with possible attention or learning problems is high-maintenance parenting—just as is parenting a child with asthma and allergies. When you are inside this world, the perceived convenience created by the nation's energy systems and agricultural systems, both of which rely on low-cost, neurotoxic substances, looks very different.

Even if you are not living inside the world of attention and learning disorders, your child almost certainly is—at school.

As more precious educational resources and instructional time are directed toward accommodating the increasing number of children with special needs, the community of the classroom is profoundly affected. While the results may not always be negative, we should acknowledge that they are real.

To all appearances, our own school brings a remarkable combination of compassion and pedagogical skill to the classroom management of children whose minds are wired in alternative ways—whatever the cause. As a result, my two children may have much less exposure to socioeconomic diversity than I did, but they have a much greater exposure to—and understanding of—a diversity of learning abilities and disabilities. While hanging around the parking lot during school dismissal recently, I overheard the following conversation:

> Child one: *So does he have Asperger's or is he being mean to you?*
> Child two: *Asperger's.*
> They nod, shrug, and get on the bus.

And during a potty break at a highway rest stop, while chaperoning a field trip, I was a participant in the following scene:

> Group of boys to me: *Hey, Faith's mom, can you help Kenny? He's scared of the toilets.*

Indeed, Kenny was frozen to a bench, rocking rhythmically, his eyes squeezed shut.

> Group of boys to Kenny: *Hey, dude. Faith's mom is going to stay with you. Don't freak out.*

So I sat with Kenny on a bench outside of a cinder-block men's room filled with roaring, auto-flush toilets while the others took their turns inside. And as I held his hand, I could only marvel at the tender mercies of his ten-year-old comrades. In my elementary school, any boy frightened by toilets—with or without a special-needs designation—would have been tormented to within an inch of his life. But here was a kid whose classmates offered protective services. Who seemed to intuit what was required, even when Kenny himself was rendered mute by psychic trauma. Ditto for the special-needs boy whose revulsion of vegetables could trigger hour-long, head-banging meltdowns. The whole class had learned to keep their carrot sticks out of sight.

I've noticed that these lessons in charity seem to carry on, even outside the world of school. Example: When I observed a teenager disrupting a children's game at a community event in Ithaca, I intervened. The boy ignored me and continued behaving aggressively. I looked around for assistance. Finally, Elijah pulled me aside—*Mom, it's okay. He's just autistic.*

The development of the brain is so drawn out and indeterminate it makes the procrastinating lungs look downright efficient by comparison. Beginning its development twenty-five days after the egg and sperm unite, the brain adds the last, refining touches to its frontal lobe about twenty-five years later.

The brain also rivals the lung in the art of packing a suitcase. Both organs specialize in cramming many feet of functional surface area into a confined, bony space. The growing brain accomplishes its task by rolling up and folding over. It also crumples. At first, the brain is a smooth lump, but, starting at about the sixth month of pregnancy, it begins to form worm-like mounds called *gyri* separated from each other by creases called *sulci*. By growing wrinkles, the brain can triple in size during the first year of life.

At the cellular level, biologists use the language of electricity and computer science to describe the process of brain development. The brain's fibers truly are wired together to form a living electrical web, and they really do conduct currents. That much of the story is not metaphor. The brain has circuitry. The electrician, however, is the ever-shifting relationship between genes and the environment. As foremost neuroscientist Lise Eliot describes the process:

> The brain itself is literally molded by experience: Every sight, sound, and thought leaves an imprint on specific

neural circuits, modifying the way future sights, sounds, and thoughts will be registered.

By contrast, when we talk about programming the brain—invoking the language of computer science—we are speaking metaphorically. Unlike a computer, which can be programmed and reprogrammed indefinitely and at any time, the brain passes through rigidly sequential, relentlessly directional stages of development in which early processes limit and govern the activity of the later ones. The options for reprogramming become progressively more restricted, as the early events create the architecture—the hardware, if you will—for later events. Thus, if mistakes are introduced up front, everything that comes after that point is permanently altered.

Children's brains are also missing antitrespassing defenses that help protect the brains of adults. For example, if you are old enough to read this, you probably have an enzyme called PON1 that can break down organophosphate pesticides. PON1 activity, however, is very low in babies—rendering newborns as much as 65 to 130 times more vulnerable to pesticide exposure than adults. (And some children are born with a variation of PON1 that is not very effective. So even as they grow up, they remain vulnerable.) In all these ways, a child's brain is less like a computer and more like a vulnerable ecosystem with no real equivalent in the world of grown-ups.

Like a five-act play, there are five stages of brain development, and while they are universal, they don't occur simultaneously in all areas of the brain. The centers of vision and hearing, for example, finish all five stages sooner than the region of language acquisition.

Acts I and II are brief. First is neurogenesis, the actual creation of the brain cells, which takes place between weeks three and eighteen in pregnancy. The second stage is the migration of those neurons within the skull. Through these long-distance journeys, all the brain structures form: the brain stem, cerebellum, various lobes of the cerebral cortex, and so on. This migration is highly choreographed. Under the guidance of various helpers, the individual neurons glide along, like dancers in a Russian ballet, to precise locations on stage. Among those offering stage direction and assistance to the moving neurons are thyroid hormones (disrupted by PCBs and, possibly, brominated flame retardants) and a crew of helpers called glial cells (disrupted by pesticides).

Along the way, neurons establish points of communication with various other members of the troupe. The creation of these connections—synapses—is the third stage. Synaptogenesis takes off in earnest after birth and continues into the second year of life. You might imagine that this phase—the actual wiring together of the electrical web—would be meticulous and precise. It's not. It is fast and furious. In the time span between birth and age two, about a quadrillion different synapses are established—an unfathomable number. Brain development is no longer a ballet. It's an all-night rave. It's a hyperextended excess of connectivity.

In fact, it is excessive. And, thus, the fourth step is the pruning back of about half of the connections. Between early childhood and adolescence, synaptogenesis actually runs backward, and 20 billion synapses per day—according to Lise Eliot—are disconnected. This two-step process of frenzied, lavish overproduction followed by a mass retrenchment seems so inefficient that generations of neurobiologists and evolutionists have pondered the point of it all.

Their best guess is that synaptic pruning allows for learning to shape the brain. The connections that are trimmed away are the ones that are rarely electrically activated. What is gained at the end of the process is a more coherent brain that is better adapted to its environment. What is lost is plasticity—the ability to adapt to some other environment. (As we shall see in Chapter 9, the hormones of puberty resculpt the brain and accelerate the process of synaptic pruning.)

Finally, the neurons undergo myelination: They are wrapped with increasingly thicker insulating layers of fat, which are provided to them by glial cells. This swaddling prevents electrical leakage and speeds neuronal signals along. Myelination, which overlaps in time with synapatogenesis and synaptic pruning, begins in the back and lower reaches of the brain and advances, slowly, to the front and outer layers. It is the biological basis for maturity. Differential myelination is why the brain can direct breathing (governed by the brain stem) before it can execute a plan or exercise good judgment (overseen by the frontal lobe). The brain stem is myelinated by birth; two decades later, the cerebral cortex, which is the site of higher-order thought, is still getting dressed. Along the way, thanks to progressive myelination, we learn to roll over, sit up, hold a spoon, crawl, walk, jump on one foot, read, memorize, imagine, calculate, scheme, rhapsodize, and respond to social cues.

Like a star athlete with a team of coaches, each electrically snappy neuron is surrounded by many myelin-making glial cells. In addition, the glial cells offer nutrition and first aid in case of injury. This assistance is critical because neurons don't routinely regenerate, as do blood cells or bone cells. Blood cells quickly replace themselves, as do, far more slowly, the honeycombed cells of your skeleton. Thus, you don't really inhabit the same body that you were born into—or even the same one you occupied a decade ago. However, you still have the same brain. Your neurons—the ones that began forming when you were the size of a paper clip—last a lifetime. Although adult brains can, under some circumstances, sprout new neurons, the brain cells you were born with are, by and large, the same ones you have now. (Learning and experience, of course, continue to tinker with their wiring over a lifetime.)

During the migration stage of brain development, the loyal glial cells also play another role: They form bridges for the neurons to travel along—from their birthplace deep in the brain to the brain's outer edges. Treacherously, glial cells are the preferential targets of assassination by organophosphate pesticides. This fact explains a couple of things. First, because glial cells are present in all parts of the brain, it explains why the pesticide chlorpyrifos can undermine such a wide range of unfolding skills, from children's ability to pay attention to the speed at which they can tap their fingers and from visual memory to verbal comprehension. Second, it also explains why you can't extrapolate to children the results of studies conducted on adults. Low-level chlorpyrifos exposure certainly disturbs the brain functioning of adults, too, and can lead to tremors, anxiety, and rapid heart rate. But adults possess stationary neurons, wrapped up in fat, whereas fetuses, infants, and children have mobile, half-naked brain cells. A chemical exposure that attacks the cells that enable their migration has bigger, most devastating consequences.

The ability to pay attention—to stay on task, dig in, and inhibit impulses long enough to achieve a goal or figure something out—is a glorious thing. So is its opposite: the capacity to find inspiration in whatever presents itself. The contrary skills of attention and wonder seem to me two sides of the same golden coin. Together, they constitute what my mother might call *inner resources*. At the very least, they are a hedge against boredom.

While pregnant with Faith, I met a remarkable three-year-old who seemed to have both skills in great abundance. He was in the audience for one of my college lectures—his mother had parked his stroller next to her chair in the auditorium—and he quietly amused himself for an entire presentation (the requisite fifty minutes) and the Q and A session that followed (fifteen more minutes). Sometimes he played with puzzles and books his mom put before him—and looked to be deeply concentrating—and other times he just gazed around inquisitively at whatever caught his eye.

I caught up with his mother after the event—she was a student at the community college campus where I was the speaker—to let her know how impressed I was. And, beaming, she told me her story. A teenager when her son was born, she knew that she wanted to go to college but had no money for childcare on top of tuition. So she waited until her baby was a toddler and then started taking him every Sunday morning to Mass—*because church is free*—and, in the back pew, began the work of teaching him how to sit still and be quiet.

At first, he could manage only ten minutes. But each week, with lots of praise, library books, and secret hand signals, he learned to keep it together for a little longer. And when he could sit through an entire High Mass service, she signed up for classes. And here they were.

This account was an epiphany for me—not quite yet a mother—because it offered an alternative to the two usual choices: hiring babysitters or bringing a disruptive child into an adult setting. And so I set out to copy her methods—minus the Catholic Church—with mostly happy results. Although there have been a few *what was I thinking* episodes, each of my kids could, by the time they were four, reliably sit through a lecture.

And then through a three-hour public health workshop.

And then through a four-hour oncology seminar.

This ability allowed us to travel together. Other skills useful to travelers soon followed: The ability to eat what you're served. The ability to fold up and go to sleep when you're tired. The necessity of patience and manners.

I like to think that the decision Jeff and I made in 1999 to not replace the stolen television enabled some of this—although, in an age of Internet-enabled, game-playing camera phones, it sounds almost quaint to say so. Basically, our kids are unplugged. Which is to say, our house offers almost no access to electronic media. Computers represent tools for grown-ups to accomplish work-related projects; they are not family entertainment centers nor portals to the wider universe.

I don't know how early-life immersion in a digital world influences pediatric brain development. I don't think anyone knows yet. Given the absence of data, hundreds of years of time-honored cultural practices just seem to me a better bet—and none of them draw from the energy grid. To develop an attention span: read a book, play a musical instrument, practice a sport. To cultivate wonder: go outside and check on the whereabouts of the sandbox toad.

Whenever I mention my digitally unmediated childrearing practices in mixed company, I'm likely to get one of three responses. First, someone will offer expressions of awe for even attempting to parent without video games, as though I had announced an intent to forsake coffee or host a foreign exchange student. But children who invent their own games, make their own lunch, stage plays, and possess a decent repertoire of knock-knock jokes are in need of less of my attention, not more. They also, Jeff says, have better scissors skills. And as far as I can see, life is more peaceful without battles over screen time. (It's possible my house is messier.)

Alternatively, someone is sure to warn me that, by disallowing my kids broadband access, I'm denying them the skills they need for the twenty-first century. In fact, I'm trying for the opposite. It was a hiker with keen observational skills who discovered the first signs of white-nose syndrome in bats and so alerted the world to an unfolding wildlife catastrophe. It was the humble Secchi disk thrown from the sides of boats by researchers who knew how to pilot them that revealed that 40 percent of our oceanic plankton has vanished; satellite monitoring had missed it. It is people with farming, cooking, and canning skills who are reinvigorating rural economies through the creation of local, organic agricultural systems.

My thoughts here are also influenced by anecdotal accounts from colleagues who teach field biology. Increasingly, they say, college students seem unfamiliar with the basic features of the natural world—including what kind of clothes and footwear are appropriate for a trip to, say, a bog. Knowledge that many of us picked up from scout camps, or from just knocking around outside as a kid, is no longer universally shared. It is news to many otherwise bright undergraduates that moths come from caterpillars and beetles from grubs. Flowers turn into fruit via insect pollination. The moon waxes from right to left. Poison ivy has three leaves. One ecologist I spoke with recently said that his students who grew up playing video games in the basement seem to have a hard time noticing things out in the field, such as the small movements of birds. *Do they lack peripheral vision?* he wondered.

From all this, I infer that ecological literacy requires tuning children's synapses to the rhythms of the natural world rather than to virtual ones. Nature is a subtle place, with less dead-ahead action than a digital game. Observing it requires an eye trained to notice the almost undetectable—the half footprint in the mud, an absence of bees—and a willingness to wait and see what happens. Some facility with knots and fire is also useful. Habituation to electronic media—for comfort, amusement, escape from boredom, or because parents want it quiet in the backseat—seems unaligned with those goals.

Hey, let's play I Spy.

The third response I get, mostly from people my age or older, are nostalgic stories about childhoods spent roaming the earth, unsupervised by adults, flying on bicycles over hill and dale until the call for supper. The best part of being a kid, they remind me by way of offering their approval, is living in an adventure-filled, adult-free world out in the woods or in an abandoned lot, organizing their own baseball games, negotiating the rules.

In fact, that's not the world my kids inhabit, if only because there are no packs of roaming kids anymore. They're all engaged in organized activities refereed by grown-ups. Or they are inside downloading something. In fact, it's hard to overstate the radical transformation of childhood that has taken place over the course of a single generation. Since the 1970s, the amount of time that children spend in unstructured, outside play has, according to surveys, declined by half and is now exceeded by the time spent inside vehicles. And when they are not being transported—or while they are—children are interacting with screens. In 2010, U.S. youth ages eight to eighteen spent an average of 7.5 hours each day involved with some form of electronic media. (Up from 6.5 hours in 2005.)

So, yes, in other people's houses, my kids have access to computer games. And, no, my kids do not enjoy the kind of freedom to roam that I once had. In fact, I'm a little testy about the oft-repeated accusation that mothers these days are irrationally paranoid about stranger abduction, which, as we are forever reminded, is no more frequent now than years ago. The fact is that years ago the outdoor world was full of children, and there was safety in those numbers. You can't just tell a kid to go outside and play when no one else is out there. If nothing else, it's lonely.

In my attempt to counteract the allure of online role-playing games where many children gather, the childlessness of nature is the big problem.

Exposure to air pollution—whether indoors or out—harms children's cognitive development. Specifically, it reduces intelligence. Thus, the combustion of fossil fuels not only creates a climate problem, it renders our children less able to solve it—or any other problem.

Polycyclic aromatic hydrocarbons (PAHs)—the sooty chemicals released from tailpipes and power plants—are the leading culprit. In studies of young laboratory animals, PAHs behave as neurodevelopmental toxicants, and they appear to have the same effects in children—although the mechanism by which they impede brain growth is not yet known. As part of their series of long-term studies, researchers at the Columbia Center for Children's Environmental Health demonstrated in 2006 that three-year-olds who had higher levels of exposures to PAHs during pregnancy—as measured by their levels in umbilical cord blood—scored lower on cognitive tests when compared to lesser exposed counterparts.

Subsequent studies of 400 mother–child pairs in New York City and in Kraków, Poland revealed similar findings. In both countries, five-year-olds exposed to above-average levels of PAHs scored, on average, about four points lower on standardized tests of reasoning ability and intelligence.

Research conducted in China before and after a coal-fired power plant was shuttered reveals similar results—but with a happy twist: two-year-olds with elevated prenatal exposures to PAHs showed poorer neurobehavioral development when compared to less exposed toddlers. However, after the power plant shut down two years later, differences among an identical cohort of two-year-olds disappeared. When coal-burning emissions were eliminated, concentrations of PAHs in umbilical cord blood declined, and children's development benefited. Mothers whose babies were born after the coal plant closed had smarter children.

Coal extinguishes intelligence by a second route: through its release of mercury into the atmosphere, which then finds its way into the brain cells of children through the medium of fish.

From the earth to the air to the water to the dinner table. And so into blood and neurons.

When coal is extracted from its carboniferous tunnels and burned in power plants, the mercury it contains vaporizes. Once airborne, the newly liberated metal can wander the skies for up to a year, traveling thousands of miles. Eventually, it falls with raindrops and snowflakes back to the earth's surface. (Urban smog encourages this homecoming.) Bacteria then convert mercury into the potent—and persistent—brain poison, methylmercury. Concentrating as it moves through the food chain, methylmercury can magnify its powers many times in aquatic systems, where food chains are long. Our main source of exposure, then, is seafood—especially large, predatory fish.

When consumed by a pregnant woman or a young child, fish ushers molecules of methlymercury into a developing body and into a developing brain. Here it can cause many kinds of damage. Methylmercury stunts the growth of baby glial cells, disrupts the transmission of dopamine, and damages cells in the hippocampus—a seahorse-shaped structure that serves as an office of memory. Altogether, sufficient prenatal exposure to methylmercury is associated with loss of IQ, learning disabilities, forgetfulness, attention deficits, as well as balance and coordination problems.

In 2003, the Centers for Disease Control quantified the problem: One of every twelve U.S. women of reproductive age has blood mercury levels above that known to be safe. Accordingly, every year, more than 300,000 infants are born at risk for mercury-induced cognitive impairment. In a 2005 study, no doubt designed to reach the ears of politicians deaf to all but monetized impacts, researchers at Mount Sinai School of Medicine's

Center for Children's Health and Environment put a price on these findings. Noting that the diminishment of IQ leads to reduced economic productivity, these authors estimated the societal cost associated with mercury emissions from U.S. power plants at about $1.3 billion a year from loss of intelligence alone. Coal is not cheap.

In August 2009, the U.S. Geological Survey released the results of a comprehensive study that sampled streams across the nation—including the isolated and the pristine. Mercury was found in every single fish caught in them. By 2008, all fifty states had issued advisories warning their residents, especially women of reproductive age, and their children, to limit their consumption of certain freshwater fish. Indeed, mercury is the reason for 80 percent of all fish advisories issued for lakes and streams. Some ocean fish, including tuna, are also sufficiently contaminated to trigger recommendations for consumption limits. Indeed, tuna, all by itself, contributes more than one-third of the total methylmercury exposure from seafood. (Tuna salad: the new lead paint.)

Divorcing women and children from fish is, of course, an imperfect solution to the problem of mercury contamination, as it deprives them of the not inconsiderable health benefits of eating fish. A better solution would be to divorce ourselves from coal—and pursue a full-bore affair with renewable energy.

The burning of coal is not the only industrial practice driving the creation of methylmercury, but it is the single biggest one. All by themselves, coal-fired power plants contribute about 40 percent of industrial mercury emissions. Medical and municipal waste incinerators also play a role, but their input has fallen dramatically with the phase-out of mercury from many consumer products—and the closing of many incinerators. For these improvements, we have state laws and local ordinances to thank—and the never-give-up efforts of environmental advocates who pushed lawmakers to act. Other significant contributors to atmospheric mercury are cement kilns and chlor-alkali plants that employ mercury to pull salt apart for chlorine (in preparation for the manufacture of PVC).

If you are wondering what cement might have to do with mercury, the answer is, well, coal. To make cement, kilns burn coal to cook limestone rocks. Both coal and limestone contain mercury. Furthermore, the powdery gray ashes left over from burning coal in power plants are allowed as a raw material for cement-making. (The name for this practice is *beneficial reuse*.) The tighter the pollution control devices in the smokestacks of coal plants—and they are getting tighter—the greater the concentration of mercury in the ashes. New national rules have recently been established to deal with the high-and-getting-higher mercury emissions from cement plants.

Coal ash is also poured into building materials and consumer products, including cosmetics, countertops, wallboard, carpet backing, and, yes, flooring tile. In these ways, coal—and whatever's in it—finds its way into our homes by more routes than electricity.

For all their shared faith in the transformative power of education, my parents were less convinced that harp lessons blazed a path to a bright future. So when I came home from a fourth grade field trip and announced a fervent desire to learn the harp, my mom offered an alterative idea. Why not piano lessons? We didn't have a piano, but I could practice at church. And if I liked it and kept at it, she and dad would get me a piano.

I liked it, and I kept at it, and true to their word, my parents bought me a piano—and cheerfully paid for the next *ten years* of weekly lessons, driving me to music contests, attending every recital. My father even signed up for adult lessons at the community college so he could learn, too. His devotion to my musical education strikes me as extraordinary now. Especially considering that I entirely lacked talent (for which I compensated by prodigious use of the pedal).

Nevertheless, I still like it, and the first purchase I made when we moved into our house was a refurbished upright piano. To commemorate Faith's first recital, my mother sent a package of old songbooks and sheet music that she had scooped from the bench of my own childhood piano, where they had undoubtedly sat for more than thirty years. Faith immediately seized on *The Red Book*, one of my first lesson books, and began to sight read. Her favorite was "Tune of the Tuna Fish (copyright 1945), which introduces the key of F—

> Tuna fish! Tuna fish! Sing a tune of tuna fish!
> Tuna fish! Tuna fish! It's a favorite dish.
> Everybody likes it so. From New York to Kokomo.
> Tuna fish! Tuna fish! It's a favorite dish.

After we belted out the song a few times together, Faith asked, *Mama, what's a tuna fish? Have I ever had one?*

In fact, she hadn't.

A few weeks later, at a potluck picnic, someone offered Faith a tuna salad sandwich. She loved it. On the ride home, she announced that she would like tuna sandwiches for her school lunches. She wants to eat one *every day*. And bursts into song. *Everybody likes it so. From New York to Kokomo.*

The U.S. Food and Drug Administration has guidelines for monthly tuna consumption limits specific for pregnant women and children—as well as for nursing mothers and women who might become pregnant. There is considerable debate about whether these current restrictions are sufficiently protective, and different kinds of tuna—albacore versus light—have different limits.

I have no idea what the guidelines say, and I have no intention of finding out. Whatever they are, they're not useful to me. In my experience, children who discover a new food to their liking want it served at every meal from here to Sunday and are not content to eat it once a month, or even once a week. Occasional samplings are not part of their ritualized dietary ethos.

How do you explain to a kid with a newfound taste for tuna that she'll have to wait a week before she can have her favorite dish again? Do you tell her that she's already consumed her weekly quota of a known brain poison, as determined by the federal government? Or do you make up some other excuse? And am I to maintain a tuna consumption calendar to keep track of our ounces per month per person?

In the end, I did talk with Faith about the problem with tuna—while also reassuring her not to worry about the potluck sandwich. I said that keeping mercury out of fish required generating electricity in some way other than burning coal and that I was working hard on that project.

Soon after, we went hiking in the woods near the day camp she had attended earlier in the season. She summarized for me the history of the old stone building where snakes and turtles are housed in one wing and bunk beds fill the other. It was originally built, she explained, as a *preventorium*. Children whose parents were sick with tuberculosis stayed there so they wouldn't get sick, too.

I saw an opening. *You know, we don't have to worry about tuberculosis anymore. We fixed that problem.*

She said she knew that.

The top of the hill offered a view across Cayuga Lake. On the far bank floated the vaporous emissions from New York State Electric and Gas Corporation's Cayuga Plant, whose coal-burning stacks were plainly visible against a cloudless sky. In the year that Faith was born, the Cayuga facility released 323 pounds of mercury.

That's where the mercury comes from that gets inside the fish. Someday, we'll fix that problem, too.

Then they can do something else with the coal building, Mama.

CHAPTER NINE
Eggs (and Sperm)

WHEN ELIJAH was a very little boy, his hair was long and blond.

And here two story lines diverge.

In my version, I thought he didn't want a haircut, and I wasn't interested in forcing him to have one. In his version, he thought that I wouldn't let him have a haircut.

For the sake of the narrative, let's go with Elijah's account, on the grounds that mothers shouldn't publish tales about their children when the facts are in dispute. So, during the years in which I was obliviously ignoring my son's desire for a haircut (*Why didn't you just ask me?*), he was often perceived as a girl.

I didn't fully realize just *how* often, however, until, having discovered our mutual miscommunication, we paid a visit to the barbershop.

Thereafter, no one called him sweetheart or honey anymore. It was all buddy, pal, guy, or dude. The sing-songy voice previously used to deliver a greeting—by everyone from the waitress to the lumberyard guy—dropped an octave. And the *Hey there, Mr. Man* came with demands for an NBA-style slap-me-five.

A more perceptive mother might have noticed the transgender mistake much sooner and taken steps to correct it. But my other child is a daughter. I was accustomed to hearing a beaming, melodic *good morning, sweetie* when an adult addressed one of my kids. I thought—oblivious again—that's just how adults talked to small children. I didn't realize there were different salutations for girls and boys.

In hindsight, I don't know why I was so surprised at this discovery. Everything else about childhood is an exercise in extreme gender stereotyping, the likes of which one seldom encounters in adult life. The most conservative of men—the upholders of tradition and the defenders of unfettered capitalism—sit next to me on airplanes dressed in shirts of pale pink, a shade that not even the most unconventional parents would consider for their infant sons.

Along with color limits, the rules for decorative elements on little boys' clothes are unwaveringly strict. Bees are allowable motifs—but not butterflies. Camouflage-style vines are okay, as are palm trees and anything jungle-y, but flowers are out, and fruits are questionable. A pumpkin hat for a boy is fine, but not a strawberry hat. Bananas are okay, especially if gorillas are holding them. Peapods and root vegetables are fine. Cherries, not fine. Suns, moons, comets: *ja*. Rainbows, *nein*. Mammals are permitted. Birds are taboo. Unless it's *archaeopteryx*. All dinosaurs are acceptable. At one point, my personal mission was to find, in the bins of gently used children's clothing at our local consignment shop, something in Elijah's size that was not festooned with weaponry, a combustion engine, or a large, predatory animal. I rarely succeeded.

Toys, especially the figurative kind, seem even more polarized, as though they had been soaked in great vats of steroidal hormones before hitting the retail shelves. Over here, futuristic, muscle-bound robo-guys with no discernible ecological connections. They're all about power. Over there, coy, glittery princess fairies who occupy monarchal kingdoms with pollination systems (flowers and butterflies) but where the rules of democracy and science don't apply. For girls, it's all magical thinking.

I've heard many parents of boy/girl families say that child-rearing has convinced them that sex differences in behavior and learning are hardwired. If by this image, we mean "innately predisposed," I'm not so convinced. My kids enjoyed childhoods that were, for all intents and purposes, commercial free, and in the early years, I can't say that I noticed a lot of differences. In fact, Faith and Elijah followed strikingly similar pathways of development: both sat up on their six-month birthdays, spoke their first words before they crawled, and didn't walk until fifteen months. Until Faith was five and Elijah two, when we lived in relative isolation in the cabin—with woods and spongy swaths of cattails for a playground—my kids showed roughly equivalent interests in books, balls, dolls, music, art, trains, frogs, dinosaurs, dance, and outdoor exploration. And it was Faith, not Elijah, who got in trouble for shoving during nursery school sing-alongs.

Maybe, of the two, Elijah showed a greater early interest in throwing and catching—and that explains his greater interest in sports now. Maybe Faith possessed the deeper congenital need to squish things—and that explains her greater interest in baking projects. That's possible. But it's equally likely that their various proclivities were differentially encouraged (or ignored) by teachers, peers, and probably—in ways that we can't even see—by Jeff and me. Certainly, Faith's early curiosity about cooking was nurtured along by a family friend of ours who was willing to spend hours with her in the kitchen. Faith also seems to possess an unusually keen palate, with a remarkable ability to discern subtle tastes and smells. Is this an inborn olfactory talent? Or did two years of breastfeeding, with its banquet of different flavors, help create it?

Because we are sensitive to accusations of preferential treatment, bias, or unfairness, these accidental factors—especially the ones involving the influence of birth order—tend to get left out of the stories that parents create to explain the origins of their children's temperaments, learning styles, and talents. Instead, we cobble together a narrative from selective memories that serves to highlight the foreordained nature of it all. I do this, too. When, after a soccer game, someone told me that Elijah seemed to have a natural love of the game, I heard myself say, "Well, you know, his very first word was *ball*."

At which point Faith chimed in, not without contempt, "Actually, Elijah's first word was *wa-wa.*"

She was right. I'd forgotten the days of *wa-wa.* This is why everyone needs an older sister to demythologize the childhood stories created by one's mother.

If, on the other hand, by asserting that sex differences are hardwired, we simply mean, "organized as if by permanent electrical connection," I wouldn't argue. Developmental pathways laid down in early childhood—however the wiring occurs—can become destiny. As neuroscientist Lise Eliot points out, infant brains are so malleable that any inborn differences between boys and girls—and she believes they are real but small—can easily become magnified when reinforced by gender stereotypes.

Here's one more tale from the family annals at our house: When Elijah was three, he found a handsaw that Jeff had left, in a violation of all child safety standards ever drafted, lying on the porch. Without the permission or knowledge of anyone, Elijah spirited the tool away, stashing it under the mattress of a futon couch in my office. (This part of the story we reconstructed after the fact.) A few days later, while I was in the kitchen working on dinner, Elijah announced with a sigh that he was feeling a little tired and wished to lie down for a while. Could he go in my office?

Okay, said the oblivious mother, who, a full five minutes later, thought, *Wait, that doesn't sound right,* and decided to check on him. He was not there. The mattress of the futon couch was askew, and the door to the backyard was wide open. Stepping outside, I spied my son high up in the magnolia tree.

Sawing a limb.

Not, thankfully, the one he was sitting on.

Here was the dilemma. A gotcha reaction on my part would likely trigger a panic reaction on his part. I could easily imagine the various horrible consequences of that. But ignoring the scene wasn't an option either. So, employing some stealthy moves of my own, I crept to within catching distance but stayed out of sight. Then I waited. A few sawed branches fell around me. And then, finally, down came Elijah. Slowly. Shimmying along the gnarly trunk, limb by limb, holding on to the tree with one hand, clutching a big shiny saw with the other. During his descent, he hummed to himself. I held my breath. Once his feet were back on earth, he surveyed the scene. And there was mom!

What followed was a long moment of blinking and staring followed by laughter. Then came the words: *See, Mama, I found a saw. . . .*

So in the family treasury of stories—which is how we come to understand who we are—what is the message of this one? Is it about Elijah's innate attraction to tools and weapons? Or to secrecy and adventure? Is it evidence of a male attraction to danger? Is it about how Faith had *never ever* done anything like that when she was his age and how boys are so different from girls when it comes to impulse control?

Or is it about two semi-exhausted, semi-distracted parents who unwittingly created the opportunity for a premeditated heist in ways that had not happened with our daughter because 1. we weren't yet engaged in home construction projects when she was small; and 2. we had only one kid, not two, to keep track of when she was his age?

Endocrinology concerns itself with the hormonal messages that fly, like fleet-footed Mercury, through the bloodstream from the Mount Olympus of the glands to the lowly cells of the body. Those bearing receptors that can bind a particular hormone are the intended readers of its message.

Actually, glands both send and receive hormonal messages, and some brain cells act like glands—which is why, all together, the brain's neurons and the body's glands are collectively known as the neuroendocrine system. Inside the center of the brain, for instance, sits the throne-shaped hypothalamus, supreme governor of sexual maturation. You cannot grow up without a hypothalamus. More specifically, you cannot grow up without the assistance of its gonad-stimulating neurons (of which there are only about 1,000). At about the end of the first decade of life for girls—a little later for boys—these sleeping neurons wake up and begin drizzling into the blood gonadotrophin-releasing hormone, and, when received and amplified by the pituitary gland, that's the sound of a gavel coming down: Childhood is adjourned; puberty is now coming into session.

But new research shows that the hypothalamus itself is studded with receptors of all sorts. Some of these respond to hormonal messages produced by glands located elsewhere in the body. Some respond to chemical cues—enzymes and neurotransmitters—produced by neurons and glial cells located elsewhere in the brain. And, in turn, these far-flung brain cells are the receivers of messages streaming in from everywhere.

Some of these incoming signals encourage the slumbering neurons of the hypothalamus to remain quiescent. (Melatonin, which conveys information about light levels in the environment, has this effect.) And some of them encourage arousal. (Leptin, a hormone that reports to the brain on the status of the body's fat reserves, has this effect but, all by itself, is not able to trigger puberty. Kisspeptin, a protein produced by neurons in the forebrain, appears to be a necessary conspirator.) When the balance of permissive signals finally exceeds the balance of inhibitory ones, the gonadotrophin-releasing neurons of the hypothalamus break dormancy and begin directing the work of pubertal onset. Sitting, as it does, downstream of a whole nexus of chemical cues, the hypothalamus turns out to be less a prime mover than a messenger itself.

Adolescent puberty, then, is the end result of a big game of pass it on. It's more like a classroom exchange of Valentines than the ringing of a preset alarm clock. Furthermore, it's not even a unique event.

There is also *infant puberty*. Those two words come as a surprise to many people, including the parents of infants. But it's real, and it's normal. Both boys and girls come into this world with wide-awake hypothalamic neurons, actively secreting gonadotropin-releasing hormone. In response, the testicles or ovaries of newborns produce sex hormones—testosterone or estrogen. New parents often notice the startling effects. Red, puffy genitals. Acne. Swollen nipples. These traits recede and disappear entirely by nine months. (Newborn boys experience a peak of testosterone at three months.)

During infant puberty, sex hormones help organize the neurological system—as they will a second time during adolescent puberty. In boys, the release of gonadotropin-releasing hormone in infancy may play a role in masculinizing the brain and almost certainly assists with testicular maturation. In both sexes, infant puberty seems to prepare the brain for its job in adult life as the regulator of sex hormones.

After a few brief months, this whole system is switched off and remains latent until adolescent puberty. And that's the endocrinological definition of being a kid: childhood is a hormonal hiatus between two puberties, made possible by a temporary inhibition of the hypothalamus—through an unknown mechanism.

The chemicals known as endocrine disruptors—like the chemicals known as neurotoxicants—have the power to alter developmental pathways during childhood. About 200 endocrine disruptors have been identified so far. They include certain pesticides, plastics, detergents, flame retardants, air pollutants, dioxins, PCBs, and heavy metals, including cadmium and lead.

Because the brain itself is organized under the influence of hormones, endocrine disruptors and neurotoxicants are overlapping identities. Any chemical, for example, that disrupts the activity of thyroid hormones—which act as wilderness guides for trekking fetal neurons—can bring ruin to the brain. But for our purposes here, we focus on that subset of endocrine disruptors with the potential to interfere with the hormones that oversee the creation of the reproductive system.

Endocrine disruptors, like neurotoxicants, are not screened or regulated as a group. No single federal agency is tasked with the job of monitoring all chemicals with the power to mimic or otherwise interfere with the flow of hormones. Of the many chemicals flagged as possible disruptors of sexual development, three are currently receiving sustained attention by researchers and are the subjects of intense, contested debate: bisphenol A, phthalates, and atrazine. The first two are ingredients of plastic, and the third is an herbicide. All are produced in high volume, and human exposure to all three is widespread—indeed, both bisphenol A and phthalates are considered ubiquitous in the environment and in people.

More specifically, 93 percent of Americans have measurable levels of bisphenol A in their urine, according to the Centers for Disease Control, which keeps track of these things. Researchers at the Columbia Center for Children's Environmental Health—the folks who put air monitors on pregnant women—found phthalates in virtually all personal air samples, as mentioned in Chapter 1. The Centers for Disease Control also found universal exposure to phthalates among Americans of all ages, with the highest levels in children. Its presence in human amniotic fluid and umbilical cord blood indicates that prenatal exposures are more than hypothetical. And, according to researchers at the National Institute for Environmental Health Sciences, 60 percent of Americans are exposed, mostly through drinking water, to atrazine, the second most common pesticide used in the United States.

At more than 8 billion pounds per year, bisphenol A has one of the largest productions of any chemical in commerce. It is a single chemical compound, and its primary purpose is for the manufacture of plastic: Bisphenol A molecules strung together make the polymer called *polycarbonate*. Bisphenol A is also used in epoxy resins that coat the inside of food and beverage cans, as dental sealant, and in printer inks and coated paper, such as that used for cash register receipts.

As an endocrine disruptor, bisphenol A is an estrogen mimic and behaves like a forged document. In the mammalian body, it can attach to the estrogen receptor and so elicit the sorts of cellular responses that are triggered by real estrogen (albeit far more weakly). Its talent at faking out the endocrine system is not a new discovery. Indeed, the affinity of bisphenol A for estrogen receptors was demonstrated in 1936. But before it could be widely marketed as a pharmaceutical hormone, a stronger synthetic estrogen took its place—the infamous DES, diethylstilbestrol. Stripped of its career as a synthetic hormone, bisphenol A languished in chemistry labs until it was repurposed as the building block for polycarbonate.

Thus, its innate ability to act like an estrogen is not in dispute. It's known to shed from plastic bottles and can linings into the food and beverages contained therein, especially when heated—as when a bottle of breast milk or formula is warmed in a pan of water. No one disputes that fact either. Where the contested accusations fly is over exposure estimates and the risks of harm from those exposures, especially for infants and children. Uncertainties swirl around questions such as, How quickly does the liver send bisphenol A to the kidneys?; and, Is there enough to cause harm? Various state and federal agencies are pondering these data gaps. Meanwhile, in 2010, over the objections of the American Chemistry Council, the Canadian government declared bisphenol A a toxic substance.

Phthalates and atrazine bring a more furtive approach to endocrine disruption. Their methods of inflicting damage are not through mimicry. Phthalates, as you'll recall, are a big family of chemicals with lots of uses. They turn up in the cosmetics department (as ingredients in perfumes, lotions, aftershave, nail polish, shampoo), in the home improvement section (vinyl flooring, shower curtains, wallpaper, garden hoses), and in children's outerwear (e.g., Curious George raincoats), but, most covertly, in our food. Some, but not all, of these many phthalates are reproductive toxicants. Instead of copying sex hormones, phthalates cut their supply lines, most notably during the production of testosterone and insulin-like factor 3. (In spite of its unsexy name, insulin-like factor 3 is as much an elixir of manhood as its famous friend, testosterone. More on both these hormones momentarily.) The end result, in many animal studies, is demasculinization. And in rats, exposure to phthalates during prenatal life can trigger testicular cancer in later life.

In spite of elegant research that has revealed the vulnerability of the developing male reproductive tract to damage from phthalates, many uncertainties remain. Marmoset testicles, for example, appear less sensitive to phthalate-induced injury than rat testicles owing to hardwired differences in the speed at which they can metabolize phthalate esters. Are human baby boys more like rats? Or marmosets? And, given that we are exposed to mixtures of phthalates of many different molecular weights, how do we tease apart the effects of one from the other? And given that we are almost all exposed on a daily basis to both phthalates and bisphenol A, how might they interact within the bodies of pregnant women? Within the testicles of baby boys? In 2008, Congress banned the use of six phthalates in nipples, pacifiers, teething rings, and toys that might be mouthed. This legislation, while welcome, does not address the exposures during pregnancy.

As for atrazine, its game is to trick the body into making more of its own estrogen. We are all, male or female, yin-yang creatures with mixtures of male and female sex hormones coursing through our bloodstreams. Our genetic sex determines the balance of the two. To make a sex hormone, you start with cholesterol and, like twisting a sausage-shaped balloon into a poodle, amend the molecule to make testosterone. A few more manipulations and, voilà, estrogen. In laboratory studies, atrazine enhances the production of an enzyme called aromatase, which is used by the body to convert testosterone into estrogen. Aromatase is the balloon twister. The more aromatase, the faster the conversion. The end result is higher estrogen levels. Less yang, more yin. (Aromatase inhibitors, by contrast, are chemotherapeutic agents used as a treatment for some estrogen-positive breast cancers.)

In experiments with developing tadpoles, atrazine has feminizing effects—of the most dramatic sort. Male tadpoles exposed at key moments in development turn into fully functional, egg-laying female frogs. In female lab animals, atrazine also interferes with hormones from the pituitary gland, whose job it is to deliver messages from the hypothalamus to the ovaries. Given these effects—and observations from other labs about atrazine's ability to alter mammary gland architecture in female rodents— questions have been raised about atrazine's potential to alter the pathways of breast development in girls. There are many uncertainties. What is the message of frogs and rats for humans? Is there a safe threshold level below which exposure effects are negligible? And what should parents of young children do between now and when scientists secure conclusive answers?

And who gets to decide when conclusiveness has been secured?

At this writing, the EPA is, not for the first time, reviewing the health effects of low-level atrazine exposure with an eye toward understanding its possible developmental effects. In 2006, in spite of the remaining uncertainties, atrazine was banned for use in the European Union.

Attending scientific meetings on environmental threats to reproductive health would surely be an unsettling experience for any mother of young children, biologist or not. From the program of one such conference, here are the titles of some of the presentations I saw: "Prenatal Exposures and Male Reproductive Disorders," "Urogenital Birth Defects in Newborns," "Postnatal Life Events That Determine Adult Male Reproductive Function," "Early Postnatal Environmental Contaminant Exposures and Reproductive Health Effects in the Female," "Male Mediated Developmental Toxicity," and "Alterations in Puberty."

Surely, what makes these topics so profoundly disturbing is the juxtaposition of something public, noxious, and invasive (chemical contaminants) with something that verily defines the words *private, innocent,* and *off-limits* (the reproductive organs of infants and children). We are talking here about threats to that part of my children that I am charged, above all else, with safeguarding: their sexuality, their fertility, their connection to future generations, and thus to the abiding, ongoingness of life itself. These are the body parts about which the necessary motherly refrain is *This is private. This is just for you. No one else is allowed to touch you there.*

Chemical trespass plus children's genitalia: It's a violation almost unbearable to contemplate.

And yet, at the same time, there is something, for me, almost liberating about these PowerPoint presentations, with all their micrographs and diagrams and statistical confidence intervals. And that's because, in their exploration of the evidence linking exposure to endocrine-disrupting chemicals with impaired reproductive health and fertility, they create a forum about an issue for which we have no other language. The data bear witness. They offer vocabulary words that, however stilted or icky, allow us to see that the development of our children's reproductive tracts (icky words) is jeopardized by insufficiently protective chemical policies. And protection is what I am charged as a parent with providing.

"Altered Semen Quality in Relation to Urinary Concentrations of Phthalate Monoester and Oxidative Metabolites." "Phthalate Ester-induced Gubernacular Lesions Are Associated with Reduced INSL3 Gene Expression in the Fetal Rat Testis."

These are terrible words.

Yet, they are better than silence.

You can go to medical and scientific meetings about asthma, and hear from the experts the latest evidence on its causes and triggers—its possible relationships to diesel exhaust, phthalates, climate change, or pet dander. And you can then go home and talk about the rising rates of childhood asthma with whomever you please. You can attend any number of seminars and workshops on autism and the changing diagnostic criteria for attention disorders. And then you can join advocacy groups, such as the Learning and Developmental Disorders Initiative, and continue the conversation. By contrast, it's very difficult to talk openly about, for example, gubernacular lesions. Outside of endocrinology seminars, there is precious little discourse on the topic.

The gubernaculum lowers the testicle into the scrotal sac during fetal development. Most of it regresses, but part of it turns into a ligament that secures the testicle within the scrotum.

If anyone touches you there, tell me. It's not your fault. Don't keep it a secret.

A few years ago, I wrote a commissioned report on the falling age of puberty in girls and had the opportunity to present the findings at public meetings, including a Congressional briefing. I had never before entertained so many questions posed in whispers. On the way to the bathroom, in the hallway, while I was putting on my coat and hailing a cab, audience members approached me, one by one and whispered to me their stories and questions. This included at least one EPA official whose adopted seven-year-old daughter was developing breasts and what, he whispered into my ear, did I know about early puberty among foreign-born adoptees? (There's a small body of literature on this very question, actually.)

When I sat down with a reporter in Toronto, she began our interview by announcing flatly, *My four-year-old son has one ball.* What startled me was not the content of the sentence but the fact that she wasn't whispering.

Until week five of pregnancy, boys and girls have the same genital tract. Then, the genes associated with the embryo's genetic sex awaken and spur the production of hormones that guide the reproductive tract along one of two very different directions, male or female. Of the two, the development of the male tract is more intricate and involves considerably more steps. Whereas the female version hews closely to the original template, the male genitalia, when finished, is an elaborate piece of engineering that barely resembles the five-week-old unisex prototype that was its starting point.

Testosterone plays a starring role in much of the design work and subcontracts the rest. After triggering the production of a second hormone to direct the development of the prostate gland, the penis, and the scrotum, testosterone goes to work overseeing the construction of the transportation system along which—someday—sperm will travel. Thus, a continuous pipeline is laid from the interior of each testicle to the tip of the penis. But this ductwork doesn't take the direct route. Instead, it heads north, into the pelvic cavity, loops over the pubic bone, skirts along the bladder, finally turns south and shoots forward again, making a complete loop-the-loop. This circular arrangement allows various glands to contribute the liquid portion of the semen. Meanwhile, inside the testicles, squadrons of cells begin creating a well-organized center for sperm production, much of which will be devoted to nurturing the sperm cells after their creation. Like glial cells to neurons, a whole support team is needed for every sperm cell. Although not required until after puberty, the infrastructure for spermatogenesis is laid down during fetal life.

The testicles begin their development high inside the body, near the kidneys. Installing them within their permanent home outside the body involves a two-part, carefully orchestrated procession. During part one, a suite of hormones, including all-important insulin-like factor 3, guides the testicles from the kidneys to a base camp in the pelvis. Then, near the end of pregnancy, like a mountain climb run backwards, the testicles begin their final descent into the scrotal sacs.

The aforementioned gubernaculum is the rappelling rope.

Quite a lot of male hormones are required to carry out all of the above. Exposures to chemicals that interfere with the production of these hormones, block their messages, or destroy the cells where they are made can jeopardize the whole process. The result can be a developmental disruption with the (icky and stilted) name, *testicular dysgenesis syndrome*. It can manifest as a range of smaller and larger alterations, including a collection of four conditions that sometimes occur in isolation but more often cluster and which seem to have a common cause: namely, abnormal development of the testicles in prenatal life due to insufficient levels of male hormones to choreograph the show. Two of these conditions—hypospadias and cryptorchidism—are birth defects visible in the delivery room; another two—testicular cancer and low sperm count—do not show up until adulthood.

Hypospadias occurs when the opening of the penis is located somewhere along the shaft rather than at the tip. Its incidence in some geographic areas has more than doubled since the 1970s. As with autism, some, but not all, of this apparent increase is attributable to greater awareness and changing diagnostic criteria.

Cryptorchidism is the medical term for undescended testicle. It is a word I rather like. *Hidden orchid* is the image conjured up by its etymology . . . except that orchids were named for their resemblance to testicles and not the other way around. But the poetry of the word belies the unhappiness of the condition. Its prevalence in some parts of the industrial world appears to be increasing, and, along with hypospadias, cryptorchidism is a risk factor for low sperm count and testicular cancer. In men diagnosed with an undescended testicle, the risk of developing a later testicular tumor increases significantly, a pattern that suggests that testicular cancer has its roots in fetal development.

Testicular cancer is now the leading cancer among young men. Its incidence in the United States has doubled since the 1960s. During the same period, sperm counts and semen quality have declined, along with testosterone levels. Due to changing methodologies for ascertaining sperm quantity and quality, many uncertainties surround these historical time trends. However, there are some striking contemporary geographic patterns that persist even with rigorously standardized data collection. These are correlative patterns that offer no proof but do provide clues for further inquiry. Men from rural Missouri, for example, have half the number of moving sperm than do men in urban Minnesota. They also have higher pesticide exposures. In Sweden, a group of men whose mothers had higher PCB levels had elevated rates of testicular cancer. In Taiwan, the sons of mothers who had unknowingly consumed PCB-contaminated rice oil while pregnant grew up into men with smaller penises and larger numbers of sperm abnormalities, as compared to unexposed counterparts.

Testicular dysgenesis syndrome has its origins in many places. Family history—and, thus, heredity—contributes to the risk for cryptorchidism, as does preterm birth, low birth weight, alcohol consumption during pregnancy, and maternal smoking. But, in addition to the indirect evidence for environmental harm gathered from the above human studies, an impressive and growing body of evidence from the lab bench and from the field strongly implicates endocrine-disrupting chemicals as actors in the story.

Wildlife observations of male animals offer striking parallels to patterns seen in human men. In a series of now-famous field studies, for example, zoologists in Florida demonstrated that the failure of alligator reproduction in Lake Apopka is attributable to infertility among the males who had abnormally small penises, low testosterone levels, and high estrogens. They were further able to show that the contamination of the lake with pesticide was the likely cause of the reproductive anomalies among male alligators. Uncontaminated alligator eggs painted with pesticide—at concentrations replicating that found in the water—produced male alligators with reproductive deficits identical to those seen in males born in the lake.

In the club of the demasculinated, alligators are not alone. Field biologists report reproductive anomalies in male mammals as well. They involve species up and down the food chain—including the big, fierce guys that are often emblazoned on clothes for little boys. In addition to pesticide-exposed panthers with undescended testicles, they include frogs with eggs inside their testicles and many species of hermaphroditic fish, from Mississippi sturgeon to Canada whitefish.

From experimental studies in lab animals, researchers have shown that insufficient testosterone can induce all four different manifestations of testicular dysgenesis. As described above, testosterone has many executive functions in male reproductive development—from formation of the urethra to oversight of testicular descent and, in adult life, management of sperm production. Within the developing testicle, testosterone is produced by a specialized group called *Leydig cells*. These cells are also the main targets of phthalate interference. When they gain entry to the inner chambers of the Leydig cells, phthalates alter the production of enzymes and proteins that guide the transport of cholesterol—testosterone's starting point. Phthalates also disable genes that oversee the production of insulin-like factor 3, which, together with testosterone, is the GIS system for the traveling testicle during its journey to the bottom of the scrotum.

Along with these findings, lab bench researchers also discovered that inadequate testosterone during the time the fetal testes are developing can result in an externally visible mark: short-ened anogenital distance. This is a new finding—and links to specific abnormalities included in the testicular dysgenesis syn-drome are just being discovered—but it looks like short male anogenital distance will turn out to be one of the earliest and easiest to identify markers of this syndrome.

Basically, a shorter anogenital distance means that a male newborn possesses a smaller perineum—which is the name for that springy expanse of skin and muscle that extends between the anus and the base of the penis (and in females, to the vagi-nal opening). All mammals possess perineums, although it's a body part you've probably not given much thought to—let alone worried if yours was bigger, smaller, thicker, thinner, perkier, or plainer than everybody else's. Exception: A woman in childbirth thinks about her perineum once every thirty seconds or so because she's attempting to push the baby past it. Delivery with-out perineal ripping is a core skill of midwifery. Obstetricians bring a less artisanal approach to perineums and are more likely to slice through one (episiotomy) to speed the baby's arrival and sew it back up after the fact. Among the people truly attuned to their anogenital distances are mothers with stitches in their perineums.

To the business at hand: In mammal species, including us, males have longer perineums—greater anogenital distances—than females (about 50 percent greater, on average). In rodents, male perineums are fully twice as long. An anogenital distance in a male that resembles that of a female is an easy-to-spot sign that testosterone production during fetal development has been inadequate. Male rodents exposed to anti-androgenic chemicals, including phthalates, have shorter perineums. The greater the disruption to testosterone production, the more femi-nized the perineum.

With the lab discovery that anogenital distance is a sensitive and verifiable measure of endocrine disruption—and is visible and measureable at birth—attention turned once again to human studies, and researchers began comparing anogenital distances among various cohorts of baby boys. So far, the findings include these: boys with shorter perineums have a higher prevalence of undescended testicles and smaller penises. Boys whose mothers had the highest body burdens of phthalates had the shortest perineums. About one in four U.S. women is believed to have phthalate levels in her body high enough to reduce the ano-genital distance in their sons.

Of the many oversights—okay, *mistakes*—I've made as a mother, one of my bigger ones involves bedtime. As in, I failed to establish one. A night owl by nature, I always resented the lights-out directive that, every evening of my own childhood, extinguished the fun just as the groove was really getting groovy. Quite possibly, that's why, as a parent, I have a hard time shouldering the job of making people go to bed.

Of my two kids, Faith is better suited to the accidental policy of *Oh, wow, look at the time; what are you still doing up?* which is what passes for rule of law in my household. Elijah, on the other hand, would probably prefer the comforting routine of nine o'clock Taps. He often has to remind me—sometimes during a very exciting part of the book!—that he needs to go to sleep now.

In saying this, I'm not intending to present myself as some kind of cool, bohemian mom. I'm now convinced—a dozen years into parenting—that shirking bedtime enforcement duties is truly not a good thing. There is a price to pay at the breakfast table when every night is a slumber party, and I would bet that a big part of the morning crabbiness quotient in our house originates from my own adolescent immaturity around going to bed at a decent hour. In a self-evaluation of my parental skills, I would give myself an A- on Encouragement of Healthy Eating Habits and a D+ on Bedtime. Had I a parenting do-over, I would handle things differently.

Having botched the whole notion of enforced bedtimes, I also seem to have missed the lesson on how to carry off a brisk yet loving tucking-in ceremony. Jeff or I still lie down with each of our kids every night and stay with them until they sink below the surface of sleep. No amount of systems theory can dress this up as an efficient practice. Nevertheless, in this one regard, I have few regrets.

Lying on a narrow bed in the dark with an elbow jutting into my ribs, I allow phone calls to go unanswered, dishes to remain unwashed, and essays to lack concluding paragraphs. With some exaggeration, I could say that the priceless benefit of sharing with my children the vulnerable moments before sleep is that I get to hear what's really on their minds—the confessions, fears, and existential questions that don't emerge in the light of day. In truth, this happens only occasionally. More typically, Faith recounts, with reportorial accuracy, the whole arc of the day's events, and Elijah provides, with fantastical flourishes, a play-by-play account of the recess hour's soccer game. In these nightly recitations, I still learn a lot about the worlds my children occupy and have created for themselves.

Of the two, Elijah's world is more mysterious to me. How is it possible—I sometimes ask myself while half listening in the dark to yet another explanation of how a player can be offside if he receives the ball on a free kick but not from a corner kick or a goal kick—that Jeff and I gave birth to this child? Neither my artist husband nor I hold the remotest interest in organized team sports. (My personal approach to all games involving flying objects basically derives from dodge ball: duck.) And yet I now stay up late (okay, *even later*) to memorize the names and faces on Elijah's stack of sports cards and study the soccer rulebook so I can impress him by asking informed questions. And, after school, I practice kicking balls, so he can practice goalkeeping.

He gives me lots of credit for these efforts. Elijah and I adore each other in big, bewildering, and complicated ways. All things treasured by my son are sources of delight for me—including those that, sans Elijah, I would scarcely notice. I suspect that this animation of formerly ignored things (who knew I had a special talent for air hockey?) is one of the unseen joys of parenting that is imperceptible to non-parents. I certainly didn't know about it.

My relationship with Faith requires less translation. Because we jog together a couple of times a week, I actually have a more straightforward athletic partnership with my daughter than with my son. And running side by side along country roads offers us plenty of time for reportage. At age twelve, Faith's ear for dialogue is as keen as when, in preschool, she referred to herself in the third person. Since I myself occupy a narrative world, I'm on familiar ground while listening to her serialized accounts.

Accordingly, my nighttime ritual with Faith is less an attempt to decode the contents of her heart than it is an enduring act of physical affection. Curled next to me in the dark, she continues her daily news updates while I finger-comb her hair and breathe her in. By contrast, in the upright, daytime world, privacy is now the watchword. The bedroom door stays shut. The forgotten towel is handed over the top of the unfurled shower curtain. The needed roll of toilet paper is passed through a crack in the bathroom door. *I'm getting dressed. Stay out.*

Is there a female analogue to testicular dysgenesis syndrome?

Not really, if only because so much female reproductive anatomy is hidden from view. Instead, researchers use another metric to identify possible environmental influences on reproductive development: age at onset of puberty. Among U.S. girls, it's falling—and at a speed far faster than genetics alone can explain and in ways that increased body fat can't entirely account for. Various lines of evidence—laid out below—point to environmental endocrine disruption as one of several contributing causes.

It would be easy to say that early puberty is to girls as testicular dysgenesis syndrome is to boys. That's not quite right, though. While both are signs of increased risk, we don't know if boys are reaching sexual maturity sooner or later than in generations past. Far fewer data are available for boys. (Some researchers do allege that pubertal age is also shifting for boys as well. Maybe. Not yet convinced, I'm awaiting the results of ongoing studies.) By contrast, we just happen to have a rich collection of pubertal data for girls that dates back a hundred years—possibly because assessing breast development and onset of menstruation in pubescent girls is a more culturally acceptable topic of study than, say, assessing penile diameter and scrotal volume of pubescent boys.

A more principled explanation would be that the well-established connection between early sexual maturation and breast cancer risk has made urgent the examination of female pubertal timing and its determinants. All things being equal, early blooming girls—those whose first periods arrive before age twelve—are more vulnerable to breast cancer after menopause and, as breast cancer patients, are more likely to be diagnosed with an aggressive tumor. Conversely, for each year menstruation is delayed, the risk of breast cancer declines by 5 to 20 percent. A first period at age 16 or over decreases breast cancer risk by 50 percent compared with a first period at age 11.

The mechanism by which early puberty makes a breast cancer diagnosis in later life more likely—and possibly more fatal—is not entirely clear, but two aspects appear to be significant. First, early puberty is associated with increased lifetime exposure to estrogens. Second, early puberty opens wide the window of time between first period and first pregnancy, an interval that is considered critical for breast cancer risk. The shorter that vulnerable time span, the better. In either case, it's clear from the breast cancer literature that identifying the causes of early puberty in girls—and intervening to eliminate them—is a meaningful place to begin a program of breast cancer prevention for our daughters.

From the fields of psychology and anthropology comes this interesting gender discrepancy: Early-maturing boys tend to be treated as leaders by their peers and teachers and are admired, whereas early-maturing girls are more likely than their late-blooming counterparts to be scorned, harassed, and subject to multiple forms of victimization, including violence. From this body of research, I conclude that the culture in which my children live projects esteem onto boys entering manhood and, onto girls becoming women, sexual objectification.

Indeed, early puberty in girls is associated with a number of startling problems, which reads like a list of every bad thing that can happen to a teenager. Girls who are the first in their cohort of friends to develop breasts report more negative feelings about themselves and suffer more from anxiety. Early-maturing girls are more likely to experience depression, eating disorders, and suicide attempts. They are more prone to early drug abuse, early smoking and alcohol use, and early sexual initiation, which itself is linked to a greater number of lifetime partners. They are over-represented in the criminal record and underrepresented at college commencement ceremonies.

In the late 1990s, a noisy debate on early puberty spilled out of the corridors of clinical practice. It began when a group of pediatric endocrinologists, protective of their patients, recommended lowering the age at which puberty should be considered precocious (an official medical condition) to spare younger girls invasive diagnostic work-ups and drastic hormonal interventions to arrest puberty's progression. They argued that the existing standards were out of date with today's trends. These standards, which had long ago set the age of precocious puberty at eight, now automatically categorized 14 percent of U.S. girls as abnormal.

Public health researchers countered that what had become the norm was not necessarily normal. Or good. And changing the definition of abnormal didn't make it so. Moreover, referring to redefinitions as "updates" made those rightfully concerned about the downward shift in pubertal timing seem somehow out of touch, like elderly aunts who still insist on handwritten RSVPs in an Internet age. Nevertheless, those advocating for change won the argument. In 1999, the cut-off age for precocious puberty was pushed back from eight to seven for white girls and from seven to six for black girls.

Narrowing the definition of *precocious* has normalized early puberty. Yet, I'm sympathetic with the compassionate impulse behind this revision. As its proponents point out, most early-blooming girls have no underlying medical problem in need of treatment. And trend is not destiny. In spite of the extra stress they endure, plenty of early-maturing girls develop into happy, high-achieving adults. Indeed, it's not clear if the very real psychosocial problems faced by early-maturing girls carry over into later life. One large-scale study finds little evidence that early-maturing girls become troubled adults, although they do apparently continue to suffer from significantly higher rates of depression. Other studies do report persistent differences: Early-maturing girls perform less well in school and are less likely to finish college.

But there are no data to show that temporarily halting precocious puberty with injections of hormones—an extreme act of endocrine disruption to be sure—spares girls the various risks that come with early maturation, including the elevated risk for breast cancer. Offering these girls social and psychological support may prove a more effective and compassionate approach than offering them syringes of synthetic anti-estrogens. As always, knowing when to intervene in the life of a child—and with what tools—is a vexing decision.

Meanwhile, caught up in an argument about the normalcy or not of third graders wearing bras, we are avoiding another question: Why are we willing to dream up radical interventions for girls growing breasts and not for our systems of chemical regulation?

In all female mammals, including humans, sexual maturation is a trait that responds to cues from the external environment (availability of food, mates, and shelter) as well as from the internal environment (presence of body fat; absence of infectious diseases). Both must be favorable for successful reproduction, which, in female mammals, requires an immense commitment of calories and nutrients as resources are diverted to the tasks of making babies, giving birth, and making milk. Among white-tail deer—a species I once studied intensely—yearlings and even spring fawns can become sexually mature by the fall mating season if food is plentiful. If it is not, a doe won't go through puberty until age two or older.

Female sexual maturation is a reactive state of affairs.

Against this backdrop, the falling age of puberty in girls—a trend that was set in motion more than a century ago and is still underway now—looks like an extension of a natural process: Girls developed the ability to reproduce at younger and younger ages in response to less disease and plentiful calories. Between the mid-nineteenth and mid-twentieth centuries, the age of pubertal advent dropped steadily—about three months every decade. This downward trend was temporarily interrupted during the Great Depression, when malnutrition increased—along with the average pubertal age of girls. When good times returned, pubertal age resumed its slow descent.

The trends of the last fifty years, however, are more complicated. The tempo of advancing puberty has gathered abrupt speed. The different milestones of puberty—breast development and first menstrual period, for example—that were once tightly coupled to each other have become less connected. And puberty is not just starting earlier; it's also unrolling more slowly and thus lasting longer. Moreover, these changes are seen in thinner girls as well as in chubbier girls. Such trends suggest that girls' endocrine systems are being subtly rewired by stimuli other than good health and sufficient food.

To interpret these patterns requires some familiarity with the whole parade of hormonally driven changes in girls that marches by during the two- to six-year period of time known as puberty.

The procession of events occurs roughly as follows. After their long nap, the gonadotropin-releasing neurons of the hypothalamus re-awaken, like so many electrical Sleeping Beauties, and activate a hormonal circuit that, in turn, arouses the ovaries, which begin secreting estrogen. The result is breast development and, eventually, onset of menstruation. A second signaling pathway, also originating in the hypothalamus, stimulates the adrenal glands atop the kidneys, which begin androgen production. The result is pubic hair, underarm hair, acne, and oily skin. These two signaling circuits appear to operate independently of each other. (For most girls, breasts appear before pubic hair.) The arrival of menstruation—menarche—is a late-stage event. Following on its heels is the arrival of ovulation, which signals the attainment of fertility and, thus, the end of puberty. Ovulation— the release of the first egg from the ovary—typically follows the first menstrual period by about a year, and the first menstrual period follows breast budding by about two years.

Along the way, under the direction of sex hormones, the pelvis widens, fat accumulates, the vagina lengthens, the uterus inflates, and the folds of the vulva blanch from red to pink. Peaking just before the first menstrual period, a pubertal growth spurt takes the girl to her final adult height, while estrogen ossifies the ends of her long bones. She is now as tall as she'll ever be. The brain is also transformed during puberty. New neuronal connections sprout and elaborate, and older pathways are pruned away. New synapses form. Others disappear. White matter increases in volume; gray matter decreases. In both males and females, pubertal resculpting of the brain's circuitry is believed to make possible the emergence of abstract thinking, values, autonomy, adult social behaviors, and the capacity to consider alternative points of view. The brain also gains efficiency, becoming faster in its processing.

But speed and the development of higher order thought does not come without a price: During the course of sexual maturation, the brain loses plasticity. As a result, after puberty, the ability to learn complex new skills declines dramatically. Playing a musical instrument. Riding a bicycle. Mastering a sport. Acquiring a new language. The prepubertal brains of children are far better equipped for these tasks than the less moldable brains of adults. Indeed, after puberty, one cannot learn to speak a foreign language without an accent. It's simply no longer possible. *C'est la vie.*

Moreover, the risk-taking parts of the brain develop sooner, under the direction of pubertal hormones, than the control centers of the brain. These facts raise for me some questions. If our daughters now have fewer years of cognitive flexibility remaining to them, is accelerated sexual development diminishing their full intellectual, musical, and athletic potential? And if female puberty is starting sooner and unscrolling over a longer number of years, are our daughters now experiencing a more prolonged disconnect between the development of risk-taking and the development of good judgment?

When Faith and I read *Little Women* together, the following sentence jumped out at me—

> Don't try to make me grow up before my time, Meg: it's hard enough to have you change all of a sudden; let me be a little girl as long as I can.

Jo—the headstrong, creative-writing tomboy—speaks these words to her older, vainer sister, Meg. At this point in the novel, Jo is fifteen. Meg is sixteen. In 1868, when *Little Women* was first published, sixteen was the approximate average age for onset of menarche among U.S. girls. By the turn of the century, average U.S. menarchal age had fallen to 14.2 years. By the mid-twentieth century, it was thirteen. Since then, the age at which U.S. girls experience their first period has continued to decline but at rates that differ markedly among racial and ethnic groups. Among U.S. white girls, the average menarchal age has fallen only slightly over the past half-century and now stands at 12.6 years. Among U.S. black girls, average menarchal age is 12.1 years, and the ongoing rate of decline is somewhat swifter, as it is among Mexican American girls.

The more troubling story, though, is what's happening with the age of breast budding, which has fallen far more rapidly than the age of menarche. This discovery came as a surprise. In the early 1960s, two British pediatricians who had studied the pubertal development of 192 girls in an English orphanage had announced that the mean onset of breast development was 11.2. (Yes, I, too, wonder about their methodologies.) Detailed and meticulously gathered, their data became the basis for what is considered normal puberty.

By the 1990s, many pediatricians suspected that most U.S. girls were experiencing breast development considerably earlier than eleven, and a large study was launched. Published in 1997, the results astonished everyone. The mean age of breast development was about 9.8 for U.S. white girls and 8.8 for black girls. About half of all girls showed signs of breast development by their tenth birthdays, with 14 percent attaining breast development between their eighth and ninth birthdays. Other researchers also discovered that onset of breast budding and onset of menarche were not as tightly coupled to each other in years past. In other words, breast development and menarche are occurring earlier and earlier in the lives of U.S. girls, but the age of breast budding is falling more rapidly than the age of menarche.

In 2009, these findings were replicated in Denmark, where, over a period of fifteen years, average age at onset of breast development fell by a full year—from 10.9 years old (in 1991–1993) to 9.9 (in 2006–2008). During that same interval, age at onset of menarche fell by only a few months. Soon after, a 2010 study in the United States documented a continued drop in pubertal onset, with even more girls starting breast development at age seven and eight than in the 1997 study. In this most recent survey, 10 percent of white girls started breast development at seven years old, along with 23 percent of black girls, 15 percent of Hispanic girls, and 2 percent of Asian girls.

It's hard not to be stunned by these data. At age seven, I lost my first tooth. I learned to write my name in cursive. I was introduced to chapter books. I hadn't started piano lessons.

The overall trend, then, is this: U.S. girls get their first periods, on average, a few months earlier than did girls forty years ago. But they get their breasts, on average, nearly two years earlier— closer to age nine than to age eleven, with a sizeable minority sprouting breasts far earlier than that. In the time span between my pubertal life and my daughter's, the childhoods of U.S. girls have been significantly shortened. And, as with testicular dysgenesis syndrome, discourse about the lived experience of these pubertal time shifts is difficult. Silence surrounds the topic.

Let me give you an example.

In 1971, I was in the sixth grade when I first noticed an achy lump behind my nipple that could only mean that I was growing breasts. According to the historical data, this discovery, at age 11.5, made my initiation into puberty a little later than average for the times. *Average* is exactly how I experienced this event in real life. I was more or less in the middle of the pack of girls who, one by one, were taken by our mothers to the undergarment outfitter—in the downtown department store with the squeaky floor—to be equipped with bras. By the end of sixth grade, most of us had made that trip.

Two years later, I awoke on a predawn December morning to a different ache, and the toilet swirled with blood. This time, I was outfitted from my mother's personal stash of feminine hygiene products. Afterwards, I pretended to fall back asleep, while mom herself sat at the foot of my bed, wordlessly patting my leg. I hoped she wouldn't leave; I wasn't ready for all this blood. She stayed, and the gray air lightened toward sunrise. At 13.3 years old, I was on the late side of average for menarche.

And as for my experience with puberty in the year 2010, as the devoted mother of a twelve-year-old daughter, the words *private entrance* bar the door of that story. And to those words, I yield.

There it is. The mother's need to protect trumps the writer's need to divulge. The timing of my daughter's puberty is a conversation I'm not having.

What is driving the declining age of puberty in U.S. girls? And how can we explain racial and ethnic differences?

Heredity clearly plays a role in setting the pace of sexual maturation. Mothers and daughters exhibit similarities in pubertal timing and tempo. Tellingly, identical twin sisters show greater correlation in pubertal onset than do fraternal twins. Nevertheless, genetics can't explain racial and ethnic differences nor the decline in age over time. Menarche, for example, occurs far earlier in U.S. black girls than among black South African girls from well-off families—or among black girls in Cameroon or Kenya. Moreover, a century ago, U.S. black girls actually entered puberty significantly *later* than U.S. white girls.

With its multitude of signaling pathways, the instrument that controls pubertal timing is, by its very intricacy, vulnerable to perturbation. Several factors appear able to alter the regulation of the hypothalamus and could thereby hasten the onset of puberty in girls. Premature birth is one. Low birth weight is another. Both are endocrine-altering events that, through unknown pathways, dramatically increase the chance that a girl will develop pubic hair before the age of seven. It's likely that alterations in insulin levels, which affect adrenal gland functioning, are the underlying mechanism. Low birth weight and prematurity are on the rise within the United States, and, as noted in earlier chapters, the risks for being born too early or too small may themselves be influenced by chemical exposures of the mother (among other important factors). Along with phthalates, PCBs, and air pollution, atrazine appears on the list of chemicals with demonstrable links to shorter pregnancy and lower birth weights.

In addition, overweight and obesity are almost surely playing a role in driving down the age of puberty in girls and may also help explain the racial disparities. Multiple lines of evidence argue for a connection. First, obesity, which has tripled in prevalence among children over the past three decades, dramatically alters levels of hormones to which the hypothalamus is known to be responsive, including insulin and leptin. Second, the trend of increasing body mass for U.S. girls coincides with the trend for earlier puberty. Third, a higher percentage of black girls are obese, and, as a group, black girls also reach puberty sooner than white girls. And, most directly, several large, carefully designed studies show that, as a group, chubby girls develop breasts sooner than lean girls.

And yet, there are equally compelling reasons to believe that the increased fatness of U.S. girls is not the whole story behind the falling age of puberty. In the 2009 Danish study, downward trends in the average age of breast budding persisted even after taking increasing body weight into account. A 2009 study of Chinese girls likewise documented a decline in age of onset for breast development for which obesity was not the complete explanation.

Within the United States, the association between body mass and early puberty, so clearly documented among white girls, is not so apparent for black girls: Even after adjusting for body mass, black girls still have earlier onset puberty, and they also have decreased insulin sensitivity (meaning that more insulin is needed to transport blood glucose into body tissues). These racial disparities suggest an alternative hypothesis: The falling age of puberty is not a direct consequence of increasing fatness itself but may be a result of increasing insulin resistance (for which type 2 diabetes is another potential consequence).

Physical inactivity—quite apart from obesity—may also be playing a role. Exercise is clearly protective against early puberty—but through pathways that are not clearly understood. Strenuous physical training may inhibit the puberty-stimulating neurons of the hypothalamus. As a group, bedridden girls have earlier than average puberties, while female athletes have later puberties. Admittedly, teasing apart the effects of thinness from the effects of exercise is difficult. Girls with anorexia tend to have delayed puberties, as do ballet dancers and runners—but so do elite swimmers and ice skaters, who are typically less lean.

Psychosocial stress is also an endocrine disruptor that acts to hasten pubertal onset—perhaps through adrenal release of cortisol to which the hypothalamus is responsive. Trauma, family dysfunction, and father–daughter relationships all seem to play a role. Conflict and stress within families are consistently associated with early puberty. So is the tragedy of sexual abuse. Absence of a biological father—with or without the presence of a stepfather or other unrelated adult male—has been linked to earlier breast development in a number of studies. No one knows why, although many have speculated. (Do girls living with biological fathers receive pheromones that inhibit puberty, perhaps as an evolutionary mechanism to prevent incest?) Conversely, the presence of siblings in the household lowers the risk of early puberty. So does household crowding: Age of puberty goes up when the number of bedrooms in the house goes down.

Evidence strongly indicates that exposures to endocrine-disrupting chemicals also are playing a role in accelerating puberty in girls. A multidisciplinary expert panel co-sponsored by two different federal agencies in 2008 concluded just that. And a close review of that evidence implicates some of the same suspects fingered in the testicular dysgenesis story.

Well-designed prospective human studies are understandably scarce. Nobody is purposefully exposing cohorts of baby girls to endocrine-disrupting substances and then monitoring their various pathways to puberty. Thus, many human studies try to make sense of accidents that involved the inadvertent exposure of children to chemical toxicants. For example, in 1973, brominated flame retardants (polybrominated biphenyls) were mistakenly added to cattle feed in Michigan. As a result, a cohort of 327 girls were exposed as fetuses or infants when their pregnant or lactating mothers ate beef and milk from the poisoned cows. Researchers enrolled these girls—now middle-aged women—in a long-term study that documented, among other things, that those exposed to the highest levels of flame retardants in early life began menstruating up to a year earlier than girls with less exposure.

Another type of human study measures background levels of hormones and hormonally active agents within the bodies of girls and women within the general population. In the urine of U.S. girls six to eight years old, for example, researchers have identified a wide spectrum of hormonally active agents, including phthalates and bisphenol A. These results confirm that exposures of young girls to ubiquitous environmental endocrine disruptors are, well, ubiquitous.

The authors of the 2009 Danish study, which documents a decline in the age of breast development among the girls of Copenhagen, did not attempt to measure endocrine disruptors in the bodies of their young subjects. They did, however, measure blood levels of two different reproductive hormones that serve as the intermediary messengers between the hypothalamus and the ovaries. Follicle-stimulating hormone and luteinizing hormone oversee the estrogen production that, in turn, is needed for breast development. The levels of these two naturally occurring hormones in the 2006–2008 cohort were no higher than in the 1991–1993 cohort. The investigators surmised that external estrogenic factors may be acting in concert with the body's own sex hormones to explain the hastening pace of breast development in Denmark.

Supporting the thin, provocative body of evidence from human studies is a fat, more conclusive body of evidence from animal studies. All together, controlled animal experiments show that exposure to environmental estrogens in early life accelerates the pace of sexual development—through a variety of mechanisms and at doses similar to background levels to which human children are routinely exposed. Precocious puberty in lab animals can be induced via exposure to synthetic estrogens either pre-natally or shortly after birth.

The evidence is especially damning for bisphenol A. In female rats, prenatal and early life exposures can trigger early onset of sexual maturation, especially breast development. At concentrations just slightly greater than what U.S. federal agencies consider safe, exposures altered hormonal signals from the hypothalamus and so reprogrammed sexual development— which is to say that while the exposures to bisphenol A were temporary, the effects were permanent, and rats so exposed entered puberty sooner.

When a progress report on risk factors for early puberty was recently released by a breast cancer organization, I received a communiqué. The final paragraph contained a list of *steps that parents can take*. The first one was *help daughters maintain healthy weight*.

Check. Faith is a regular string bean.

The second step was *model exercise*.

Check. One hour a day, six days a week, unless I've got a kid home sick from school. And Faith can already beat me in a 5K run.

The third one was *reduce exposure to environmental chemicals that may act like . . . hormones in the body, such as flame retardants, pesticides, and chemicals in plastics*.

Okay. How would I do that exactly? And how would I know that I had succeeded?

My daughter belongs to the world now, with its water cycles, air currents, and food chains. She spends her days in a school full of equipment and furniture that no doubt contain flame retardants. She rides home on a diesel-powered bus. She flies around town on her bicycle—or scooter or skateboard—to flute lessons, piano lessons, the public library, passing by pesticide-treated fields and lawns as she goes. She accepts invitations to sleepovers that involve fingernail polish. She signs up for cooking classes and brings home the leftovers in plastic containers. She scours thrift stores for funky bargains. I don't always know what they're made of. And when we lie together in the dark at the end of the day, I'm aware that under her skin somewhere lies a wide-awake hormonal network that is receiving incoming signals—from tonight's dinner; from the chemicals in her shampoo; from the agricultural practices of the county; from the coal plant across the lake; from her father, brother, and me. This much I know: Her body is the medium for a much larger message.

New research shows that bisphenol A is released from burning plastic. It drifts in the wind and attaches to inhalable particles. So don't give me any more shopping tips and Web sites to consult. Instead, give me federal regulations that assess chemicals for their ability to alter puberty before they are allowed access to the marketplace. Give me a functioning endocrine-screening program, with validated protocols, as mandated by 1996 legislation. Give me chemical reform based on the precautionary principle.

I can stand at the base of a tree in case my saw-wielding son falls out of it, but I can't place myself between 200 known or suspected endocrine-disrupting chemicals and the body of my daughter. I can't stop the wind from blowing.

To the index of things felt acutely by parents but imperceptible to everyone else, let's add *birthday cakes, the sadness of.*

I'm referring to the moment that comes after the candles are blown out and before someone stumbles over to the wall switch and flips the lights back on.

After the song is sung. Before the cake is cut.

The moment when the waxy smoke balloons up from the red-tipped wicks into the sudden darkness, and the mother of the birthday child realizes that there will never again in this house be a four-year-old boy. Or a seven-year-old girl. Or a six-year-old boy. Or a nine-year-old girl. Or an eight-year-old boy. Or an eleven-year-old girl.

Was it ever thus? Was the sadness of birthday cakes a secret known to my own parents? And to their parents, too?

To be sad about birthday cakes is to lament the speed at which the earth circles the sun, which is a helpless, silly sort of sadness. It doesn't last long. Especially now that the lights are on again and before you is a laughing child—boy or girl—in possession of manners and psychomotor skills, who is doling out cake and ice cream to the neighbor kids and hamming it up for the camera.

It is never the birthday kid, but the other one—the younger or older sibling who is trying to be a good sport about it all—who looks over and notices. *Mama, why are you crying?*

Oh, someday you'll find out.

And, unbeknownst to him or her, that very sentence was also my unspoken wish, delivered just seconds earlier, as the mighty exhale blew out the flames that number the years already gone by. It's the same wish for every cake: *Oh, let them grow up. Let them find out. Let it go on.*

CHAPTER TEN

Bicycles on Main Street
(and High-Volume Slickwater
Hydraulic Fracturing)

ON A RECENT morning in May, I perched on the steps of my front porch—with its still-spongy floorboards—and read an advance copy of a policy statement just released by the American Association of Pediatrics, which acknowledges that a child's life is influenced by the environment in which he or she lives.

I sat very still and turned the pages as silently as possible because, to my left, as if to illustrate the point, a pair of juncos was constructing a nest in the hanging pot of fuchsia above the porch swing. A flash of wings. A rustle of leaves. The flowers shivered, and the pot swung gently, as though an invisible breeze was blowing.

We've lived in this house long enough for me to learn that shade-loving fuchsia is the only flower that will agree to bloom under my porch eaves, canopied as they are by maple, spruce, and walnut trees. Of the six different hooks on which I hang potted flowers each spring, the one nearest the porch swing, for reasons known only to juncos, is the spot chosen each year for egg-laying. A slate-gray pair with a serious work ethic has faithfully nested here since Elijah was young enough to call them *jumpos*. Or maybe these are now the great-great-grandchildren of the original homesteaders. Either way, with the annual arrival of Mother's Day come the hoppy birds with architectural plans and strands of grass clamped in their beaks, and we have to abandon the porch swing until the new generation has fledged.

The report argued that pediatric health and wellbeing is predicated not only on the quality of air, water, and food but also on the physical aspects of communities themselves—the so-called built environment. As such, urban sprawl has been an exercise in design that works against children's health—which is an irony, if nothing else, since low-density, suburban developments in formerly rural areas are magnets for families with young children. Often, that's where the good schools and parks and playgrounds are.

And yet, sprawl, with its necessary dependencies on automobile travel and long-distance food transportation, increases air pollution, exacerbates asthma, promotes child obesity, and, compared with traditional small town centers, erodes the safety of playing, walking, and bicycling, thus decreasing the likelihood that kids will make it to their schools under their own steam. In 1969, 41 percent of American children walked to school. Three decades later, that figure had dwindled to 13 percent. When parents were asked what prevents their own children from contributing to this statistic, the top two reasons were, according to this report, physical distance and traffic danger.

With urban sprawl, the report's authors note, the one-hour-per-day recommendation for exercise is more likely to be met—if it's met at all—in school settings or through structured activities supervised by adults. This trend—I'm adding a footnote of my own here—transforms parents, *mostly mothers*, into chauffeurs and, along the way, further increases our collective fossil fuel dependencies. The carbon footprint of children's play has never been larger.

I looked up from the report at my shale sidewalk—marked hundreds of millions of years ago by an ancient sea that rippled over it and marked last week by the chalk drawings of my children, now half erased by rain. Connecting my front door to Main Street, three blocks away, this crooked sidewalk deserves the credit, I suddenly realized, for so much more than service as an art easel and a hopscotch court. It has played a key role in many of my parenting successes heretofore.

Although it's way too early for a parenting retrospective, here's a quick progress report on a few projects I'm happy about, interwoven with a grateful paean to the sidewalk.

One is the ongoing campaign to instill in each of my kids a sense of financial self-reliance. Saving and spending, budgeting and splurging, the remorse of buyers and misery of misers. Frequent, early experience with each of these conditions, when the stakes are small, seems like a good thing. To that end, the children living in my house receive a monthly allowance of $1 times their age. If that proves insufficient, I also accept bids on home improvement projects.

On the first Saturday of each month, Faith and Elijah, flush with cash, follow the sidewalk downtown to the community bank. It's up to each of them to decide how much to deposit and how much to keep on hand for purchases. Near the bank, farther down the sidewalk, are lots of little shops, including a used bookstore, where a kid whose money is burning a hole in a pocket can quickly become divested of a month's worth of spending money. For big purchases, especially those involving withdrawal slips and saving account numbers, Jeff and I require a 24-hour think-it-over period. (Elijah once spent one-third of his entire net worth on a Beatles anthology. Jeff and I decided that we were okay with that.) Faith now also receives a yearly clothing allowance. It's up to her to create a school wardrobe out of it. If asked, I'll offer dressing room opinions. I still provision the unexciting essentials, like rain boots, socks, and snow pants.

Sidewalks, a community bank, and a walkable downtown make my allowance scheme possible. They permit Faith and Elijah a measure of financial independence that they would not have if every trip to a bank and a store involved a commute, a drive-through window, and a shopping mall. And with no one begging me to buy things for them, I lead a less-harassed life.

Not just a lucky accident, the presence of sidewalks and pedestrian crosswalks throughout our entire village was part of a strategic, long-range plan, for which public hearings were held, revenue raised, and blueprints drafted. In those days, I was too busy with preschool children to participate in any of that civic-minded work, but I am the lucky beneficiary, as are all stroller-pushing mothers who live here.

The move from the car-dependent cabin into this village had felt like a faith healing. The sidewalks allowed me to rise and walk, rolling my small children before me. What I had in mind at the time was simply running errands. I didn't anticipate how these same sidewalks would eventually also link us to the village's various subcultures. Now they walk us to the farmers' market, the pizzeria, and the neighborhood bar with its foosball table and string band performances. And with these excursions have come lessons for my kids on how to act in public, along with opportunities to spy on a variety of grown-ups and over-hear adult conversations.

I also credit sidewalks for literacy, compliance on household chores, and the ability to live in a small house with a modest mortgage. These disparate things are all connected to the fact that the village library lies less than a mile from our house, con-nected to us by continuous slabs of shale and concrete. The kids and I make a weekly pilgrimage here each Saturday. Now that everyone has moved on to thicker chapter books, our checkout rate has declined, but back in the days of picture books, we checked out a profligate number—between thirty and fifty a week, which I then hauled home in the double stroller. At some point, this practice was declared embarrassing, and I switched to a red wagon. (Doesn't that seem a *more* eccentric choice for materials transport?)

My method was a simple one: Beginning with the authors whose names started with A (Ahlberg, *Each Peach Pear Plum*), I indiscriminately grabbed an armload of books from a single shelf and checked them out. Then I wandered off to read the newspaper, while the kids chose a few books of their own. The next Saturday, I returned the previous week's haul—read or unread—and grabbed another armload beginning with wherever on the shelf I had left off. In this way, week after week, season after season, we plowed through the library's entire collection of books. Once we finished the alphabet (Zion, *Harry the Dirty Dog*), I started over again.

Or I dipped into the nonfiction holdings, which are organized by topic. One week, we took home eleven books on Easter. Another week brought rainforest animals, followed by a week of sports, a week of creation myths, and a week of dog training manuals. The week of joke books was a bit much. But, mostly, the topics never seemed to matter. The achievement was that every Saturday, new titles appeared in the book baskets that I'd scattered in various rooms. (As for compliance on household chores, the deal was—and is—nobody goes to the library until the bathroom is cleaned.)

And these continuously resupplied baskets have contributed to peaceful cohabitation in a house where solitude is in short supply. A stack of library books in a bathroom offers a child a private world. As does a book basket on the stair landing or behind the couch. Connection to a public library, via an undisrupted network of sidewalks, allows me to live in a smaller house, shoulder a smaller mortgage, pay smaller utilities bills. Walk more. Worry about money less.

A few miles in the opposite direction is Cayuga Lake, with its swimming beach and hiking paths. There are neither sidewalks nor pedestrian trails to take us there—many of us are advocating for them—but it is bikeable, if one takes a fearless attitude towards hilly back roads and a one-mile stretch of two-lane highway. In this lake, I've taught each of my children how to swim. I also periodically sign them up for formal lessons in chlorinated pools with certified instructors, but during the fleeting weeks of summer here in Ithaca, I use the lake as the classroom.

As Elijah will gladly tell you, the middle of Cayuga Lake is 435 feet deep. (Its final fifty feet lie below sea level.) *One of the deepest lakes in North America*, he freely informs out-of-town bathers. Although not ready to swim out to the twelve-foot dock himself, Elijah is yet convinced that anyone—including his sister—

who can dive from it and into *one of the deepest lakes in North America* must be a truly skilled swimmer. Sometimes, while we're paddling around in four-foot water—safely separated from 435-foot depths by a rope of bobbing buoys—I tell him again how I often swam out to that dock while pregnant with him sideways in my belly, hoping my front crawl would coax him to turn and line his body up with mine.

And did I do it?

Not right away. But later on. When you were ready.

The story of the dock is intended to inspire him to keep practicing his own front crawl. *I once swam here for both of us. Now you try.* The landscape thus connects us to each other—I'm not just driving my kids from one supervised activity to the next—and the stories we tell about it link us to the seasons, and, for Elijah, to his own origins. We know where to find the sumac along Seneca Road that is always the first to turn color in the fall, which is also the most reliable place to find milkweed pods or catch the flash of a bluebird in July. Just beyond the sumac is the field of goldenrod that Faith once declared should be called *silverrod* from November through March. And so we call it that every time we pedal by it.

The evidence for a much older story about origins also surrounds us here. While wandering through the shoreline woods, Faith matter-of-factly showed me a stone she found with fossil tracings and then skipped it into the lake and raced ahead. Her casualness prompted in me an obvious thought: Although I am a newcomer with no extended family here, my children are native to this place. Rooted here, they take for granted its abundance of fossils, waterfalls, vineyards, apple stands, pastures, and ripply sidewalks. For them, these are the familiar features of home. The discovery of a seashell's silhouette pressed into shale is an unrare finding. The juncos in the fuchsia pot are delightful but ordinary. It's just the way things are.

Cayuga Lake is the longest of the eleven Finger Lakes, which resemble not so much fingers but the claw marks left by glaciers. Which is, in fact, their origin.

By the time the ice sheets arrived, the shale they gouged was already hundreds of millions of years old. After their retreat, the tumbling creeks that flowed into these lakes broke apart this fossiliferous rock and carried it away. Its redeposition downstream created fan-shaped deltas—rich with skipping stones—such as the one that made our swimming beach, on which we spread towels and picnic blankets.

What remained upstream of these deposits were glens and gorges. And when hiking the waterfall-crazed gorge that flanks the west side of this lake, we are connected to a past that is almost unfathomable. Nearly as high as the lake is deep, the walls of the gorge are fissured and crumbly. In fact, they are less like walls than they are the interiors of ancient rooms, as though the sides of a great, multistoried house had fallen away, offering a glimpse into the stratified chambers within, stacked one on top of another. And the deeper into the gorge you hike, the sunlight narrowing into a slit farther and farther above your head, the more you realize that, deep below your feet, layer after Devonian layer, the house goes on.

One of the layers far below is a shale formation called the Marcellus, which was created nearly 400 million years ago—before the earth knew trees—when the Acadian range eroded into a nameless sea. Its silt sank to the ocean floor, together with the remains of ancient creatures that lived and died in its shallow waters.

Under pressure, this burial ground turned to shale, forming a chalkboard the size of Florida. And the plankton and animals trapped inside became bubbles of methane (natural gas). Because eroding mountains shed many elements, this trough also captured uranium, mercury, arsenic, and lead. And so, in a bedrock layer that ranges from 2 to 200 feet thick, at a depth of zero to a mile or so below the earth's surface, extending for some 600 miles throughout West Virginia, Ohio, Pennsylvania, and New York, the Marcellus Shale's rock, methane, and heavy metals have remained locked together.

Underlain and infused by brine. Overlain by drinking water aquifers.

The Marcellus Shale is the basement foundation of New York State. Nevertheless, it comes blistering out of the ground in the little village of Marcellus—sixty miles and three lakes east of my village. That community became its namesake.

The Marcellus Shale holds the largest natural gas deposit in the United States. And so, this subterranean landscape has become ground zero for a form of energy extraction called high-volume slickwater hydraulic fracturing.

Drilling for gas by fracturing shale in vertical wells is an established practice, but, prior to the twenty-first century, capturing a multitude of unconnected gas bubbles dispersed within a horizontal formation like the Marcellus was not economically feasible. Indeed, until just a few years ago, much of the natural gas trapped in shale was considered unrecoverable because it is scattered like a fizz of bubbles in a petrified spill of champagne.

Enter horizontal wells and high-volume slickwater hydraulic fracturing. Or, to use the world's ugliest gerund: *fracking*.

Using this method, a drill bores down until it intersects with the shale, turns sideways, and continues horizontally for up to a mile or more. Steel pipe, some portions of which are encased by cement, is laid in the borehole. Explosives are detonated along the horizontal pipe to perforate it and then, under enormous pressure, a slurry of water, sand, and chemicals is forced into the rock, fracturing it and opening preexisting fractures, known as "joints." The grains of sand (or glass beads or epoxy) hold the cracks open. The chemicals—and there are hundreds for drillers to choose from—serve several purposes. Most notably, they reduce friction (thus the term *slickwater*) so that the fracking fluid can flow easily. In addition, the mixture includes acids, rust and scale inhibitors, and pesticides to kill microbes. Sometimes it includes gelling agents, petroleum distillates, glycol ethers, formaldehyde, and toluene.

And up the borehole flows the gas. The cocktail of water and chemicals forced into the fractured shale also flows back up. But not all of it. Some 40 to 70 percent stays behind in the rubble.

Fracturing deep bedrock is not a gentle process. Exerting up to 10,000 pounds of pressure per square inch, fracking has been compared to smashing a windshield with a baseball bat. A single fracking operation requires an access road, 2 to 8 million gallons of fresh water, between 10,000 and 40,000 gallons of chemicals, and at least 1,000 diesel truck trips. Between 34,000 and 95,000 wells are envisioned for New York State, with 77,000 likely.

Here, then, is an interesting mathematical word problem, of the kind presented to third graders to help them practice learning the place values of very large numbers. At least 34,000 wells are planned. Add to 34,000 four zeroes and multiply by a number between 1 and 4 to estimate the volume of chemicals that will be pumped into the ground. Add to 34,000 six or seven zeroes to estimate the volume of fresh water that will be used. Divide that product by 2 to determine, roughly, how many gallons of toxic flowback water will come back out of the hole and require disposal somewhere else. Use that same number to approximate how many gallons of water will remain buried in the fractured bedrock. To 34,000, add three zeroes to determine the number of diesel truck trips that will be added to the roads.

The results of all this multiplication is why, in New York State, at the time I am writing these words, there is a temporary moratorium on high-volume slickwater hydrofracturing. We are thinking hard about all those zeroes. Just to the south, our sister state of Pennsylvania—also located atop the Marcellus—has made a different decision. Since the first drilling permit was issued there in 2004, fracking is going full tilt, with thousands of new permits being issued at gold-rush speed. We are watching closely the unfolding results.

One thing is clear already—just as there is no such thing as a little bit pregnant, neither is there any such thing as a little bit of fracking. Due to the economies of scale and required infrastructure, fracking is an all or nothing, shock and awe operation. Either the drillers come into an area and plaster it with well pads—or they don't come.

Natural gas is many things. To a homeowner, it is the blue flame beneath the teakettle and the whoosh of the furnace kicking on. To a geologist, it is the vaporous form of petroleum. To an industrial chemist, it's a feedstock and a starting point for many petrochemicals, such as fertilizer and PVC plastic. To a climatologist, natural gas is the Dr. Jekyll and Mr. Hyde of fossil fuels: When burned it generates only about half the greenhouse gases of coal, but when it escapes into the atmosphere as unburned methane, it's one of the most powerful greenhouse gases of them all. Although shorter lived than carbon dioxide, natural gas can be twenty times more powerful at trapping heat.

Because high-volume slickwater fracking is not, at this writing, being practiced in New York State, there are many unknowns and a great debate about how to think about the things we don't know about. But at least five things are known with certainty.

The first is that, within our state, the land that lies above the Marcellus Shale is full of farms and vineyards. It's the state's foodshed and wine-growing country. It also contains some of the largest unbroken forest canopy in the Northeast. Fracking, thus, represents the industrialization of a rural landscape. If it goes forward, it will usher in the biggest ecological change since the original forests here were cleared. More than shale will be fractured.

Drilling for gas in farm country requires the construction of well pads, roads, and pipeline in fields and pastures. Drilling for gas in forests additionally requires cutting trees and clearing land. These activities fragment wetlands and woodlands and diminish their ability to provide habitat for wildlife, filter rainwater, and prevent flooding. Even before the actual fracking begins, the erosion and run-off from deforestation and road construction will invariably send sediment into streams, threatening invertebrates, amphibians, and fish.

Less obviously but no less real, fracking threatens migratory birds, some of whom are already suffering catastrophic declines in their population. Forest fragmentation provides avenues of access for the predators of bird nests—namely, squirrels, skunks, crows, and cowbirds. According to Audubon Pennsylvania, the bird species now threatened by fracking activities include the hermit thrush, ovenbird, veery, winter wren, and seven species of warblers. They also include scarlet tanagers and wood thrushes. Seventeen percent of the entire world's population of scarlet tanagers reproduce in the forests of Pennsylvania. Twelve percent of the world's wood thrushes lay their eggs there.

Sure thing number two: Fracking brings urban-style air pollution to the rural countryside. Even if accidents never happen, the business of drilling fills the air with ozone-making, smog-producing combustion byproducts of the type linked in adults to cancer and heart disease and, in children, to lowered IQ, preterm birth, asthma, and stunted lung development. What's more, the airborne contaminants from gas drilling travel long distances—up to 200 miles. The air emissions from shale drilling in the Haynesville shale in Texas and Louisiana, for example, are projected to raise ozone levels dramatically not only in the communities near the gas patch but in counties far away from the wellheads. That is to say, the health costs of drilling will be borne by children living in areas where no one is benefiting financially from land leases. These emissions accompany every stage of the gas extraction process from clearing timber to building well pads and hauling away toxic flowback fluid. Drilling, fracking, and gas production require generators, pumps, compressors, drilling rigs, condensate tanks, and endless fleets of diesel trucks. These trucks, many of which will be hauling toxic chemicals, will fill our rural roads and two-lane highways.

Like the one my kids and I bicycle along to get to the lake.

Like the one we call *Main Street* when it passes through our village. The one with sidewalks and crossing guards that help kids walk to school.

Sure thing number three: Accidents happen. Especially during industrial activities involving explosive vapors. (See *Illiopolis, evacuation of.*) Gas pockets explode. Blowout protectors fail. Chemicals spill. Trucks hauling toxic liquids crash. Holding ponds and waste pits leak. Sludge tank walls collapse. In Pennsylvania, in less than three years of fracking, 1,500 environmental violations have occurred, including one that involved an exploded well that blasted a fire hydrant stream of poisonous frack fluid for sixteen hours. In hundreds of cases, petroleum products, fracking chemicals, or flowback fluids have ended up in creeks, streams, or groundwater.

Sure thing number four: Fracking makes water disappear. This is worth thinking hard about. Usually, when we say that water is being "wasted"—because, say, some misinformed soul opens a carwash in the middle of the desert or insists on watering the lawn during droughts—we don't mean that the water itself exits the water cycle. (A quick review: Rain recharges groundwater, groundwater is the mother of rivers, rivers flow to oceans, oceans evaporate and become clouds, and clouds carry rain.) Wasting water may draw down a reservoir or deplete an aquifer, but the water itself is merely transferred to a different place on the proverbial wheel, which is forever turning.

When you brush your teeth or wash your car, water flows, via plumbing, from a municipal well to a sewage treatment plant—and then heads down the river to somewhere else. It's not really gone (although the resulting depletion of aquifers is nonetheless a hugely serious problem). When a farmer pumps groundwater to irrigate a field, the moisture enters the root hairs of crop plants, exits the stomata of leaves, rises as vapor to the clouds, and comes back down as rain—if not falling on the same watershed from whence it came, then maybe falling in another one a couple of time zones away—or on all the ships at sea. When fresh water is added to the oceans in this manner, sea levels rise—there is evidence for this—and huge amounts of energy are required to turn salt water back to fresh for drinking. Nevertheless, if destructive groundwater pumping were stopped, the hydrological cycle would eventually—over a long time— recharge the world's groundwater aquifers.

Conversely, what happens to water during fracking is different from what happens to water when you leave the tap running while brushing your teeth. When a single well is fracked, several million gallons of fresh water are removed from lakes, streams, or groundwater aquifers and are entombed in deep geological strata, up to a mile or more below the water table. Once there, this water is, very likely, removed from the water cycle permanently.

As in, forever. It will no longer swirl with tadpoles or ripple with fish. It will no longer ascend into clouds, freeze into snowflakes, melt into rivulets, cascade over rocks, turn with the tide, soak into soil, rise through roots, or flow from your tap. It will not become blood, tears, sweat, urine, milk, sap, nectar, yolk, honey, or the juice of a fruit. It will never again float a leaf boat, swell a bud, quench a thirst, fill a swamp, spill over an edge, slosh, dribble, spray, trickle, splash, drip, or glisten. Never again fog, mist, frost, ice, dew, or rain. It's gone.

Not that you would want it to come back. It's poisonous now.

Sure thing number five: Sooner or later, the gas will run out.

Beyond these certainties, lie questions.

Can fracking contaminate groundwater? The answer depends, in part, on what is meant by fracking. Over and over during public meetings on hydrofracturing, I've heard drilling proponents claim that no confirmed cases of drinking water contamination have ever occurred from fracking. If by this, they mean only the short period of time during which the explosions and pressurized injections of chemicals and water split the shale, they may be right. But if fracking refers to the whole operation—from clearing the land to evaporation from holding ponds—there are many documented cases of water contamination with compounds associated with gas development—including the carcinogen, benzene. Their likely sources are mostly located on the surface of the earth: from pits that hold flowback fluid, from leaking tanks, from leaking wellheads, from spills.

And everything depends upon the integrity of the steel-and-cement-encased wells themselves, which bisect aquifers on their way down to the shale below. Their external surfaces have direct and ongoing contact with groundwater. And their internal surfaces have direct and intermittent contact with the highly toxic substances that are poured, under intense pressure, down the hole—and with the more toxic substances that are forced back up. When drilling through drinking water, there is no room for error. The drinking water that overlies the Marcellus Shale services 4 million people. Can the external surface of *every* well bore that pierces a water supply *always* be trusted to remain an unbreachable barrier to the substances flowing up and down its interior surface? A single cracked well bore can bring eternal, irremediable ruin to groundwater.

And if this happened, how would we know?

In this, the story of Pavillion, Wyoming, population 166, is particularly instructive. With 211 active gas wells, 50 inactive wells, and 37 pits that formerly held drilling fluids, Pavillion is a veritable gas patch. A few years ago, residents noticed that their water looked different. It was yellow, cloudy, and oily. It bubbled. It smelled like chemicals. And some people felt sick. If you think that doesn't sound good, you'd be right. A joint investigation by the EPA and the Agency on Toxic Substances and Disease Registry found petrochemicals in twenty water wells. These included diesel fuel, benzene, cyclohexane, methane, propane, and ethane, along with traces of arsenic and a pesticide used to keep microbes from growing in the pipeline. The EPA recommended that Pavillion residents not drink their water. Turning on fans while showering—to avoid possible explosions caused by methane accumulation in indoor air—was also said to be useful.

But what was the source of the drinking water contamination? If you think the answer to this question is easy, you'd be wrong. For two reasons, investigators couldn't reach a final conclusion. First, agricultural practices in the area are of a type that can also contaminate groundwater. And second, in Pavillion—as almost everywhere else in the world—the subterranean world of groundwater flow is uncharted and mysterious.

There is a very limited understanding of the groundwater flow . . . in the Pavillion area. The lack of this important information limits the ability to predict the velocity and direction of contaminant movement. . . . The hydrology of the site has not been adequately characterized. The relationships between groundwater contamination and the private wells and how the well water could be impacted in the future are not well understood. . . . It is not possible to determine if these data are suggestive of past or future contaminant levels in well water.

Likewise in Pennsylvania, uncertainties are compounded by the underground chaos left over from coal mining operations. So, even when contamination of water occurs after fracking operations, it's difficult to use the tools of science to connect the two causally. Which leads to another question: If, because of geological ignorance and confounding variables, we are systematically unable to identify fracking-induced drinking water contamination when it does occur, how do we document evidence of harm? And how should we proceed if we cannot?

From here, it gets more frustrating.

Granted the environmental equivalent of diplomatic immunity, fracking enjoys special exemptions from a whole raft of regulations—Clean Water Act, Clean Air Act, Superfund Act, National Environmental Policy Act—that govern other types of industrial activities. Fracking gets a pass on federal right-to-know laws as well, which means that natural gas operations do not report their air and water emissions under the Toxics Release Inventory. And a special amendment to the 2005 Energy Policy Act specifically grants fracking exclusion from the Safe Drinking Water Act, which authorizes the EPA to regulate all injection of toxic chemicals into the ground. Thus, no federal requirement compels a drilling company to disclose the specific chemical formulation of its fracking fluids. The ingredients can remain a trade secret. The result is that researchers not only lack basic knowledge about the behavior of groundwater in areas where fracking is occurring and going to occur, they also don't know what chemicals to test for when there is a suspected problem. In any case, without federal oversight, there are few researchers.

The best generic analysis of fracking fluid ingredients—pieced together through old-fashioned sleuthing—comes from biologist Theo Colborn and her research team at TEDX (The Endocrine Disruptor Exchange). They report that of the 300-odd chemicals that are presumed ingredients of fracking fluid, 40 percent are endocrine disrupters and a third are suspected carcinogens. A third are developmental toxicants. Over 60 percent can harm the brain and nervous system. But without full disclosure of fracking chemicals, documenting water contamination linked to fracking is tough.

And there's a *furthermore*. The chemicals found in fracking fluid are mysterious because the formula is proprietary. But the chemicals found in the flowback water are an even bigger cipher. When this fluid comes back out of the well, it carries with it the same chemicals it carried in—together with substances that, for the past 400 million years, had been safely locked up a mile below us, estranged from the surface world of living creatures. At the very least, these include brine, various radioactive elements, and heavy metals. In many areas, they include volatile poisons, such as benzene, toluene, and xylene. Freed at last from their subterranean chambers in the shale, the naturally occurring substances combine in the flowback fluid with all the undisclosed chemical additives.

What should we do with this lethal fluid—a million or more gallons of it for every wellhead? The technology to treat it—to transform it back to drinkable water—does not exist. And, even if it did, where would we put all the noxious, radioactive substances we capture from it? The trend is toward recycling it, but flowback water gets brinier and more poisonous with every reuse. This makes accidents and spills more dangerous. And, at some point, this highly concentrated, highly toxic liquid still has to be disposed of. More typically, flowback fluid sloshes around in open evaporation pits and ponds and is sometimes hauled away in fleets of soot-emitting diesel trucks to be forced, under pressure, down underground disposal wells. Sometimes it is run through municipal sewage-treatment plants, which release the toxicants downstream. And it is sometimes clandestinely dumped. Or to express this perplexity using the official language:

> Potential disposal options for wastewater and other wastes containing radioactivity are currently unclear.

And what about the fluid left underground? Could it escape from the now-shattered shale—like bats out of Pandora's box— rise through rock and somehow find its way into drinking water? Many geologists dismiss the idea that groundwater could be contaminated from below. And that's because aquifers do not lie directly above the shale layer that is fractured for gas; they are separated from it by many layers of rock and by hundreds, if not thousands, of feet.

Truly? There are no vertical passages, faults, fissures, hairline cracks, or channels anywhere underground that volatile hydro-carbons might find in the warm, fractured depths of the Marcellus Shale? And so worm its way upward into a drinking water aquifer? Not even through one of the thousands of unmapped, abandoned wells—embedded in the earth like so many cocktail straws—from long-ago gas explorations?

Can shale safely contain toxic chemicals forever? And, with no extra planet to use as a laboratory, how would we test that question?

Fracking's economic story also holds many questions, although they are often drowned out by breathless assertions. Leasing your land to a gas company can get you out of debt. It can save the farm. It can allow you to retire. The going rate is $5,000 an acre with between 12 and 20 percent royalties. The Marcellus Shale could be worth a trillion dollars. It can provide enough natural gas to supply the nation's consumption for two years. Or eleven years. Or twenty years. Or a hundred years.

In fact, no one knows how much gas is recoverable or how often the wells will need to be refracked to stimulate their pro-duction. But these uncertainties are posed as bookmaker's odds rather than as questions for thought. Tellingly, a methane-containing formation under the ground is called a *shale play*. It's something worth betting on. It's a gamble. Play the odds.

Companies are placing their bets and laying big money on the table. Halliburton, the baron of fossil fuels, has set up shop in Williamsport, Pennsylvania, just a few hours drive from my house, with plans to stay. Of the five big U.S. shale plays, the Marcellus is considered the safe bet. It's close to eastern markets. The cost of production is low. Environmental, health, and social costs not included.

Which raises questions that have no answers. No study of the cumulative impact of fracking on public health or agriculture has ever been conducted.

The language of fracking is rich in metaphors, which conceal uncertainties under an aura of inevitability. The Marcellus Shale is "the Saudi Arabia of natural gas," says *Intelligent Investor Report*. To the *Toronto Globe and Mail*, it is a "prolific monster" with the potential to "rearrange the continent's energy flow." Shale gas is said to be a "paradigm-shifting innovation," as though fossils were a new computer operating system.

Most often, shale gas derived from fracking is called the "bridge" to the future, on the grounds that, until something better comes along, it's the cleanest burning of all the fossil fuels: Per unit of energy generated, methane generates about half the carbon dioxide emissions of coal, But setting fire to natural gas at the electric power plant is not the beginning of its carbon emissions. Leaks and innumerable gallons of diesel precede its arrival there. And yet, to date, no comprehensive, life-cycle analysis of natural gas obtained from high-volume slickwater hydrofracturing has been undertaken. The full range of its greenhouse gas emissions is, thus, an open question. A preliminary assessment conducted at Cornell University in 2010 suggests that a complete consideration of all emissions—including leakage of methane gas—might make shale-derived natural gas look much less benign. This first-pass analysis concludes that natural gas drilled out of the Marcellus Shale has a global greenhouse effect (per unit of energy generated) roughly on par with coal obtained from mountain-top removal—a shameful form of energy extraction that even its proponents do not refer to as a bridge to anywhere. Blowing up mountains for coal and blowing up bedrock for shale gas begin to feel like two sides of the same counterfeit coin, two versions of the same ecological crime.

Here's another unanswered question: By tapping a new, abundant source of fossil fuel—however big or small its carbon emissions—are we not creating economic disincentives to investments in renewable energy and lowering the costs for all the things that natural gas is used to make: pesticides, anhydrous ammonia, PVC flooring? Where is this bridge leading us?

Nevertheless, under the banner of clean and green, the natural gas rush is on. Energy companies are looking at shale plays in Poland and Turkey. Fracking is underway in Canada. But nowhere has the technology been as rapidly deployed as in the United States. We are literally shattering the bedrock of our nation in order to bring methane out of the earth, consuming enormous quantities of precious fresh water to do so, without any clear knowledge of the health or environmental consequences.

Fracking destroys more than bedrock. It fractures our food systems, our communities, and our families. Even as my son exults in his newfound soccer triumphs, plans are being laid to turn his deeply loved home into a staging ground for fossil fuel extraction and so threaten his hair-trigger respiratory system— along with our community's interlaced system of streams and groundwater and our interlaced systems of local organic farms, vineyards, and dairies. Even as my daughter ponders a future career in marine biology, the world's plankton stocks and coral reefs are disappearing, casualties of fossil-fuel combustion.

Ultimately, the environmental crisis is a parenting crisis. It undermines my ability to carry out two fundamental duties: to protect my children from harm and to plan for their future. My responsibility as a mother thus extends beyond push mowers and clotheslines to the transformation of the nation's energy systems along renewable lines. Fine. With joy and resolve—and accepting the full severity of the situation vis-à-vis the world's oxygen-making plankton—I hereby devote myself to the task. When I watch my children breathing in their sleep, it doesn't feel like a choice. Happily, I'm in good company. And I have this quote for inspiration:

> Recent studies indicate the U.S. and world could rely 100 percent on green energy sources within 20 years if we dedicate ourselves to that course.

The language of addiction is often invoked in conversations about our economy's desperate relationship with fossil fuels. Up to a point, that's a useful way to frame the problem—especially when pondering the irrationalities of shale gas drilling and our own enabling behaviors that drive the need for it. Consider the drunk who has already cashed out his kids' college fund, hocked the family heirlooms, burned the furniture, and terrified the dog. He's beginning to grasp that he has a problem. He's also running out of whiskey. He flirts with the idea of Alcoholics Anonymous. But wait! He suddenly discovers a fully loaded wine cellar buried deep beneath the basement of his house. Falling in love with his own cleverness, he begins to lay plans to blow up the foundation to get at it. Overhearing the news, his alarmed family members hold an emergency meeting. What will they decide to do? Stay out of his way and pretend there is no problem? Help him get the wine and then regulate its consumption? Insist on overseeing the detonation of the basement? Or bar the way to the cellar steps?

The addiction analogy begins to fall apart, however, with the realization that, although their corporate charters may grant them legal personhood, energy companies are not people. There is no psychological bottom for them to hit. They are not motivated by life-changing epiphanies or by admissions that life has now become unbearable.

And when you live in a community that is targeted for fossil fuel extraction, when Halliburton and Exxon are massing at the border, the situation feels less like an Al-Anon meeting and more like foreign takeover. The word *plundering* comes to mind. In my own county, 40 percent of the land acreage is leased to gas drillers. In the township where I live, 37 percent is leased. In the village where I live—with its sidewalks, string bands, and alphabetically ordered library shelves—14 percent of the land is already under lease. As with any occupation, the populace is divided. Some people just wish the whole thing would go away. Some people find fracking such a vile and preposterous idea that they don't believe it will really happen—not here in the land of heirloom wheat fields and organic vineyards. Others, hoping for personal gain or believing that occupation is inevitable, plan to ride the tiger. From a position of collaboration, they can insist on greater oversight. So they believe. And there are others, open in their declarations of opposition, who seek to bar the door to the cellar steps.

"The shale army has arrived," said a representative from an energy company to a Canadian newspaper. "Resistance is futile."

The second half of that statement, anyway, is not true. Unless you believe that sitting at a segregated lunch counter or standing before a line of tanks in Tiananmen Square is just a waste of time. At a meeting at the fire station, located at the end of my street, the various candidates for mayor and village board all declared their unified opposition to fracking. They spoke about ways to deny the drilling companies access to our sewage treatment plant. They spoke about ways to decrease our own demand for natural gas and fossil fuels. Questions were raised about the role of local government. What protections could it offer our landscape, farms, and local economy? Given our collective commitment to sustainable development, could the township, for example, enact a total ban on fracking?

Shortly thereafter, a meeting about fracking at the village library drew a standing room only crowd. The conversation included a lively discussion about a community on nearby Keuka Lake that had successfully turned away fracking waste trucked in from Pennsylvania, which was headed to an old well for disposal. An older man in the audience declared passionately, "We have to be ready to lie down in front of the trucks."

On the way home, walking on an unbroken sidewalk made of shale above an as-of-yet unshattered bedrock made of shale, Elijah handed his stack of books to me—a week's worth of Magic Treehouse stories—and slipped his hand in mine. During the meeting, he had been so engrossed in books that I wondered how much he had overheard.

We shouldn't wreck this place down, right, Mom?

Resources and Sources

Further Resources

350.org   www.350.org
A movement working to unite the world around solutions to the climate cri-sis, beginning with the scientific premise that 350 parts per million (of carbon dioxide in the atmosphere) is the outer safe limit for humanity. (Current level: 388.)

Beyond Pesticides   www.beyondpesticides.org
A coalition that works to eliminate toxic pesticides in homes, schools, lawns, and the food supply.

CoalSwarm   www.sourcewatch.org
A portal on the collaborative encyclopedia SourceWatch that provides infor-mation on the coal industry to citizen groups working to promote a transition away from coal. SourceWatch also provides information on gas drilling and fracking.

**Collaborative on Health and the Environment
    www.healthandenvironment.org**
An international partnership committed to strengthening the scientific and public dialogue on the impact of environmental factors on human health and catalyzing initiatives to address these concerns. The Web site provides links to the following projects and working groups:

    Initiative on Children's Environmental Health: An alliance of health professional and child advocates devoted to the promotion of children's health through the prevention of environmental exposures.

    Learning and Developmental Disabilities Initiative: A partnership of disabilities organizations, researchers, and environmental health groups work-ing to understand the impact of environmental pollutants on neurological health.

Working Group on Asthma and the Environment: A consortium of scientists, health-affected groups, and community organizations working to understand the link between air quality and respiratory health.

Columbia Center for Children's Environmental Health
 www.ccceh.hs.columbia.edu
A research center that works to protect children's health by conducting scientific studies linking common pollutants—including air contaminants, pesticides, mercury, and endocrine disruptors—to children's health.

Earthworks   www.earthworksaction.org
A research and advocacy organization working on behalf of communities confronting minerals and mining operations, including, through its Oil and Gas Accountability Project, drilling and hydraulic fracturing.

TEDX (The Endocrine Disruption Exchange)
 www.endocrinedisruption.com
A research organization that compiles scientific evidence on the health and environmental problems caused by exposure to chemicals that interfere with hormones, including those used in natural gas drilling.

Environmental Health News   www.environmentalhealthnews.org
A foundation-funded journalism organization that offers free subscriptions to a daily news digest and maintains searchable archives on a variety of environmental health topics, including air quality, asthma, autism, cancer, children's health, climate change, coal, fracking for natural gas, green chemistry, male reproduction, pesticides, phthalates, and water issues.

Environmental Working Group   www.ewg.org
Research and advocacy organization that has undertaken studies of product safety, drinking water contaminants, and body burdens of toxic chemicals, including those found in umbilical cord blood.

Food Studies Institute   www.foodstudies.org
An educational organization that promotes hands-on, curriculum-based experiences with nutritious foods as a way to transform the school lunch program and to teach food preparation skills to children and their parents.

Green Chemistry and Green Engineering
Fields of research that seek to eliminate hazardous chemicals at the design stage—and detoxify high school and university chemistry labs:

> **Berkeley Center for Green Chemistry   http://bcgc.berkeley.edu**

> **Center for Green Chemistry and Green Engineering at Yale**
> **www.greenchemistry.yale.edu**

> **Green Chemistry Education Network   www.cmetim.ning.com**

> **U Mass Boston Center for Green Chemistry**
> **www.greenchemistry.umb.edu**

Warner Babcock Institute for Green Chemistry
www.warnerbabcock.com

Healthy Building Network   www.healthybuilding.net
The conscience of the green architecture movement and green building certification standards (Leadership in Energy and Environmental Design, LEED), HBN seeks to transform the building materials market. An outspoken critic of PVC plastic and arsenic-treated wood.

Health Care Without Harm   www.noharm.org
A global movement that seeks to detoxify the healthcare industry through elimination of mercury and PVC and other toxicants used in hospitals. Promotes safer chemicals, medical recycling, healthy food systems, green building, and green purchasing.

Healthy Schools Network   www.healthyschools.org
An advocacy group that documents and publicizes school environmental problems while shaping education, health, and environmental policies that promote healthy, non-toxic learning environments.

Institute for Agriculture and Trade Policy   www.iatp.org
A research organization that advocates for family farmers, fair trade policies, rural wind power, and an end to the overuse of antibiotics in agriculture.

LocalHarvest, Inc.   www.localharvest.org
A nationwide, informational resource site for community supported agriculture, farmers' markets, and sources of local, organic food.

Making Our Milk Safe   www.safemilk.org
A group founded by breastfeeding mothers seeking to keep breast milk safe by eliminating industrial pollutants that contaminate it. A project of the Center for Environmental Health (www.ceh.org).

Midwives Alliance of North America   www.mana.org
Membership organization that represents the profession of midwifery, which practices a model of care that monitors all aspects of the mother's environment—physical, psychological, and social—during the childbearing cycle and seeks to minimize technological intervention in childbirth.

National Center for Safe Routes to School   www.saferoutesinfo.org
National program helping parents and communities to design routes to school free of traffic and other hazards so children can safely walk and bicycle to school.

National Farm to School Network   www.farmtoschool.org
The voice of the farm to school movement, supporting efforts to bring food from local farms to school cafeterias nationwide as a way supporting community-based food systems while improving children's health.

Pesticide Action Network   www.panna.org
An international citizens' network that advances alternatives to pesticides. Maintains a searchable database.

Physicians for Social Responsibility   www.psr.org
The medical and public health voice of the environmental health movement, working on the climate crisis, the spread of nuclear weapons, and the degradation of the environment.

Project Laundry List   www.laundrylist.org
The central organizing force of the "right to dry" movement. Helps individuals reduce energy consumption and save money through the cultural rediscovery of clotheslines.

Safer Chemicals, Healthy Families   www.saferchemicals.org
A nationwide effort by a coalition of organizations and individuals, including parents, to reform U.S. chemicals policy.

Science and Environmental Health Network   www.sehn.org
The leading voice for the precautionary principle in the United States. Encourages the practice of science in the public interest and advocates for the idea of legal guardianship for future generations.

Women's Voices for the Earth   www.womensvoices.org
A national organization that works to eliminate toxic chemicals that affect women's health, including cosmetics and cleaning products.

IN CANADA:
Canadian Partnership for Children's Health and Environment
   www.healthyenvironmentforkids.ca
An alliance of twelve organizations with expertise in issues related to children, public health, and the environment.

Source Notes

AUTHOR'S NOTE: A full citation is provided for each source the first time it appears in the notes for each chapter. Thereafter, only author and title are given. Whenever I was aware of them, I also provide citations for (accurately written) articles and books in the popular literature for readers with less appetite for peer-reviewed technical papers.

FOREWORD

xi   **Elijah Lovejoy:** P. Simon, *Freedom's Champion: Elijah Lovejoy* (Carbondale, IL: Southern Illinois University Press, 1994). Quotes are found on pages 67, 113, 114. See also the delightful children's biography by Jennifer Phillips, *Elijah Lovejoy's Fight for Freedom* (Shoreline, WA: Nose in a Book Publishing, 2010).

xiii   **The oceans' plankton:** M.R. Lewis and B. Worm, "Global Phytoplankton Decline Over the Past Century," *Nature* 466 (2010): 591–96; D.A. Siegel and B.A. Franz, "Oceanography: Century of Phytoplankton Change," *Nature* 466 (2010): 569, 571.

xiii   **Quote from Rachel Carson:** R. Carson, *Silent Spring* (New York: Houghton Mifflin, 1962), p. 13.

xiv   **Species extinctions:** W.F. Frick et al., "An Emerging Disease Causes Regional Population Collapse of a Common North American Bat Species," *Science* 329 (2010): 679–82; J. Schipper et al., "The Status of the World's Land and Marine Mammals: Diversity, Threat, and Knowledge," *Science* 322 (2008): 225–30.

xiv   **Current trends in pediatric disorders:** L.J. Akinbami et al., "Status of Childhood Asthma in the United States, 1980–2007," *Pediatrics* 213 (2009): S131–45; M. Altarac and E. Saroha, "Lifetime Prevalence of Learning Disability among U.S. Children," *Pediatrics* 119 (2007): S77–83; American Lung Association, *Asthma and Children Fact Sheet*, Feb. 2010, www.lungusa.org; T.E. Froehlich et al., "Prevalence, Recognition, and Treatment of Attention Deficit/Hyperactivity Disorder in a National Sample of U.S. Children," *Archives of Pediatric and Adolescent Medicine*

161 (2007): 857–64; S.G. Gilbert, *Scientific Consensus Statement on Environmental Agents Associated with Neurodevelopmental Disorders* (Learning and Developmental Disabilities Initiative, Collaborative on Health and the Environment, 2008); M. King and P. Bearman, "Diagnostic Change and the Increased Prevalence of Autism," *International Journal of Epidemiology* 38 (2009): 1224–34; M. Lloyd-Smith and B. Sheffield-Brotherton, "Children's Environmental Health: Intergenerational Equity in Action— A Civil Society Perspective," *Annals of the New York Academy of Sciences* 1140 (2008): 190–200. J.M. Perrin et al., "The Increase of Childhood Chronic Conditions in the United States," *Journal of the American Medical Association* 297 (2007): 2755–59; U.S. Centers for Disease Control, *Attention Deficit/Hyperactivity Disorder (ADHD)*, 2009, *Summary Health Statistics for U.S. Children: National Health Interview Survey, 2006*, and "Premature Birth," 2010.

xv **Conclusions of a pediatric environmental health investigation:** B. Lanphear et al., "Trials and Tribulations of Protecting Children from Environmental Hazards," *Environmental Health Perspectives* 114 (2006): 1609–12.

xvi **Well-informed futility:** Public communications scholar Gerhart Wiebe coined the term in 1973 to describe the psychological state in which viewers of television newscasts find themselves. G. Wiebe, "Mass Media and Man's Relationship to His Environment," *Journalism Quarterly* 50 (1973): 426–32. See also G. Schmidt, *Positive Ecology: Sustainability and the "Good Life"* (Burlington, VT: Ashgate Publishing, 2005).

xvii **quote by Lovejoy:** Illinois Anti-Slavery Convention, *Proceedings of the Ill. Anti-Slavery Convention: Held at Upper Alton on the Twenty-sixth, Twenty-seventh, and Twenty-eighth October 1837* (Alton, IL: Parks and Breath, 1838). This document is available through the Abraham Lincoln Historical Digitization Project, http://lincoln.lib.niu.edu.

CHAPTER ONE

2 **Freeze tolerance of wood frogs:** R.E. Lee et al., "Dynamics of Body Water During Freezing and Thawing in a Freeze-Tolerant Frog (*Rana sylvatica*)," *Journal of Thermal Biology* 17 (1992): 263–66.

2 **Thomas Road amphibian migration:** K. Block et al., "Minimizing Highway Mortality for Migrating Amphibians," student research project supervised by biologist John Confer, Ithaca College.

3 **Health effects of PCBs:** For a highly readable history of PCBs and a summary of the evidence of harm, see T. Dracos, *Biocidal: Confronting the Poisonous Legacy of PCBs* (Boston: Beacon Press, 2010); L.S. Birnbaum and D.S. Staskal-Wikoff, "5th International PCB Workshop—Summary and Implications," *Environment International* 36 (2010): 814–18; O. Boucher et al., "Prenatal Exposure to Polychlorinated Biphenyls: A Neuropsychologic Analysis," *Environmental Health Perspectives* 117 (2009): 7–16; J.G. Brody et al., "Environmental Pollutants and Breast Cancer Risk: Epidemiologic Studies," *Cancer* 109 (2007, S12): 2667–711; L.S. Engel et al., "Polychlorinated Biphenyls and Non-Hodgkin Lymphoma," *Cancer Epidemiology, Biomarkers & Prevention* 16 (2007): 373–76; K.P. Stillerman et al., "Environmental Exposures and Adverse Pregnancy Outcomes: A Review of the Science," *Reproductive Sciences* 15 (2008): 631–50.

3   **History of PCBs in the Hudson:** E.A. Garvey, "PCBs in the Hudson River: Role of Sediments," *Clearwaters* 32 (2002); E. Kolbert, "The River: Will the EPA Finally Make G.E. Clean Up Its PCBs?" *New Yorker*, 4 Dec. 2000.

4   **EPA investigation:** U.S. Environmental Protection Agency, "Hudson River PCBs: Actions Prior to EPA's February 2002 Record of Decision (ROD)," www.epa.gov/hudson/.

4   **PCBs seeping through shale bedrock:** More specifically, PCBs that migrated through the bedrock fractures entered the river when an old wooden gate collapsed at a nearby abandoned paper mill, allowing water to flow through a tunnel cut through the bedrock. D. Cargill, "The General Electric Superfraud—Why the Hudson River Will Never Run Clean," *Harpers*, Dec. 2009, pp. 41–51; New York State Department of Environmental Conservation, *Record of Decision: GE Hudson Falls Plant Site Operable Units No. 2A-2D, Village of Hudson Falls, Town of Kingsbury, Washington County, New York, Site Number 5-58-013*, March 2004; J.H. Guswa et al., "An Innovative Approach for Hydraulic Containment of PCB Contamination in Fractured Bedrock," National Groundwater Association, Sept. 2005.

4   **Citizen advocacy:** For example, Environmental Advocates of New York, Friends of a Clean Hudson, Riverkeeper, and sloop Clearwater.

4   **PCBs and thyroid hormone in pregnancy:** J. Chevrier et al., "PCBs, Organochlorines, and Thyroid Function during Pregnancy," *American Journal of Epidemiology* 168 (2008): 298–310.

7   **Now famous intelligence briefing:** "Bin Laden Determined to Strike in U.S.," President's Daily Briefing item, 6 Aug. 2001, declassified and approved for release 10 April 2004.

10   **The obscure origins of human labor:** R.M. Karmel, "The Onset of Human Parturition," *Archives of Gynecology and Obstetrics* 281 (2010): 975–82.

10   **Epidemiological studies of environmental influences on onset of labor:** These are reviewed in the report, *Shaping Our Legacy: Reproductive Health and the Environment* (Program on Reproductive Health and the Environment, National Center of Excellence in Women's Health, University of California, San Francisco, 2008).

11   **Phthalates and early labor:** R.M. Whyatt et al., "Prenatal Di-(2-Ethylhexyl) Phthalate Exposure and Length of Gestation Among an Inner-city Cohort," *Pediatrics* 124 (2009): e1213– e1220; R.M. Whyatt et al., "Maternal Prenatal Urinary Concentrations of D-(2-Ethylhexyl) Phthalate in Relation to the Timing of Labor: Results from a Birth Cohort Study of Inner-city Mothers and Newborns," *Epidemiology* 19 (2008): S220.

12   **Air pollution and early labor:** J.R. Barrett, "Delivering New Data: Local Traffic Pollution and Pregnancy Outcomes," *Environmental Health Perspectives* 117 (2009): A505; M. Brauer et al., "A Cohort Study of Traffic-Related Air Pollution Impacts on Birth Outcomes," *Environmental Health Perspectives* 116 (2008): 680–86; L.A. Darrow et al., "Ambient Air Pollution and Preterm Birth: A Time-Series Analysis," *Epidemiology* 20 (2009): 989–98; J.D. Parker et al., "Preterm Birth after the Utah Valley Steel Mill Closure: A Natural Experiment," *Epidemiology* 19 (2008): 820–23;

R. Slama et al., "Meeting Report: Atmospheric Pollution and Human Reproduction," *Environmental Health Perspectives* 116 (2008): 791–98.

12 **Barometric pressure and onset of labor:** E.A. King, "Association between Significant Decrease in Barometric Pressure and Onset of Labor," *Journal of Nurse Midwifery* 42 (1997): 32–34. See also, O. Akutagawa et al., "Spontaneous Delivery Is Related to Barometric Pressure," *Archives of Gynecology and Obstetrics* 275 (2007): 249–54; E. Hirsch et al., "Meteorological Factors and Timing of the Initiating Event of Human Parturition," *International Journal of Biometeorology* 6 June 2010 [epub ahead of print]; K.L. Noller et al., "The Effect of Changes in Atmospheric Pressure on the Occurrence of the Spontaneous Onset of Labor in Term Pregnancies," *American Journal of Obstetrics and Gynecology* 174 (1996): 1192–97.

13 **Consequences of preterm birth:** R.L. Goldenberg et al., "Preterm Birth 1: Epidemiology and Causes of Preterm Birth," *The Lancet* 371 (2008): 75–84; Institute of Medicine, *Preterm Birth: Causes, Consequences, and Prevention* (Washington, D.C.: National Academies Press, 2006).

16 **Breastfeeding:** M. Bartick and A. Reinhold, "The Burden of Suboptimal Breastfeeding in the United States: A Pediatric Cost Analysis," *Pediatrics* 125 (2010): 1048–56; A. Stuebe, "The Risks of Not Breastfeeding for Mothers and Infants," *Reviews in Obstetrics and Gynecology* 2 (2009): 222–231. I reviewed the value of breastfeeding in S. Steingraber, *Having Faith: An Ecologist's Journey to Motherhood* (New York: Berkley, 2003).

21 **World Trade Center attacks:** R. M. Brackbill et al., "Surveillance for World Trade Center Disaster Health Effects Among Survivors of Collapsed and Damaged Buildings," *Morbidity and Mortality Weekly Report* 55 (2006): 1–8; S. Rayne et al., "Polychlorinated Dioxins and Furans from the World Trade Center Attacks in Exterior Window Films from Lower Manhattan in New York City," *Environmental Science & Technology* 39 (2005): 1995–2003.

21 **Effects of the WTC disaster on children:** J. Reibman et al., "The World Trade Center Residents' Respiratory Health Study: New-Onset Respiratory Symptoms and Pulmonary Function," *Environmental Health Perspectives* 113 (2005): 406–11; J. Reibman et al., "The World Trade Center Residents' Respiratory Health Study: New-Onset Respiratory Symptoms and Pulmonary Function," *Environmental Health Perspectives* 113 (2005): 406–411; A.M. Szema et al., "Post-9/11: High Asthma Rates Among Children in Chinatown, New York," *Allergy and Asthma Proceedings* 30 (2009): 605–11; A.M. Szema et al., "Clinical Deterioration in Pediatric Asthmatic Patients After September 11, 2001," *Journal of Allergy and Clinical Immunology* 113 (2004): 420–26; P.A. Thomas et al., "Respiratory and Other Health Effects Reported in Children Exposed to the World Trade Center Disaster of 11 September 2001," *Environmental Health Perspectives* 116 (2008): 1383–90.

22 **Effects of the WTC disaster on pregnant women:** P. J. Landrigan et al., "Impact of September 11 World Trade Center Disaster on Children and Pregnant Women," *Mount Sinai Journal of Medicine* 75 (2008): 129–34; S.A. Lederman et al., "The Effects of the World Trade Center Event on Birth Outcomes among Term Deliveries at Three Lower Manhattan Hospitals," *Environmental Health Perspectives* 112 (2004): 1772–78; F. P. Perera, "Relationship Between Polycyclic Aromatic Hydrocarbon-DNA Adducts, Environmental Tobacco Smoke, and Child Development in the

World Trade Center Cohort," *Environmental Health Perspectives* 115 (2007): 1497–1502; F. P. Perera et al., "DNA Damage from Polycyclic Aromatic Hydrocarbons Measured by Benzo[a]pyrene-DNA Adducts in Mothers and Newborns from Northern Manhattan, the World Trade Center Area, Poland, and China," *Environmental Health Perspectives* 114 (2005): 709–14; F.P. Perera et al., "Relationships Among Polycyclic Aromatic Hydrocarbon-DNA Adducts, Proximity to the World Trade Center, and Effects on Fetal Growth," *Environmental Health Perspectives* 113 (2005): 1062–67.

22   **Effects of the WTC disaster and other disasters on sex ratio:** T.A. Bruckner et al., "Male Fetal Loss in the U.S. Following the Terrorist Attacks of September 11, 2001," *BMC Public Health* 10 (2010): 273–81; R. Catalano et al., "Exogenous Shocks to the Human Sex Ratio: The Case of September 11, 2001 in New York City," *Human Reproduction* 21 (2006): 3127–31; R. Catalano et al., "Sex Ratios in California Following the Terrorist Attacks of September 11, 2001," *Human Reproduction* 20 (2005): 1221–27; R. Catalano and T. Hartig, "Communal Bereavement and the Incidence of Very Low Birthweight in Sweden," *Journal of Health and Social Behavior* 42 (2001): 333–41.

In addition to national disasters, the phenomenon of the disappearing male is also associated with maternal exposure to certain chemical contaminants, including PCBs. D.L. Davis et al., "Declines in Sex Ratio at Birth and Fetal Deaths in Japan, and in U.S. Whites but Not African Americans," *Environmental Health Perspectives* 115 (2007): 941–46; I. del Rio Gomez, "Number of Boys Born to Men Exposed to Polychlorinated Biphenyls," *The Lancet* 360 (2002): 143–44; M.G. Weisskopf et al., "Decreased Sex Ratio Following Maternal Exposure to Polychlorinated Biphenyls from Contaminated Great Lakes Sport-Caught Fish: A Retrospective Cohort Study," *Environmental Health* 2 (2003): 2.

CHAPTER TWO

27   **Differences between children and adults:** P.J. Landrigan and A. Garg, "Chronic Effects of Toxic Environmental Exposures on Children's Health," *Clinical Toxicology* 40 (2002): 449–56.

28   **Mouth breathing:** D. Mukerjee, "Assessment of Risk from Multimedia Exposures of Children to Environmental Chemicals," *Air Waste Management Association* 48 (1998): 483–501.

28   **Models not representative of children:** The EPA's policy, since 1995, is to consider the risks to infants and children explicitly as part of its risk assessments on environmental hazards. However, this policy does not constitute a rule and is not "intended, nor can it be relied upon, to create any rights enforceable by any party." It's also not retroactive. U.S. Environmental Protection Agency, "Policy on Evaluating Health Risks to Children," Nov. 1995.

28   **Reference man:** President's Cancer Panel, *Reducing Environmental Cancer Risk: What We Can Do Now*, 2008–2009 Annual Report (Washington, D.C.: National Cancer Institute, April 2009).

29   **Toxic Substances Control Act does not regulate on the basis of developmental effects:** B. Borrell, "America Pushes to Overhaul Chemical Safety Law," *Nature* 463 (2010): 599. See also R. Steinzor et al., "Corrective Lenses for IRIS: Additional Reforms to Improve EPA's Integrated

Risk Information System (Center for Progressive Reform, white paper #1009, Oct. 2010) and U.S. Government Accountability Office, *Chemical Assessments: Low Productivity and New Intra-agency Review Process Limit the Usefulness and Credibility of EPA's Integrated Risk Information System*, GAO-08-440, 2008.

29   **National Children's Study:** Involving the National Institutes of Health, the Centers for Disease Control and Prevention, and the Environmental Protection Agency, the National Children's Study will examine the role of the environment in the development and health of children from birth through age 21. www.nationalchildrensstudy.gov.

29   **Toxic Chemicals Safety Act of 2010:** This bill, H.R. 5820, sponsored by Representative Bobby Rush of Illinois, is at this writing, in the first step of the legislative process. www.govtrack.us/congress/.

30   **9.5 hand-to-mouth interactions per hour:** U.S. EPA, FIFRA Scientific Advisory Panel, *Final Exposure Document*, 23–25 Oct. 2001.

31   **Toxicology of arsenic:** C.O. Abernathy, "Health Effects and Risk Assessment of Arsenic," *Journal of Nutrition* 133 (2003, S1): 1536S–1538S; U.S. Agency for Toxic Substances and Disease Registry, *Toxicological Profile of Arsenic* (Atlanta, U.S. Department of Public Health and Human Services, 2007).

32   **Geology of arsenic:** J. Hauserman, "Arsenic Arrives in 'Toxic Trade'—The United States Is the World's No. 1 Arsenic Consumer, and China Provides Most of It," *St. Petersburg Times*, 29 Dec. 2001.

32   **Organic vs. inorganic arsenic:** U.S. Agency for Toxic Substances and Disease Registry, *Toxicological Profile for Arsenic*, August 2007. Complicating this picture, however, is recent research that reports high toxicity within particular organic (trivalent methylated) species. E. Dopp et al., "Cellular Uptake, Subcellular Distribution and Toxicity of Arsenic Compounds in Methylating and Non-Methylating Cells," *Environmental Research* 110 (2010): 435–42.

32   **Arsenic and glucocorticoids:** R.C. Kaltreider et al., "Arsenic Alters the Function of Glucocorticoid Receptor as a Transcription Factor," *Environmental Health Perspectives* 109 (2001): 245–51. In addition, arsenic is a potent endocrine disruptor and interferes with the signals of steroidal hormones critical in development. J.C. Davey et al., "Arsenic as an Endocrine Disruptor: Arsenic Disrupts Retinoic Acid Receptor- and Thyroid Hormone Receptor-Mediated Gene Regulation and Thyroid Hormone-Mediated Amphibian Tail Metamorphosis," *Environmental Health Perspectives* 116 (2008): 165–72.

33   **2010 review of arsenic and methylation:** X. Ren et al., "An Emerging Role of Epigenetic Dysregulation in Arsenic Toxicity and Carcinogenesis," *Environmental Health Perspectives*, 119 (2011): 11–19. See also, A.S. Andrew et al., "Decreased DNA Repair Gene Expression Among Individuals Exposed to Arsenic in United States Drinking Water," *International Journal of Cancer* 104 (2003): 263–68.

33   **Special vulnerabilities of children to arsenic:** B. Fangstrom et al., "Impaired Arsenic Metabolism in Children During Weaning," *Toxicology and Applied Pharmacology* 239 (2009): 208–14; National Research Council, *Arsenic in Drinking Water* (Washington, D.C.: National Academy of Sciences, 1999).

34   **Sonti Kamesam and the history of pressure-treated wood:** This history

is summarized in my report, S. Steingraber, "Late Lessons from Pressure-Treated Wood," parts I and II, *Rachel's Environment & Health News* #784 and #785, Feb. 2004. For an engaging architectural history of the faux-redwood deck, "the platform shoe" of suburbia, see M. Dolan, in *The American Porch: An Informal Life History of an Informal Place* (Guilford, CT: Globe Pequot Press, 2002). See also, P.A. Copper, "Future of Wood Preservation in Canada: Disposal Issues," paper presented at the 20th Annual Canadian Wood Preservation Association Conference, Vancouver, BC; C. Cox, "Chromated Copper Arsenate," *Journal of Pesticide Reform* 11 (1991): 2–6; Environmental Working Group and Healthy Building Network, *The Poisonwood Rivals: A Report on the Dangers of Touching Arsenic Treated Wood*, Nov. 2001; J. DiGangi, "Timeline History of the Wood Treatment Industry" (Environmental Health Fund, 2001); S. Fields, "Caution—Children at Play; How Dangerous Is CCA?" *Environmental Health Perspectives* 109 (2001): A262–69; I. Lerner, "Potential Litigation Creates Concern for Wood Preservatives," *Chemical Market Reporter*, 14 Oct. 2002, p. 14; C. Rist, "Arsenic and Old Wood," *This Old House*, March 1998, pp. 118–25; U.S. EPA, "Cancellation of Residential Uses of CCA-Treated Wood: Questions and Answers," 20 March 2003; U.S. EPA, *Guidance for the Reregistration of Wood Preservative Products Containing Chromated and Non-Chromated Arsenicals as the Active Ingredient*, EPA Office of Pesticide Programs, 1988.

36　**Fact sheets in retail stores:** Sometimes the information contained in these "fact sheets" contradicted that contained in the Material Safety Data Sheets provided to the employees of these same retail stores (Environmental Working Group and Healthy Building Network, *The Poisonwood Rivals*).

36　**Signs of harm during the 1980s:** W. Takahashi et al., "Urinary Arsenic, Chromium, and Copper Levels in Workers Exposed to Arsenic-Based Wood Preservatives," *Archives of Environmental Health* 38 (1983): 209–14; H.A. Peters et al., "Seasonal Arsenic Exposure from Burning Chromium-Copper-Arsenate Treated Wood," *Journal of the American Medical Association* 251 (1984): 2393–96.

36　**1992 Code of Federal Regulations:** Title 40 Code of Federal Regulations, Part 261.4(b).

37　**Consequences of disposal of CCA wood:** K.A. O'Connell, "Poison Planks," *Waste Age*, 1 Oct. 2003.

37　**Production increased 14-fold between 1970–1995:** Fields, "Caution—Children at Play."

37　**Stilwell's discovery:** D.E. Stilwell and K.D. Gorny, "Contamination of Soil with Copper, Chromium, and Arsenic Under Decks Built from Pressure-Treated Wood." *Bulletin of Environmental Contamination and Toxicology* 58 (1997): 22–29.

38　**Hauserman's reporting:** J. Hauserman, "The Poison in Your Backyard," *St. Petersburg Times*, 11 March 2001; "Treated Wood Industry Fights Back," *St. Petersburg Times*, 2 July 2001. Also, M. Dunne, interview with Julie Hauserman in *SEJournal*, Society of Environmental Journalists, winter 2001.

38　**2001 report on arsenic in drinking water:** National Research Council, *Arsenic in Drinking Water: 2001 Update* (Washington, D.C.: National Academies Press, Sept. 2001).

40 **Mouthing frequencies:** N.S. Tulve et al., "Frequency of Mouthing Behavior in Young Children," *Journal of Exposure Analysis and Environmental Epidemiology* 12 (2002): 259–64.

41 **Lifetime cancer risk in two weeks:** R. Sharp and B. Walker, *Poisoned Playgrounds: Arsenic in "Pressure-Treated Wood"* (Washington, D.C.: Environmental Working Group, 2001).

41 **Decks plus play equipment equals high risk:** The updated risk analysis—mostly unchanged from the 2003 draft analysis—of this subset of children is provided in EPA's final report. J. Chen et al., *A Probablistic Risk Assessment for Children Who Contact CCA-Treated Playsets and Decks, Final Report* (U.S. EPA, April 2008).

42 **Feb. 2002 announcement:** EPA, "Whitman Announces Transition from Consumer Use of Treated Wood Containing Arsenic," press release, 12 Feb. 2002. See also, S. Gray and J. Houlihan, *All Hands on Deck* (Washington, D.C.: Environmental Working Group, August 2002).

44 **EPA report released in 2008:** Chen, *A Probabilistic Risk Assessment for Children Who Contact CCA-Treated Playsets and Decks.*

46 **Well-informed futility:** G. Wiebe, "Mass Media and Man's Relationship to His Environment," *Journalism Quarterly* 50 (1973): 426–32. See also G. Schmidt, *Positive Ecology: Sustainability and the "Good Life"* (Burlington, VT: Ashgate Publishing, 2005).

47 **Peter Sandman:** "Climate Change Risk Communication: The Problem of Psychological Denial," www.psandman.com.

47 **550 million pounds of arsenic in pressure-treated wood:** D.A. Belluck et al., "Widespread Arsenic Contamination of Soils in Residential Areas and Public Spaces: An Emerging Regulatory or Medical Crisis?" *International Journal of Toxicology* 22 (2003): 109–128.

47 **Danger ignored when in playgrounds:** D.A. Belluck et al., "Widespread Arsenic Contamination of Soils in Residential Areas and Public Spaces: An Emerging Regulatory or Medical Crisis?" *International Journal of Toxicology* 22 (2003): 109–128; Sharp and Walker, *Poisoned Playgrounds.*

48 **Norway:** J. Pelley, "Playground Arsenic Dose Relatively Low," *Environmental Science & Technology* 44 (2010): 3650.

49 **CCA wood in landfills:** H. Solo-Gabriele et al., "Disposal of CCA-Treated Wood: An Evaluation of Existing and Alternative Management Options," Florida Center for Solid and Hazardous Waste Management, State University System of Florida, 1999.

51 **Technical report on 1993 ban on arsenical pesticides:** U.S. EPA, International Pesticide Notice. *EPA Cancels the Last Agricultural Use of Arsenic Acid in the United States*, Office of Pesticide Programs, 1993. See also G. Kidd, "CCA-Treated Lumber Poses Danger from Arsenic and Chromium," *Pesticides and You* 21 (2001): 13–15. www.beyondpesticides.org.

52 **2003 draft of EPA risk assessment:** U.S. EPA, *A Probabilistic Risk Assessment for Children Who Contact CCA-Treated Playsets and Decks* (Washington, D.C.: EPA, Nov. 2003).

53 **CPSC report:** U.S. Consumer Products Safety Commission, "Questions and Answers: CCA-Treated Wood," Feb. 2002. See also, CPSC, "Petition to Ban Chromated Copper Arsenate (CCA)-Treated Wood in Playground Equipment (Petition HP 01-3), Feb. 2003; CPSC, "CPSC Denies Petition to Ban CCA Pressure-Treated Wood Playground Equipment," 4 Nov. 2003, press release.

53 **2004 refinement of mathematical model:** R.M. Parsons and H.M. Solo-Gabriele, "Children's Exposure to Arsenic from CCA-Treated Wooden Decks and Playground Structures," *Risk Analysis* 24 (2004): 51–64. See also, E. Kwon et al., "Arsenic on the Hands of Children after Playing in Playgrounds," *Environmental Health Perspectives* 112 (2004): 1375–80; S.L Shalat et al., "A Pilot Study of Children's Exposure to CCA-Treated Wood from Playground Equipment," *Science of the Total Environment* 15 (2006): 80–88.

53 **Results of sealant study:** "A Set of Scientific Issues Being Considered by the Environmental Protection Agency Regarding: Studies Evaluating the Impact of Surface Coatings on the Level of Dislodgeable Arsenic, Chromium and Copper from Chromated Copper Arsenate (CCA)-Treated Wood," SAP Minutes No. 2007-02, FIFRA Scientific Advisory Panel Meeting, 15–16 Nov. 2006.

53 **2007 challenge:** L.M. Barraj et al., "The SHEDS-Wood Model: Incorporation of Observational Data to Estimate Exposure to Arsenic for Children Playing on CCA-Treated Wood Structures," *Environmental Health Perspectives* 117 (2007): 781–86.

53 **2008 final risk assessment:** Chen, *Chromated Copper Arsenate.*

53 **2010 study in New Orleans:** H.W. Mielke et al., "Soil Arsenic Surveys of New Orleans: Localized Hazards in Children's Play Areas," *Environmental Geochemistry and Health* 32 (2010): 431–40; J. Raloff, "Toxic Playgrounds," *Science News*, 23 Nov. 2009.

55 **Tips from Pediatrics for Parents:** C.O. Adler, "Toxins in Playground," *Pediatrics for Parents* 23 (2007): 11–12.

55 **Quote from updated profile:** U.S. Agency for Toxic Substances and Disease Registry, *Toxicological Profile for Arsenic.*

CHAPTER THREE

58 **Playground on PBS:** E. Campi, "CU Ecologist Talks Chemicals, Illness on PBS," *Ithaca Journal*, 10 May 2002.

60 **Design of CSA farms:** E. Henderson and R. Van En, *Sharing the Harvest: A Citizen's Guide to Community Supported Agriculture* (White River Junction, VT: Chelsea Green, 2007); A Wallace, "The Origins of CSA: Japan's Seikyou Movement Then and Now," Maine Organic Farmers and Gardeners Association. www.mofga.org. A somewhat different set of historical origins for CSAs is described by Steven McFadden, "Community Farms in the 21st Century: Poised for Another Wave of Growth?: History of Community Supported Agriculture, Part I," Rodale Institute. www.newfarm.rodaleinstitute.org.

62 **Three-fourths of organic food has no residues:** B.P. Baker et al., "Pesticide Residues in Conventional, Integrated Pest Management (IPM)-grown and Organic Foods: Insights from Three U.S. Data Sets," *Food Additives and Contaminants* 19 (2002): 427–446.

62 **Children fed organic food:** D. Wessels et al., "Use of Biomarkers to Indicate Exposure of Children to Organophosphate Pesticides: Implications for Longitudinal Study of Children's Environmental Health," *Environmental Health Perspectives* 111 (2003): 1939–46; C.L. Curl et al., "Organophosphorus Pesticide Exposures of Urban and Suburban Pre-school Children with Organic and Conventional Diets," *Environmental Health Perspectives* (2003): 377–82; C. Lu et al., "Organic Diets Significantly

Lower Children's Dietary Exposure to Organophosphorus Pesticides," *Environmental Health Perspectives* 114 (2006): 260–63.

63   **Kids like highly contaminated foods:** C. Lu et al., "Assessing Children's Dietary Pesticide Exposure—Direct Measurement of Pesticide Residues in 24-Hour Duplicate Food Samples," *Environmental Health Perspectives* 118 (2010): 1625–30. See also, Samuel Fromartz's comparison of strawberries grown organically with those grown with methyl bromide in *Organic, Inc.: Natural Food and How They Grew* (Orlando, FL: Harcourt, 2006), pp. 32–68.

63   **Pesticides on strawberries:** USDA researchers found that only 5.9 percent of commercial strawberries (samples washed before testing) were free of pesticides. Agricultural Marketing Services, *Pesticide Data Program Annual Summary, Calendar Year 2008*, U.S. Department of Agriculture, Dec. 2009. www.ams.usda.gov.

63   **The Food Quality Protection Act:** D. Payne-Sturges et al., "Evaluating Cumulative Organophosphorus Pesticide Body Burden of Children: A National Case Study," *Environmental Science & Technology* 43 (2009): 7924–30; M.L. Phillips, "Registering Skepticism: Does the EPA Pesticide Review Protect Children?" *Environmental Health Perspectives* 114 (2006): A592–95. The ten-fold safety standard is a default. The EPA can reduce or eliminate it entirely if it has data suggesting that children are no more sensitive than adults to particular pesticides (Karl Tupper, personal communication).

65   **Pesticides and ADHD:** M.F. Bouchard et al., "Attention-Deficit/Hyperactivity Disorder and Urinary Metabolites of Organophosphate Pesticides," *Pediatrics* 125 (2010): 1270–77.

66   **Eating habits tend to flourish:** Wendy Wolfe, Ph.D., personal communication.

66   **Antonia Demas:** A. Demas, "School Meals: A Nutritional and Environmental Perspective," *Perspectives in Medicine and Biology* 53 (2010): 249–56.

66   **Go, grow, and glow foods:** Brown adopted these terms from D. Werner, *Helping Health Workers Learn* (Palo Alto, CA: Hesperian Foundation, 1988).

68   **McDonald's branding:** T.N. Robinson et al., "Effects of Fast Food Branding on Young Children's Taste Preferences," *Archives of Pediatric Adolescent Medicine* 161 (2007): 792–97.

69   **Farmworker children:** A. Bradman et al., "Pesticides and the Metabolites in the Homes and Urine of Farmworker Children Living in the Salinas Valley, CA," *Journal of Exposure Science and Environmental Epidemiology* 17 (2007): 331–49; G.D. Coronado et al., "Organophosphate Pesticide Exposure and Work in Pome Fruit: Evidence for the Take-Home Pesticide Pathway," *Environmental Health Perspectives* 114 (2006): 999–1006.

70   **Amphibians and pesticides:** T.B. Hayes et al., "Atrazine Induces Complete Feminization and Chemical Castration in Male African Clawed Frogs (*Xenopus laevis*)," *Proceedings of the National Academy of Sciences* 107 (2010): 4612–17; T. Hayes et al., "Herbicides: Feminization of Male Frogs in the Wild," *Nature* 419 (2002): 895-96; A. Marco et al., "Sensitivity to Nitrate and Nitrite in Pond-Breeding Amphibians from the Pacific Northwest, USA," *Environmental Toxicology and Chemistry* 18

(1999): 2836–39. See also, *The Syngenta Corporation and Atrazine: The Cost to the Land, People and Democracy* (Land Stewardship Project and Pesticide Action Network North America, Jan. 2010).

71   **Antibiotic use in food animals:** Ali Kahn, assistant surgeon general, testimony before the Subcommittee on Health, Committee on Energy and Commerce, U.S. House of Representatives; 14 July 2010; K.M. Shea and the Committee on Environmental Health and Committee on Infectious Disease, "Nontherapeutic Use of Antimicrobial Agents in Animal Agriculture: Implications for Pediatrics" [American Academy of Pediatrics Technical Report], *Pediatrics* 114 (2004): 862–68; K.M. Shea, "Antibiotic Resistance: What Is the Impact of Agricultural Uses of Antibiotics on Children's Health?" *Pediatrics* 112 (2003): 253–58; E.K. Silbergeld et al., "Industrial Food Animal Production, Antimicrobial Resistance, and Human Health," *Annual Review of Public Health* 29 (2008): 151–69.

73   **Kale story:** S. Steingraber, "The Environmental Life of Children," in R. Cavoukian and S. Olfman (eds.), *Child Honoring: How to Turn This World Around,* (Westport, CT: Praeger, 2006), 107–115.

75   **Methyl bromide use in tomatoes and strawberry production:** M. Cone, "U.S. Has Been Stockpiling Banned Pesticide," *Los Angeles Times,* 15 Sept. 2006; A. Coombs, "Methyl Bromide Still Finds Its Way into U.S. Fields," *San Francisco Chronicle,* 24 Nov. 2007.

75   **Methyl bromide as a neurotoxicant:** R.S. Yang, "Toxicology of Methyl Bromide," *Reviews of Environmental Contamination and Toxicology* 142 (1995): 65–85.

75   **Methyl bromide as ozone destroyer:** L.O. Ruzo, "Physical, Chemical, and Environmental Properties of Selected Chemical Alternatives to the Pre-Plant Use of Methyl Bromide as a Soil Fumigant," *Pest Management Science* 62 (2006): 99–113.

75   **Whales with blistering sunburns:** L.M. Martinez-Levasseur et al., "Acute Sun Damage and Photoprotective Responses in Whales," *Proceedings of the Royal Society B*: November 2010 [epub ahead of print]. See also, A. Coghlan, *New Scientist,* 10 November 2010.

76   **Petition to U.N. Ozone Secretariat:** A good historical summary can be found in E.N. Rosskopf et al., "Alternatives to Methyl Bromide: A Florida Perspective," American Phytopathology Society, *APSnet,* June 2005.

76   **Toxicology of methyl iodide:** B.E. Erickson, "Methyl Iodide Saga Continues—EPA Gives Green Light to Soil Fumigant, but California Is Still Assessing Risks," *Chemical and Engineering News* 86 (2008): 28–30; Ruzo, "Physical, Chemical and Environmental Properties"; U.S. EPA, *Pesticide Fact Sheet: Iodomethane* (U.S. EPA, Office of Prevention, Pesticides and Toxic Substances, 2007).

77   **Tomatoes in the *Guinness Book of World Records:*** C. H. Wilber, *How to Grow World Record Tomatoes* (Metaire, LA: Acres U.S.A., 1999).

77   **Separation of animals from animal feed:** National Research Council, *Toward Sustainable Agricultural Systems in the United States.*

CHAPTER FOUR

84   **USDA Food Plans:** These figures are for July 2003. Updated figures can be found at the USDA's Center for Nutrition Policy and Promotion: www.cnpp/usda.gov. In June 2010, the average weekly cost of food at

home for a family of four with young children ranged from $134.50 (the thrifty plan) to $227.80 (the liberal plan). A. Carlson et al., *The Low-Cost, Moderate-Cost, and Liberal Food Plans, 2007* (U.S. Department of Agriculture, Center for Nutrition Policy and Promotion, CNPP-20, 2007).

85   **Higher cost of organic food:** My conclusions are described in "The Ecology of Pizza (Or Why Organic Food Is a Bargain)" Organic Trade Association, fall 2003. www.theorganicreport.org. See also, E. Simon, "Why Does Organic Food Cost More Than Conventional?" Associated Press, 22 Aug. 2005.

86   **Trends in organic agriculture:** Anonymous, "Organic Trade Association; U.S. Organic Sales Grow by a Whopping 17.1 Percent in 2008," *Food Business Week*, 21 May 2009, p. 47; C. Greene et al., "America's Organic Farmers Face Issues and Opportunities," *Amber Waves* 8 (2010): 34–39; Organic Trade Association, "Industry Statistics and Projected Growth" (Greenfield, MA: OTA, June 2010) and *Organic Trade Association's 2010 Organic Industry Survey* (both available at www.ota.com).

86   **Obstacles to organic farming:** C. Dimitri and L. Oberholtzer, "Marketing U.S. Organic Foods: Recent Trends From Farms to Consumers," USDA Economic Research Service, *Economic Information Bulletin* No. (EIB-58), Sept. 2009.

87   **The results of my pizza study:** Steingraber, "The Ecology of Pizza."

89   **My own tomatoes priced out at $2.14:** A bushel of organic utility tomatoes, purchased direct from the farmer, cost me $35. From one bushel, I could put up eighteen quarts of tomatoes. That meant $1.94 went into each jar, along with 8 cents-worth of citric acid. The jars were sealed with lids that cost 12 cents each. Since I reused the glass jars and metal rims and owned outright the big enamel canning pot and the special pinching device for lifting the tomato-filled glass jars from their boiling bath, I had no additional packaging expenses beyond the lids.

89   **Ding Dongs versus apples:** On August 5, 2010, a twelve-pack of Hostess Ding Dongs sold for $3.99 at the Tops Supermarket in Ithaca, New York. A dozen Red Delicious apples—about three pounds—cost $5.37.

90   **Hidden economic costs of conventional agriculture:** F. Kirschenmann, "Scale—Does it Matter?" in A. Kimbrell (ed.), *Fatal Harvest: The Tragedy of Industrial Agriculture* (Washington, D.C.: Foundation for Deep Ecology and Island Press, 2002), pp. 91–97.

90   **Half the cost of pesticides is included in the price:** In 1992, the indirect costs were $8 billion per year. D. Pimentel, "Environmental and Economic Costs of the Recommended Application of Pesticides," *Bioscience* 42 (1992): 750–60.

90   **Kimbrell's comments:** *Fatal Harvest: The Tragedy of Industrial Agriculture* (Washington, D.C.: Foundation for Deep Ecology and Island Press, 2002), p. 55.

91   **History of wheat:** W. Ebeling, *The Fruited Plain: The Story of American Agriculture* (Berkeley, CA: University of California Press, 1979); I.G. Malkina-Pykh and Y.A. Pykh, *Sustainable Food and Agriculture* (Southampton, UK: WIT Press, 2003).

91   **Wheat production:** G. Vocke, *Wheat Year in Review (Domestic): 2008 U.S. Wheat Production Rose in Response to High Prices*, USDA Electronic Outlook Report, WHS-2009 (Washington, D.C., USDA Economic Research Service, Jan. 2010). National Agricultural Statistics Service, "Agri-

culture Chemical Use, Wheat 2009," USDA; "Wheat's Role in the U.S. Diet Has Changed Over the Decades," USDA, Economic Research Service, March 2009 briefing; "Wheat: Background," USDA, Economic Research Service, March 2009 briefing.

92 **No national pesticide registry:** California is the sole state that does require comprehensive pesticide use reporting in agriculture. The results are made publicly available through the Department of Pesticide Regulations (www.cdpr.ca.gov), but I find the Pesticide Action Network database easier to use (http://pesticideinfo.org).

92 **Pesticides used on wheat:** These are compiled by the USDA's National Agriculture Statistics Survey: www.nass.usda.gov.

92 **2,4-D and birth defects in wheat-growing areas:** D.M. Schreinemachers, "Birth Malformations and Other Adverse Perinatal Outcomes in Four U.S. Wheat-Producing States," *Environmental Health Perspectives* 111(2003): 1159–64. See also, P.D. Winchester et al., "Agrichemicals in Surface Water and Birth Defects in the United States," *Acta Paediatrica* 98 (2009): 664–69.

93 **Chlorpyrifos linked to cognitive deficits and autism:** P.J. Landrigan, "What Causes Autism? Exploring the Environmental Contribution," *Current Opinion in Pediatrics* 22 (2010): 219–25.

93 **Nitrogen calculation for wheat:** I used two USDA databases for these calculations. The wheat acreage data was drawn from *Crop Production 2009 Summary* (USDA National Agricultural Statistics Service, CrPr2-1(10), Jan. 2010), and the nitrogen application data was drawn from a fertilizer table created from the 2009 wheat chemical use data in the NASS Quick Stats database. Thanks to USDA agricultural statistician Theresa Varner for her assistance in helping me locate these data. See USDA Economics, Statistics, and Market Information System, http://usda.mannlib.cornell.edu.

94 **History of olive oil:** J. Harwood and R. Aparicio, eds., *Handbook of Olive Oil: Analysis and Properties* (Gaithersburg, MD: Aspen, 2000), pp. 18-20; A.K. Kiritsakis, *Olive Oil* (Champaign, IL: American Oil Chemists, 1990), pp. 6–7.

94 **Pesticides in olive oil:** M.A. Aramendia et al., "Determination of Herbicide Residues in Olive Oil by Gas Chromatography-Tandem Mass Spectrometry," *Food Chemistry* 105 (2007): 855–61; C. Ferrer et al., "Determination of Pesticide Residues in Olives and Olive Oil by Matrix Solid-Phase Dispersion Followed by Gas Chromatography/Mass Spectrometry and Liquid Chromatography/Tandem Mass Spectrometry," *Journal of Chromatography* 1069 (2005): 183–94; A.M. Tsatsakis and I. Tsakiris, "Fenthion, Dimethoate and Other Pesticides in Olive Oils of Organic and Conventional Cultivation," in V.R. Preedy and R.R. Watson (eds.), *Olives and Olive Oil in Health and Disease Prevention* (Burlington, MA: Elsevier, 2010), pp. 415–424.

94 **Organic olive oil requires less energy:** C.I. Guzman and A.M. Alonso, "A Comparison of Energy Use in Conventional and Organic Olive Oil Production in Spain," *Agricultural Systems* 98 (2008): 167–76.

95 **History of tomato cultivation and sex:** J. Benton Jones, Jr., *Tomato Plant Culture* (Boca Raton, FL: CRC Press, 1999).

96 **Field tomatoes vs. fresh-market tomatoes:** Agricultural Marketing Resource Center, "Tomatoes." www.agmrc.org.

96   **Pesticide use on California tomatoes and in Fresno County:** Califor-
     nia pesticide use data from Pesticide Action Network's pesticide data-
     base: www.pesticideinfo.org.

97   **Botany of garlic:** S. Crawford, *A Garlic Testament: Seasons on a Small New
     Mexico Farm* (Albuquerque: University of New Mexico Press, 1998).

98   **Pesticide use on garlic:** Pesticide use data from Pesticide Action Net-
     work's pesticide database: www.pesticideinfo.org.

99   **Cheese and the Black Plague:** J. Thirsk, *Alternative Agriculture: A History
     from the Black Death to the Present Day* (New York: Oxford University
     Press, 1997), pp. 7–20.

99   **Dairy farming during the economic collapse:** W.D. McBridge and
     C. Greene, "Organic Sector Evolves to Meet Changing Demand," *Amber
     Waves*, March 2010; W.D. McBride and C. Greene, "Costs of Organic
     Milk Production on U.S. Dairy Farms," *Review of Agricultural Economics*
     31 (2009): 793–813. See also "Dairy Industry Report," *Natural Foods
     Merchandiser*, 2010.

101 **Organic farming cannot feed the world:** R. Paarlberg, "Attention
     Whole Foods Shoppers," Foreign Policy May/June 2010.

102 **Stanford study:** J.A. Burney et al., "Greenhouse Gas Mitigation by Agri-
     cultural Intensification," *Proceedings of the National Academy of Sciences*
     107 (2010): 12052–57.

102 **22 billion pounds of nitrogen fertilizer each year:** This is a 2008 sta-
     tistic from the Food and Agriculture Organization of the United
     Nations. http://faostat.fao.org.

103 **The wrong question:** A. Lappé, "Don't Panic, Go Organic," *Foreign Pol-
     icy* 29 April, 2010; H. Whittman et al., *Food Sovereignty: Reconnecting
     Food, Nature, and Community* (Oakland, CA: Food First Books, 2010).

103 **Terra Brockman:** "Embrace Organic Foods for Personal Community
     Health," Op-Ed, *Peoria Journal Star*, 28 Feb. 2010. www.terrabrockman
     .com.

103 **2010 health ranking study:** University of Wisconsin Population Health
     Institute and Robert Wood Johnson Foundation, *County Health Rank-
     ings*, Illinois 2010 Health Outcomes Map, Feb. 2010. www.countyhealth
     rankings.org.

104 **Organic farming can feed the world:** I. Perfecto and J. Vandermeer,
     "The Agroecological Matrix as Alternative to the Land-Sparing/Agricul-
     tural Intensification Model," *Proceedings of the National Academy of Sci-
     ences* 107 (2010): 5786–91.

104 **Haber process:** This extraordinary invention of German-Jewish scientist
     Fritz Haber is summarized by Diarmuid Jeffrey in *Hell's Cartel: IG Farben
     and the Making of Hitler's War Machine* (New York: Henry Holt, 2008).
     See also www.haberchemistry.tripod.com.

105 **University of Michigan study:** C. Badgley et al., "Organic Agriculture
     and the Global Food Supply," *Renewable Agriculture and Food Systems* 22
     (2007): 86–108.

105 **Wisconsin study:** J. L. Posner et al., "Organic and Conventional Produc-
     tion Systems in the Wisconsin Integrated Cropping Systems Trials: I.
     Productivity 1990–2002," *Agronomy Journal* 100 (2008): 253–60.

106 **Organic fields more resilient:** Fred Kirschenmann, Iowa State Univer-
     sity, personal communication.

106   **Organic fields evolve:** P. Mader et al., "Soil Fertility and Biodiversity in Organic Farming," *Science* 296 (2002): 1694–97; E. Stokstad, "Organic Farms Reap Many Benefits," *Science* 296 (2002): 1589.

106   **Potato study:** D.W. Crowder et al., "Organic Agriculture Promotes Evenness and Natural Pest Control," *Nature* 466 (2010): 109–13. See also D.G. Hole et al., "Does Organic Farming Benefit Biodiversity?" *Biological Conservation* 112 (2005): 113–30 and M. Hansen, *Escape from the Pesticide Treadmill: Alternatives to Pesticides in Developing Countries* (San Francisco: Consumers Union and Pesticide Action Network, 1988).

108   **Ecosystem services worth $33 trillion:** R. Custanza et al., "The Value of the World's Ecosystem Services and Natural Capital," *Nature* 387 (1997): 252–60.

108   **Quote about ecosystem services:** H.S. Sandu et al., "Organic Agriculture and Ecosystem Services," *Environmental Science and Policy* 13 (2010): 1–7.

109   **Northeast wheat farming:** T. Frisch, "A Short History of Wheat," *The Valley Table* 44 (2008): www.valleytable.com; E. Yowell, "NYC Food Detective: What Wheat Where?" Food Systems Network, NYC, 2010. See also www.growseeds.org/now.html.

CHAPTER FIVE

116   **Trickiness of burns:** R. Ravage, *Burn Unit: Saving Lives After the Flames* (Cambridge: Da Capo Press, 2004).

117   **Burned children:** Ravage, *Burn Unit.*

118   **Briefing paper on PVC:** S. Steingraber, *Update on the Environmental Health Impacts of Polyvinyl Chloride (PVC) as a Building Material: Evidence from 2000–2004—A Commentary for the U.S. Green Building Council*, April 2004.

119   **The Chemical Security Act of 2003:** www.theorator.com/bills108/s157 .html.

119   **Life history of PVC begins with chlorine gas:** As reviewed in my briefing paper. Steingraber, "Update on the Environmental Health Effects of Polyvinyl Chloride."

120   **Genuine linoleum is now imported:** T. Lent et al., *Resilient Flooring and Chemical Hazards: A Comparative Analysis of Vinyl and Other Alternatives for Health Care*, Health Care Research Collaboratives, April 2009.

120   **DEHP is ubiquitous in people:** Centers for Disease Control, *Third National Report on Human Exposures to Environmental Chemicals* (Atlanta, 2005).

120   **DEHP is an endocrine disruptor:** B. Kolarik, "The Association Between Phthalates in Dust and Allergic Diseases Among Bulgarian Children," *Environmental Health Perspectives* 116 (2008): 98–103.

121   **Chemicals prone to detonation:** T. Karasik, *Toxic Warfare*, Rand Corporation (for the U.S. Air Force), Santa Monica, CA, 2002.

121   **Databases pulled from the Web:** In 1990, Congress required industries that use extremely hazardous chemicals to disclose worst-case accident scenarios as part of developing their risk management plans. In the wake of 9/11, portions of these plans were removed from the EPA's Web site. The Right-to-Know Network, "Secrecy in the Bush Administration Obstructs Communities' Right to Know." 1 June 2004. www.rtknet.org.

122 **PVC plants are leaky places:** The ability of vinyl chloride monomer to drift beyond the fenceline is reviewed in Steingraber, "Update on the Environmental Health Effects of Polyvinyl Chloride."

122 **Research findings of Wilma Subra:** W. Subra, "Environmental Impacts in Communities Adjacent to PVC Production Facilities," presentation to the USGBC Stakeholders Meeting, 18 Feb. 2004.

123 **History of Formosa Plastics:** B. Bradway, "Munitions on the Prairie," *Doubletake* 26, Oct. 2003; B. Bradway, "Ill Wind: The Chemical Plant Next Door," *E Magazine*, Sept. 2002; G. Hall, "The History of Illiopolis," Illiopolis Business Association, 1984; D.M. Hines, "Boom Times—War Planted 9,700 Jobs Amid the Farms West of Illiopolis," *The State Journal-Register*, 13 Aug. 1995.

123 **Toxics emissions from Formosa:** Data from the Toxics Release Inventory, www.rtk.net.

123 **Detonation of the plant:** Illinois EPA, "Formosa Plastics Corporation Site," Fact Sheet 2, June 2004.

123 **U.S. Chemical Safety and Hazard Investigation Board:** "CSB Chairman Merritt Declares Full Root-Cause Investigation of Fatal Accident at Formosa Plastics PVC Plant in Illinois," press release, 25 April 2004.

124 **2007 final report:** U.S. Chemical Safety and Hazard Investigation Board, Investigation Report: *Vinyl Chloride Monomer Explosion (5 Dead, 3 Injured, and Community Evacuated), Formosa Plastics Corp. Illiopolis, Illinois, 23 April 2004* (U.S. CSHIB, 2007). For a timeline of events: L.A. Long et al., "Vinyl Chloride Monomer Explosion," *Process Safety Progress* 27 (2008): 72–79. See also, "After Five Fatalities, OSHA Fines Formosa Plastics $300,000," *Occupational Hazards* 67 (2005): 11–13.

131 **The results:** U.S. Agency for Toxic Substances and Disease Registry, *Health Consultation, Formosa Plastics Plant Explosion Cantrell Road and Old U.S. Route 36 Illiopolis, Sangamon County, Illinois*, EPA Facility ID: ILD005158548, 8 September 2005.

132 **Argonne National Laboratory report:** D.F. Brown et al., *A National Risk Assessment for Selected Hazardous Materials Transportation*, Argonne National Laboratory, Dec. 2000. See also L.M. Branscomb et al., *Rail Transportation of Toxic Inhalation Hazards: Policy Responses to the Safety and Security Externality*, Belfer Center Discussion Paper #2010-01, Harvard Kennedy School, Feb. 2010.

132 **2010 train derailment in Binghamton:** J. Hunter, "Railroads' Guarded Secret: Not Even HAZMAT Teams Are Told What's on Train," *Elmira Star Gazette*, 15 Aug. 2010.

132 **Terrorism and the chemical infrastructure report:** National Research Council, *Terrorism and the Chemical Infrastructure: Protecting People and Reducing Vulnerabilities* (Washington, D.C.: National Academies Press, 2006).

133 *Homeland Security Newswire:* "Chemical Industry Welcomes Extension of Current Chemical Facilities Security Measure," *Homeland Security Newswire*, 29 July 2010. See also L. Gilbert and E. Hitchcock, *Chemical Insecurity: America's Most Dangerous Companies and the Multimillion Dollar Campaign Against Common Sense Solutions* (U.S. PIRG, Aug. 2010).

136 **The fall-out shelter:** A. Szasz, *Shopping Our Way to Safety: How We Changed from Protecting the Environment to Protecting Ourselves* (Minneapolis: University of Minnesota Press, 2007).

CHAPTER SIX

139   **Surface area of the lung:** K.E. Pinkerton and F. H.Y. Green, "Normal Aging of the Lung," in R. Harding et al. (eds.), *The Lung: Development, Aging, and the Environment* (San Diego: Academic Press, 2004), 215–33.

140   **Respiration and emotions:** G.J. Tortora and S.R. Grabowski, *Principles of Anatomy and Physiology*, 7th ed. (New York, HarperCollins, 1993).

140   **Description of lung development:** S.E. McGowan and J.M. Snyder, "Development of Alveoli," and S. Orgeig et al., "Development of the Pulmonary Surfactant System," in R. Harding et al. (eds.), *The Lung: Development, Aging, and the Environment* (San Diego: Academic Press, 2004), pp. 55–73; 149–161.

143   **The taxonomy of asthma:** In the United States, about half of asthma cases are atopic. S.J. Arbes, "Asthma Cases Attributable to Atopy: Results from the Third National Health and Nutrition Examination Survey," *Journal of Allergy and Clinical Immunology* 120 (2007): 1139–45; H.Y. Kim et al., "The Many Paths to Asthma: Phenotype Shaped by Innate and Adaptive Immunity," *Nature Immunology* 11 (2010): 577–84.

144   **Environmental agents augment allergic reactions:** F.H.Y. Green and K.E. Pinkerton, "Environmental Determinants of Lung Aging," in R. Harding et al. (eds.), *The Lung: Development, Aging, and the Environment* (San Diego: Academic Press, 2004), 377–95; C.R. Mendelson, ed., *Endocrinology of the Lung: Development and Surfactant Synthesis* (Totowa, New Jersey: Humana Press, 2000).

144   **Early-life experiences alter immune functioning in ways that contribute to asthma:** R.J. Wright, "Perinatal Stress and Early Life Programming of Lung Structure and Function," *Biological Psychology*, Jan. 10, 2010 [epub ahead of print].

144   **Environmental chemicals can alter lung architecture:** M.D. Miller and M.A. Marty, "Impact of Environmental Chemicals on Lung Development," *Environmental Health Perspectives* 118 (2010): 1155–1164.

144   **Phthalates incite inflammatory responses:** C.G. Bornehag and E. Nanberg, "Phthalate Exposure and Asthma in Children," *International Journal of Andrology* 33 (2010): 333–45; K.E. Rakkestad et al., "Mono(2-ethylhexl) phthalate induces both Pro- and Anti-Inflammatory Responses in Rat Alveolar Macrophages through CrossTalk between p38, the Lipoxygenase Pathway and PPAR-alpha," *Inhalation Toxicology* 22(2010): 140–50. Phthalates come in several varieties. In addition to DEHP, the type that keeps vinyl shower curtains from cracking, other phthalates are used, for example, to carry fragrance in air fresheners and scented body lotions. For a highly readable summary, see R. Smith and B. Lourie, *Slow Death by Rubber Duck: How the Toxic Chemistry of Everyday Products Affects Our Health* (Toronto: Knopf Canada, 2009).

145   **Asthma and indoor air:** C.G. Bornehag et al., "The Association Between Asthma and Allergic Symptoms in Children and Phthalates in House Dust: A Nested Case-Control Study," *Environmental Health Perspectives* 112 (2004): 1393–97; J. J. K. Jaakkola and T.L. Knight, "The Role of Exposure to Phthalates from Polyvinyl Chloride Products in the Development of Asthma and Allergies: A Systemic Review and Meta-analysis," *Environmental Health Perspectives* 116 (2008): 845–53; M.J. Mendell, "Indoor Residential Chemical Emissions as Risk Factors for Respiratory and Allergic Effects in Children: A Review," *Indoor Air* 17 (2007): 259–77.

149 **Number of children with asthma:** In 2006, asthma killed 3,613 people of which 131 were children. American Lung Association, *Asthma and Children Fact Sheet,* Feb. 2010. www.lungusa.org

149 **Missed school days:** Centers for Disease Control, National Center for Chronic Disease Prevention and Health Promotion, "Healthy Youth! Health Topics: Asthma," Aug. 2009.

149 **The high cost of asthma:** American Lung Association, *Asthma and Children Fact Sheet.*

149 **30 percent of cases:** P.J. Landrigan et al., "Environmental Pollutants and Disease in American Children: Estimates of Morbidity, Mortality, and Costs for Lead Poisoning, Asthma, Cancer, and Developmental Disabilities," *Environmental Health Perspectives* (110) 2002: 721–28.

150 **Asthma time trends:** T. Woodruff et al., "Trends in Environmentally Related Childhood Illnesses," *Pediatrics* 113 (2004): 1133–40.

150 **Exposure to airborne contaminants:** An alternative—but not mutually exclusive—hypothesis posits that increasing hygiene plays a role in raising the risk for asthma. J. Douwes et al., "Protective Effects of Farming on Allergies and Asthma: Have We Learned Anything since 1873?" *Expert Review of Clinical Immunology* 5 (2009): 213–19.

150 **Improvement of air:** American Lung Association, *State of the Air, 2009;* F.P. Perera, "Children Are Likely to Suffer Most from Our Fossil Fuel Addiction," *Environmental Health Perspectives* 116 (2008): 987–90.

150 **Diesel exhaust:** Political scientist John Wargo discusses at length the threat of diesel exhaust for children in his chapter "The Trouble with Diesel," in *Green Intelligence: Creating Environments that Protect Human Health* (New Haven, CT: Yale University Press, 2009).

151 **Asthma and pesticides:** B.J. Proskocil et al., "Organophosphate Pesticides Decrease M2 Muscarinic Receptor Function in Guinea Pig Airway Nerves via Indirect Mechanisms," *PLoS ONE* 5 (2010): e10562.

151 **Demographics of asthma:** N. Laster et al., "Barriers to Asthma Management Among Urban Families: Caregiver and Child Perspectives," *Journal of Asthma* 46 (2009): 731–39; Y.Y. Meng, "Outdoor Air Pollution and Uncontrolled Asthma in the San Joaquin Valley, California," *Journal of Epidemiology and Community Health* 64 (2010): 142–47; N.C. Schleicher et al., "Asthma Mortality Rates Among California Youths," *Journal of Asthma* 37 (2000): 259–65; G.M. Solomon, "Asthma and the Environment," Collaborative on Health and Environment, April 2003. www.healthandenvironment.org.

151 **Rise of preterm births:** M. Davidoff et al., "Changes in the Gestational Age Distribution among U.S. Singleton Births: Impact on Rates of Late Preterm Birth, 1992–2002," *Seminars in Perinatology* 30 (2006): 8–15.

151 **Rise of obesity:** C. Ogden and M. Carroll, "Prevalence of Obesity Among Children and Adolescents: United States, Trends 1963–1965 through 2007–2008," National Center for Health Statistics, Centers for Disease Control, June 2010. www.cdc.gov/nchs/.

151 **Who gets asthma:** Obesity itself is an inflammatory state. C.M. Visness et al., "Association of Childhood Obesity with Atopic and Nonatopic Asthma: Results from the National Health and Nutrition Examination Survey 1999–2006," *Journal of Asthma* 47 (2010): 822–29.

151 **Link between outdoor air pollution and child asthma:** American Lung Association, *State of the Air, 2010* (Washington, D.C.: ALA, 2010);

L. Braback and B. Forsberg, "Does Traffic Exhaust Contribute to the Development of Asthma and Allergic Sensitization in Children?: Findings from Recent Cohort Studies," *Environmental Health* 8 (2009): 17–28; F.D. Gilliland and R. McConnell, "Effects of Air Pollution on Lung Function Development and Asthma Occurrence," in R. Harding et al. (eds.), *The Lung: Development, Aging, and the Environment* (San Diego: Academic Press, 2004), pp. 333–43; M. Jerrett et al., "Traffic-Related Air Pollution and Asthma Onset in Children: A Prospective Study with Individual Measurement," *Environmental Health Perspectives* 116 (2008): 1433–38; Kim, J.J. et al., "Residential Traffic and Children's Respiratory Health," *Environmental Health Perspectives* (2008): 1274–79; T.C. Lewis et al., "Air Pollution–Associated Changes in Lung Function Among Asthmatic Children in Detroit," *Environmental Health Perspectives* 113 (2005): 1068–1075; S. Lin et al., "Chronic Exposure to Ambient Ozone and Asthma Hospital Admissions among Children," *Environmental Health Perspectives* 116 (2008): 1725–30; L. Liu et al., "Acute Effects of Air Pollution on Pulmonary Function, Airway Inflammation, and Oxidative Stress in Asthmatic Children," *Environmental Health Perspectives* 117 (2009): 668–74; R. McConnell et al., "Childhood Incident Asthma and Traffic-Related Air Pollution at Home and School," *Environmental Health Perspectives* 118 (2010): 1021–26; R. McConnell et al., "Traffic, Susceptibility, and Childhood Asthma," *Environmental Health Perspectives* 114 (2006): 766–72; R. McConnell et al., "Asthma in Exercising Children Exposed to Ozone," *The Lancet* 359 (2002): 386–91; F. Perera, "Relation of DNA Methylation of 5'-CpG Island of ACSL3 to Transplacental Exposure to Airborne Polycyclic Aromatic Hydrocarbons and Childhood Asthma," *PLoS One* 4 (2009): e4488; F.P. Perera, "Children Are Likely to Suffer Most from Our Fossil Fuel Addiction"; C. Potera, "The Freeway Running Through the Yard: Traffic Exhaust and Asthma in Children," *Environmental Health Perspectives* 114 (2006): A305.

152   **Air pollution exposure in pregnancy changes baby's immune system:** C.E.W. Herr et al., "Air Pollution Exposure During Critical Time Periods in Gestation and Alterations in Cord Blood Lymphocyte Distribution: A Cohort of Livebirths," *Environmental Health* 9 (2010): 46–58.

152   **Asthma and living near busy roads:** M. Brauer et al., "Air Pollution and Development of Asthma, Allergy and Infections in a Birth Cohort," *European Respiratory Journal* 29 (2007): 825–26.

153   **Cytokines:** C. Chang et al., "Cord Blood Versus Age 5 Mononuclear Cell Proliferation on IgE and Asthma," *Clinical and Molecular Allergy* 8 (2010): 11–23.

153   **Air pollution interferes with asthma drugs:** Z. Qian, "Interaction of Ambient Air Pollution with Asthma Medication on Exhaled Nitric Oxide Among Asthmatics," *Archives of Environmental and Occupational Health* 64 (2009): 168–76.

153   **Effects that are not asthma:** Rachel Miller, M.D. "Addressing Urban Air Pollution and Climate Change," panel presentation before the conference "Translating Science to Policy: Protecting Children's' Environmental Health," Columbia Center for Children's Environmental Health, Columbia University, 30 March 2009; C. Potera, "Heavy Traffic Can Be a Pain in the . . . Ear?: Vehicle Emissions Linked to Otis Media,"

Environmental Health Perspectives 114 (2006): A544; S. Svanes et al., "Early Life Origins of Chronic Obstructive Pulmonary Disease," *Thorax* 65 (2010): 14–20.

For more on deaths due to exposure to particulate matter, see also California Air Resources Board, *Estimate of Premature Deaths Associated with Fine Particulate Particle Pollution (PM2.5) in California Using a U.S. Environmental Protection Agency Methodology* (California Air Resources Board, California Environmental Protection Agency, August. 2010) and M. Franklin et al., "Association between PM2.5 and All-Cause and Specific-Cause Mortality in 27 U.S. Communities," *Journal of Exposure Science and Environmental Epidemiology* 17 (2007): 279–87.

155 **The larger story:** J.F. Pearson et al., "Association Between Fine Particulate Matter and Diabetes Prevalence in the United States," *Diabetes Care* 33 (2010): 2196–201; U. Kramer et al., "Traffic-related Air Pollution and Incident Type 2 Diabetes: Results from the SALIA Cohort Study," *Environmental Health Perspectives* 118 (2010): 1273–79. This evidence is summarized in The President's Cancer Panel, *Reducing Environmental Cancer Risk: What We Can Do Now* (Washington, D.C.: National Cancer Institute, April 2010) as well as in S. Steingraber, *Living Downstream: An Ecologist's Personal Investigation of Cancer and the Environment* (Cambridge, MA: Da Capo Press, 2010), pp. 171–85.

155 **Air pollution's effect on the heart:** R.D. Brook et al., "Particulate Matter Air Pollution and Cardiovascular Disease," *Circulation* 121 (2010): 2331–78; H. Pham et al., "Central Neuroplasticity and Decreased Heart Rate Variability Following Particulate Matter Exposure in Mice," *Environmental Health Perspectives* 117 (2009): 1448–53.

156 **Testimony:** Albert A. Rizzo, M.D., American Lung Association, "S. 2995 The Clean Air Act Amendments of 2010," testimony before the Committee on Environment and Public Works and the Subcommittee on Clean Air and Nuclear Safety, U.S. Senate, 4 March 2010.

157 **Switzerland:** S.H. Downs et al., "Reduced Exposure to PM10 and Attenuated Age-Related Decline in Lung Function," *New England Journal of Medicine* 357 (2007): 2338–47.

157 **China study:** Frederica Perera, Dr.P.H., "Impact of Prenatal Exposure to Air Pollution from Traffic and Other Fossil Fuel Combustion Sources on Children's Health and Development," presentation before the conference, "Translating Science to Policy: Protecting Children's Environmental Health," Columbia University, March 2009.

157 **Atlanta:** M.S. Friedman et al., "Impact of Changes in Transportation and Commuting Behaviors During the 1996 Summer Olympic Games in Atlanta on Air Quality and Childhood Asthma," *Journal of the American Medical Association* 285 (2001): 897–905.

157 **Steel mine in Utah:** C.A. Pope, "Respiratory Disease Associated with Community Air Pollution and a Steel Mill, Utah Valley," *American Journal of Public Health* 79 (1989): 623–28.

160 **Greenhouse gas regulation:** B. Vastag, "The Long Fight Against Air Pollution," Smithsonian.com, 19 April 2010.

161 **Stanford study:** M.Z. Jacobson, "On the Causal Link Between Carbon Dioxide and Air Pollution Mortality," *Geophysical Research Letters* 35 (2008): L03809.

161 **European Respiratory Society:** J.G. Ayres et al., "Climate Change and

Respiratory Disease: European Society Position Statement," *European Respiratory Journal* 34 (2009): 295–302.

162 **Where fossil fuel emissions come from:** Perera, "Children Are Likely to Suffer Most from Our Fossil Fuel Addiction."

162 **Comments of Michel Gelobter, Ph.D.:** "Addressing Urban Air Pollution and Climate Change," panel presentation before the conference "Translating Science to Policy: Protecting Children's Environmental Health," Columbia Center for Children's Environmental Health, Columbia University, 30 March 2009.

163 **Global warming's effects on the ocean:** D.S. Arndt et al. (eds.), "State of the Climate in 2009," *Bulletin of the American Meteorological Society* 91 (2010): S1–S224. For two highly readable accounts of climate change and how we might survive it, see Dianne Dumanoski, *The End of the Long Summer: Why We Must Remake Our Civilization to Survive on a Volatile Earth* (New York: Three Rivers Press, 2009); B. McKibben, *Eaarth: Making a Life on a Tough New Planet* (New York: Henry Holt, 2010).

164 **Decline of plankton:** M.R. Lewis and B. Worm, "Global Phytoplankton Decline Over the Past Century," *Nature* 466 (2010): 591–96; D.A. Siegel and B.A. Franz, "Oceanography: Century of Phytoplankton Change," *Nature* 466 (2010): 569, 571. For a popular account, see Seth Borenstein, "Plankton, Base of Ocean Food Web, in Big Decline," Associated Press, 26 July 2010.

165 **Oxygen cycle:** Paul Falkowski, lecture, "Electrons, Life, and the Evolution of the Oxygen Cycle on Earth," Massachusetts Institute of Technology, 9 Oct. 2007.

CHAPTER SEVEN

169 **CDC guidelines:** U.S. Centers for Disease Control, "Coming in Contact with Bats," www.cdc.gov/rabies/bats/contact.index.html.

170 **Statistics on rabies:** Ten people are known to have survived rabies. U.S. Centers for Disease Control, "Bats and Human Rabies in the United States," www.cdc.gov/rabies/education/index.html.

171 **The concerted efforts of rabies prevention:** P.D. Kumar, *Biographies of Disease: Rabies* (Westport, CT: Greenwood, 2008).

171 **Climate change as a children's health issue:** Anonymous, "A Commission on Climate Change," *The Lancet* 373 (2009): 1659; L. Baker, *Feeling the Heat: Child Survival in a Changing Climate* (London: International Save the Children Alliance, 2009); Committee on Environmental Health, "Global Climate Change and Children's Health," *Pediatrics* 120 (2007): 1149–52; A. Costello et al., "Managing the Health Effects of Climate Change" [*The Lancet* and University College Long Institute for Global Health Commission], *The Lancet* 373 (2009): 1693–733; K. L. Ebi et al., "Climate Change and Child Health in the United States," *Current Problems in Pediatric Adolescent Health Care* 40 (2010): 2–18; M. Jay and M.G. Marmot, "Health and Climate Change: Will a Global Commitment Be Made at the UN Climate Change Conference in December?" *British Medical Journal* 339 (2009): 645–46; K.M. Shea and the Committee on Environmental Health, "Global Climate Change and Children's Health," *Pediatrics* 120 (2007): 1149–52; World Health Organization, "Health Impact of Climate Change Needs Attention," press release, 11 March 2009.

173   **Zach Jones:** L. Hart, "Texas Rabies Death Spurs New Concern About Bats," *Los Angeles Times*, 15 August 2006.

173   **Activists at the barricades:** They include climatologist James Hansen, director of NASA's Goddard Institute for Space Studies, journalist Bill McKibben, founder of 350.org, and Timothy deChristopher, arrested in Utah for disrupting an auction of public land for oil and gas drilling.

173   **The silence surrounding health effects:** K. Akerlof et al., "Public Perception of Climate Change as a Human Health Risk: Surveys of the United States, Canada, and Malta," *International Journal of Environmental Research and Public Health* 7 (2010): 2559–2606; E.W. Maibach et al., "Reframing Climate Change as a Public Health Issue: An Exploratory Study of Public Reactions," *BMC Public Health* 10 (2010): 2.

174   **Adults should shut up and get to work:** My evolving credo, as expressed in "The Big Talk," *Orion Magazine*, Sept.–Oct. 2008, is influenced by David Sobel: *Beyond Ecophobia: Reclaiming the Heart in Nature Education* (Great Barrington, MA: Orion Society Nature Literacy Series, 1996) and Richard Louv: *Last Child in the Woods: Saving Our Children from Nature-Deficit Disorder* (Chapel Hill, NC: Algonquin, 2008). Polar bear story is from my essay "The Environmental Life of Children," in R. Cavoukian and S. Olfman (eds.), *Child Honoring: How to Turn This World Around*, (Westport, CT: Praeger, 2006), 107–115.

176   **One in four mammal species threatened with extinction:** There are 5,487 known mammal species. According to the most recent assessment undertaken by the International Union for Conservation of Nature, 25 percent are at risk for disappearance. Of marine mammals, one in three. J. Schipper et al., "The Status of the World's Land and Marine Mammals: Diversity, Threat, and Knowledge," *Science* 322 (2008): 225–30.

176   **Climate change as de-creation:** B. McKibben, "Climate Change and the Unraveling of Creation," *Christian Century*, 8 Dec. 1999.

176   **The Eremozoic Era:** E.O. Wilson, *The Creation: An Appeal to Save Life on Earth* (New York: W.W. Norton, 2006).

177   **Children's literature on climate change:** For example, J. Bertagna, *Exodus* (New York: Walker, 2008); L. Cherry and G. Braasch, *How We Know What We Know About Our Changing Climate: Scientists and Kids Explore Global Warming* (Nevada City, CA: Dawn Publications, 2008); D. Gloiri, *The Trouble with Dragons* (NY: Bloomsbury, 2008); A. Rockwell, *Why Are the Ice Caps Melting?: The Dangers of Global Warming* (Collins, 2006). Also, Joanna Cole's latest book in the Magic School Bus series is *The Magic School Bus and the Climate Challenge* (NY: Scholastic, 2010). For a school curriculum on climate change, see B. Larson, "Making the Climate Connection," *Science & Children* 47 (2010).

178   **Because my parents are working on it:** Nancy Schimmel shared this story on the *Orion Magazine* Web site in response to "The Big Talk."

178   **Policy solutions that end subsidies for fossil fuels:** For example, M.Z. Jacobson and M.A. Delucchi, "A Path to Sustainable Energy by 2030," *Scientific American* 301 (2009): 58–65.

179   **Quote by James Hansen:** "Am I an Activist for Caring About My Children's Future? I Guess I Am," *The Guardian*, 26 August 2010.

179   **The multiplicative effect of a civil rights approach:** B. McKibben, "Multiplication Saves the Day," *Orion Magazine*, Nov.–Dec. 2008.

180 **Residences contribute 21.1 percent of emissions:** G.T. Gardner and P.C. Stern, "The Short List: The Most Effective Actions U.S. Households Can Take to Curb Climate Change," *Environment Magazine* 50 (2008): 12–24.

180 **The title of one recent paper:** T. Dietz et al., "Household Actions Can Provide a Behavioral Wedge to Rapidly Reduce U.S. Carbon Emissions," *Proceedings of the National Academy of Sciences* 106 (2009): 18452–56.

181 **The dreary list for how to green your house:** As provided by the *Wall Street Journal* in response to the news that the Russian wheat crop had burned up in a record-setting heatwave: B. Arends, "Ten Ways to Save Money by Going Green," *Wall Street Journal,* 3 September 2010.

181 **Barriers to household energy efficiency:** S.Z. Attari et al., "Public Perceptions of Energy Consumption and Savings," *Proceedings of the National Academy of Sciences* 107 (2010): 16054–59; G.T. Gardner and P.C. Stern, "The Short List: The Most Effective Actions U.S. Households Can Take to Curb Climate Change."

181 **A very rapid, 80 percent reduction:** P. Miller, "It Starts At Home," *National Geographic,* March 2009.

185 **Food and yard waste are 26 percent of solid waste:** U.S. EPA, "Composting," www.epa.gov/epawaste/conserve/rrr/composting/index.htm.

185 **Methane in landfills:** U.S. Energy Information Administration, *Emissions of Greenhouse Gases in the United States, 2008* (Washington, D.C.: U.S. Department of Energy, Dec. 2009). Compost piles do generate carbon dioxide, but because properly managed compost piles are aerated in ways that landfills are not, methane production is much lower.

186 **Systems thinking:** J. Gharajedaghi, *Systems Thinking: Managing Chaos and Complexity—A Platform for Designing Business Architecture* (Burlington, MA: Elsevier, 2006); H.T. Odum, *Ecological and General Systems: An Introduction to Systems Ecology,* rev. ed. (Boulder, CO: University Press of Colorado, 1994).

187 **Lawnmower emissions:** D. Stout, "EPA Issues New Engine Rules," *New York Times,* 4 Sept. 2009; Union of Concerned Scientists, *The Climate Friendly Gardener: A Guide to Combating Global Warming from the Ground Up,* April 2010; U.S. EPA, "Outdoor Air—Transportation: Lawn Equipment—Additional Information"; U.S. EPA, "Your Yard and Clean Air," Fact Sheet, OMS-19, May 1996. www.epa.gov.

188 **Children injured by lawn mowers:** Here is but a sampling of the published papers: J. Bayer et al., ["Stump Forming After Traumatic Foot Amputation of a Child—Description of a New Surgical Procedure and Literature Review of Lawnmower Accidents"] *Zeitschrift für Orthopädie und Unfallchirurgie* 147 (2009): 427–32; P.L. Horn and A.C. Beebe, "Lawn Mower Injuries in Pediatric Patients," *Journal of Trauma Nursing* 16 (2009): 136–41; M.E. Kurth et al., "Power Lawn Mower Decortication of Scalp and Skull," *Illinois Medical Journal* 108 (1955): 117–19; D. Vollman and G.A. Smith, "Epidemiology of Lawn-Mower Related Injuries to Children in the United States, 1990–2004," *Pediatrics* 118 (2006): 273–78.

188 **Emissions from lawn mowers in attached garages:** S. Batterman et al., "Migration of Volatile Organic Compounds from Attached Garages to Residences: A Major Exposure Source," *Environmental Research* 104 (2007): 224–40.

188 **Lawn mowers and ozone:** EPA, "Lawn Equipment—Additional Information." www.epa.gov/air/community/details/yardequip_addl_info.html.

188 **Mowing reduces the carbon-sequestering power of grass:** A. Townsend-Small and C.I. Czimczik, "Carbon Sequestration and Greenhouse Gas Emissions in Urban Turf," and "Correction to 'Carbon Sequestration and Greenhouse Gas Emissions in Urban Turf,'" *Geophysical Research Letters* 37 (2010): L02707 and L06707.

188 **Lead batteries in electric mowers:** D. Sivaraman and A.S. Linder, "A Comparative Life Cycle Analysis of Gasoline-, Battery-, and Electricity-Powered Lawn Mowers," *Environmental Engineering Science* 21 (2004): 768–85.

190 **Critics of the American lawn:** For example, N.K. Chase, *Eat Your Yard: Edible Trees, Shrubs, Vines, Herbs, and Flowers for Your Landscape* (Layton, UT: Gibbs Smith, 2010); J. Greenlee and S. Holt, *The American Meadow Garden: Creating a Natural Alternative to the Traditional Lawn* (Portland, OR: Timber Press, 2009).

191 **The absence of clothes dryers in Italy:** The editors, "Rethinking Laundry in the 21st Century," Room for Debate, *New York Times*, 25 Oct. 2009.

193 **Fires caused by clothes dryers:** National Fire Data Center, "Clothes Dryer Fires in Residential Buildings," *Topical Fire Research Series* 7 (2007): 1–7 (Emmitsburg, MD: U.S. Department of Homeland Security, U.S. Fire Administration).

193 **Statistics on clothes dryers:** K.A. Hughes, "To Fight Global Warming, Some Hang a Clothesline," *New York Times*, 12 April 2007; A. Lee, "The Right to Dry," *New Scientist* 204 (2009): 26–27; Project Laundry List, www.laundrylist.org.

194 **White-nose syndrome in bats:** V. Chaturvedi et al., "Morphological and Molecular Characteristics of Psychrophilic Fungus *Geomyces destructans* from New York Bats with White-Nose Syndrome (WNS)," *PLoS ONE* 5 (2010): e10783; W.F. Frick et al., "An Emerging Disease Causes Regional Population Collapse of a Common North American Bat Species," *Science* 329 (2010): 679–82; K. Kannan et al., "High Concentrations of Persistent Organic Pollutants Including PCBs, DDT, PBDEs, and PFOs in Little Brown Bats with White-Nose Syndrome in New York, USA," *Chemosphere* 80 (2010): 613–18; T. Kelley, "Bats Perish and No One Knows Why," *New York Times*, 25 March 2008; J. Scharr, "NY State Bat Populations May Be Extinct Within 20 Years: Report," *NBC New York*, 11 August 2010.

195 **U.S. Fish and Wildlife Service quote:** U.S. Fish and Wildlife Service, "The White-Nose Mystery: Something is Killing Our Bats," May 2010. www.fws.gov/WhiteNoseSyndrome.

CHAPTER EIGHT

201 **2006 Lancet study:** P. Gandjean and P.J. Landrigan, "Developmental Neurotoxicity of Industrial Chemicals," *The Lancet* 368 (2006): 2167–78; W. Mundy et al., "Building a Database of Developmental Neurotoxicants: Evidence from Human and Animal Studies," www.epa.gov.

201 **Polycyclic aromatic hydrocarbons:** Centers for Disease Control, "Polycyclic Aromatic Hydrocarbons," National Report on Human Exposures to Environmental Chemicals," updated tables, 2010. www.cdc.gov/exposurereport.

202   **22 percent of school spending:** These are 1999–2000 figures. U.S. Department of Education, *What Are We Spending on Special Educational Services in the United States, 1999–2000?*, American Institutes for Research, Special Education Expenditure Project, June 2004.

202   **Cost of development disabilities:** S.G. Gilbert, *Scientific Consensus Statement on Environmental Agents Associated with Neurodevelopmental Disorders* (Learning and Developmental Disabilities Initiative, Collaborative on Health and the Environment, 2008). See also, S. Gonzalez, *Mind Disrupted: How Toxic Chemicals May Change How We Think and Who We Are* (Learning and Developmental Disabilities Initiative, 2010), www.minddisrupted.org.

202   **Lifetime costs of autism:** Safer Chemicals, Healthy Families, *The Health Case for Reforming the Toxic Substances Control Act*, January 2010. M. Ganz, "The Costs of Autism," in S.O. Moldin and J.L.R. Rubenstein eds., *Understanding Autism: From Basic Neuroscience to Treatment* (Boca Raton, FL: Taylor and Francis, 2006), pp. 475–502.

203   **Trends in learning and developmental disorders:** M. Altarac and E. Saroha, "Lifetime Prevalence of Learning Disability Among U.S. Children," *Pediatrics* 119 (2007): S77–83; T.E. Froehlich et al., "Prevalence, Recognition, and Treatment of Attention-Deficit/Hyperactivity Disorder in a National Sample of U.S. Children," *Archives of Pediatric and Adolescent Medicine* 161 (2007): 857–64; S.G. Gilbert, *Scientific Consensus Statement on Environmental Agents Associated with Neurodevelopmental Disorders* (Learning and Developmental Disabilities Initiative, Collaborative on Health and the Environment, 2008); M. King and P. Bearman, "Diagnostic Change and the Increased Prevalence of Autism," *International Journal of Epidemiology* 38 (2009): 1224–34; J.M. Perrin et al., "The Increase of Childhood Chronic Conditions in the United States," *Journal of the American Medical Association* 297 (2007): 2755–59; U.S. Centers for Disease Control, *Attention Deficit/Hyperactivity Disorder (ADHD)*, 2009 and *Summary Health Statistics for U.S. Children: National Health Interview Survey, 2006.*

204   **CDC's estimate of ADHD:** S.N. Visser et al., "Increasing Prevalence of Parent-Reported Attention Deficit Hyperactivity Disorder Among Children—United States 2003 and 2007," *Morbidity and Mortality Weekly Report* 59 (2010): 1439–43.

204   **Core features of autism:** C.L. Lord and S. Spence, "Autism Spectrum Disorders: Phenotype and Diagnosis," in S.O. Moldrin and J.L.R. Rubenstein, *Understanding Autism: From Basic Neuroscience to Treatment* (Boca Raton, FL: Taylor and Francis, 2006), pp. 1–23.

205   **Changing prevalence of autism:** Centers for Disease Control, "Prevalence of Autism Spectrum Disorders—Autism and Developmental Disabilities Monitoring Network, United States, 2006." *Morbidity and Mortality Weekly Review* 58 (2009, SS10): 1–20.

205   **One-quarter of autism cases in California:** M. King and P. Bearman, "Diagnostic Change in and Increased Prevalence of Autism," *International Journal of Epidemiology* 38 (2009): 1224–34.

206   **Autism and thimerosal:** M. Aschner and S. Ceccatelli, "Are Neuropathological Conditions Relevant to Ethylmercury Exposure?" *Neurotoxicity Research* 18 (2010): 59–68.

206   **A 2010 review:** P.J. Landrigan, "What Causes Autism? Exploring the Environmental Contribution," *Current Opinion in Pediatrics* 22 (2010):

219–25. See also, E.M. Roberts et al., "Maternal Residence Near Agricultural Pesticide Applications and Autism Spectrum Disorders Among Children in the California Central Valley," *Environmental Health Perspectives* 115 (2007): 1482–89.

208   **Timing of exposures:** J.B. Herbstman et al., "Prenatal Exposure to PBDEs and Neurodevelopment," *Environmental Health Perspectives* 118 (2010): 712–19; R. Nevin, "Trends in Prenatal Lead Exposure, Mental Retardation, and Scholastic Achievement: Association or Causation?" *Environmental Research* 109 (2009): 301–10; J. Newman et al., "Analysis of PCB Congeners Related to Cognitive Functioning in Adolescents," *Neurotoxicology,* 22 May 2009; P.W. Stewart et al., "The Relationship Between Prenatal PCB Exposure and Intelligence (IQ) in 9-Year-Old Children," *Environmental Health Perspectives* 116 (2008): 1416–22.

208   **Neurotoxicants can act in concert:** T.E. Froehlich et al., "Association of Tobacco and Lead Exposures with Attention-Deficit/Hyperactivity Disorder," *Pediatrics* 124 (2009): 1054–63.

208   **Poverty and dysfunctional families:** M. Altarac and E. Saroha, "Lifetime Prevalence of Learning Disability Among U.S. Children," *Pediatrics* 119 (2007): S77–S83; T.E. Froehlich et al., "Prevalence, Recognition, and Treatment of Attention-Deficit/Hyperactivity Disorder in a National Sample of U.S. Children," *Archives of Pediatrics and Adolescent Medicine* 161 (2007): 857–64.

209   **Pesticide exposures and ADHD:** M.F. Bouchard, "Attention-deficit/ Hyperactivity Disorder and Urinary Metabolites of Organophosphate Pesticides," *Pediatrics* 125 (2010): 1270–77; V.A. Rauh et al., "Impact of Prenatal Chlorpyrifos Exposure on Neurodevelopment in the First 3 Years of Life Among Inner-City Children," *Pediatrics* 259 (2006): c1845–59.

209   **Effects of organophosphate pesticides on neurodevelopment of farm worker children:** B. Eskenazi et al., "Organophosphate Pesticide Exposure and Neurodevelopment in Young Mexican-American Children," *Environmental Health Perspectives* 115 (2007): 792–98; B. Eskenazi et al., "Association of *in Utero* Organophosphate Pesticide Exposure and Fetal Growth and Length of Gestation in an Agricultural Population," *Environmental Health Perspectives* 112 (2004): 1116–24.

211   **EPA response:** U.S. EPA, "EPA Statement Regarding Pesticides Article," 24 May 2010.

211   **Half pound of poison per child:** In 2009, the U.S. population of children aged 0–5 years old was 22.1 million. www.childstats.gov/ americaschildren/tables/pop1.asp.

211   **We live on a climate-altered planet:** B. McKibben, *Eaarth: Making Life on a Tough New Planet* (New York: Times Books, 2010).

214   **Quote from Lise Eliot:** L. Eliot, *What's Going on in There?: How the Brain and Mind Develop in the First Five Years of Life* (New York: Bantam, 1999).

215   **How the brain develops:** My description is based on the wonderfully readable book by Eliot, *What's Going on in There?*

215   **Enzymes called PON1:** B. Eskenazi et al., "PON1 and Neurodevelopment in Children from the CHAMACOS Study Exposed to Organophosphate Pesticides *in Utero,*" *Environmental Health Perspectives* (2010); C.E. Fulong et al., "PON1 Status of Farmworker Mothers and Children as a Predictor of Organophosphate Sensitivity," *Pharmacogenetics and Genomics* 16 (2006): 183–90.

215   **Stages of brain development:** My description is based on Eliot, *What's Going on in There?*

217   **Effect of organophosphate pesticides on the developing brain:** PON1 is shorthand for paraoxoase. T. Colborn, "A Case for Revisiting the Safety of Pesticides: A Closer Look at Neurodevelopment," *Environmental Health Perspectives* 114 (2006): 10–17; B. Eskanazi et al., "Developmental Changes in PON1 Enzyme Activity in Young Children and Effects of PON1 Polymorphisms," *Environmental Health Perspectives* 117 (2009): 1632–38; B. Eskanazi et al., "Exposures of Children to Organophosphate Pesticides and Their Potential Adverse Health Effects," *Environmental Health Perspectives* 107 (1999, S3): 409–19; C.E. Furlong et al., "PON1 Status of Farmworker Mothers and Children as a Predictor of Organophosphate Sensitivity"; N. Holland et al., "Paraoxonase Polymorphisms, Haplotypes, and Enzyme Activity in Latino Mothers and Newborns," *Environmental Health Perspectives* 114 (2006): 985–91; D.S. Rohlman et al., "Neuro-behavioral Performance in Preschool Children from Agricultural and Non-agricultural Communities in Oregon and North Carolina," *Neurotoxicology* 26 (2005): 589–98. My discussion in this section is also informed by the presentation of Virginia Rauh, "Prenatal Chlorpyrifos Exposure and Neurodevelopment: How Exposure to a Common Pesticide Can Damage the Developing Brain," at the Columbia University conference, "Translating Science to Policy: Protecting Children's Environmental Health," March 2009.

219   **No one knows the effects of digital media on brain development:** There is some observational evidence, however, for its negative effects on behavior and cognition. See Susan Linn, *The Case for Make Believe: Saving Play in a Commercial World* (New York: New Press, 2008). Daily television viewing from ages one to three is associated with attention problems at age seven. L.E. McCurdy et al., "Using Nature and Outdoor Activity to Improve Children's Health," *Current Problems in Pediatric and Adolescent Health Care* 40 (2010): 102–117.

221   **An average of 7.5 hours daily of media in 2010:** V.J. Rideout et al., *Generation M²: Media in the Lives of 8- to 18-Year-Olds*, Henry J. Kaiser Family Foundation study, Jan. 2010.

222   **Polycyclic aromatic hydrocarbons:** S.C. Edwards et al., "Prenatal Exposure to Airborne Polycyclic Aromatic Hydrocarbons and Children's Intelligence at Age 5 in a Prospective Cohort Study in Poland," *Environmental Health Perspectives* 118 (2010): 1326–31; F.P. Perera et al., "Prenatal Airborne Polycyclic Aromatic Hydrocarbon Exposure and Child IQ at Age 5 Years," *Pediatrics* 124 (2009): e195–e202; F.P. Perera, "Benefits of Reducing Prenatal Exposure to Coal-Burning Pollutants to Children's Neurodevelopment in China," *Environmental Health Perspectives* 116 (2008): 1396–1400; F.P. Perera et al., "Effect of Prenatal Exposure to Airborne Polycyclic Aromatic Hydrocarbons on Neurodevelopment in the First 3 Years of Life Among Inner-City Children," *Environmental Health Perspectives* 114 (2006): 1287–92.

223   **Mercury:** P. Grandjean and P.J. Landrigan, "Developmental Neurotoxicity of Industrial Chemicals, *The Lancet* 368 (2006): 2167–78; C. Johansson et al., "Neurobehavioral and Molecular Changes Induced by Methylmercury Exposures During Development," *Neurotoxicity Research* 11 (2007): 241–60; Z. Li et al., "Chemically Diverse Toxicants Converge

on Fyn and c-Cbl to Disrupt Precursor Cell Function," *PLoS Biology* 5 (2007): e35; Northeast States for Coordinated Air Use Management, *Mercury Emissions from Coal-fired Power Plant: The Case for Regulatory Action,* Oct. 2003; R. Tofighi et al., "Hippocampal Neurons Exposed to the Environmental Contaminants Methylmercury and Polychlorinated Biphenyls Undergo Cell Death via Parallel Activation of Calpains and Lysosomal Proteases," *Neurotoxicity Research,* 19 Feb. 2010 [epub ahead of print].

223 **One in twelve women:** S.E. Schober et al., "Blood Mercury Levels in U.S. Children and Women of Childbearing Age, 1999–2000," *Journal of the American Medical Association* 289 (2003): 1667–74.

224 **Societal costs associated with mercury-induced intelligence loss:** L. Trasande et al., "Mental Retardation and Prenatal Methyl Mercury Toxicity," *American Journal of Industrial Medicine* 49 (2006): 153–58; L. Trasande et al., "Public Health and Economic Consequences of Methyl Mercury Toxicity to the Developing Brain," *Environmental Health Perspectives* 113 (2005): 590–96. The $1.3 billion price tag has been disputed by analysts at the EPA's National Center for Environmental Economics: C. Griffiths et al., "A Comparison of the Monetized Impact of IQ Decrements from Mercury Emissions," *Environmental Health Perspectives* 115 (2007): 841–47. See also response to this challenge by authors: L. Trasande et al., "Methylmercury and the Developing Brain," *Environmental Health Perspectives* 115 (2007): A396.

224 **USGS reports on mercury in freshwater fish:** B.C. Scudder et al., *Mercury in Fish, Bed Sediment, and Water from Streams Across the United States, 1998–2005* (U.S. Geological Survey, Scientific Investigations Report, 2009).

224 **Fish advisories:** U.S. EPA, "Biennial National Listing of Fish Advisories, 2008," EPA-823-F-09-007, 2009. www.epa.gov/fishadvisories.

224 **Tuna accounts for one third of mercury:** Fresh and canned tuna contributes 37.4 percent of total mercury inputs to the U.S. seafood supply. E. Groth, III, "Ranking the Contributions of Commercial Fish and Shellfish Varieties to Mercury Exposure in the United States: Implications for Risk Communication," *Environmental Research* 110 (2010): 226–36.

224 **Coal plants contribute forty percent of mercury:** Northeast States for Coordinated Air Use Management, *Mercury Emissions from Coal-Fired Power Plants,* Oct. 2003

224 **Cement kilns:** The EPA estimates that U.S. cement kilns contribute 22,914 pounds of mercury to the air each year. Pointing out that this number reflects only nonhazardous waste burning kilns, the public interest law firm Earth Justice argues this figure is an underestimate. (Earth Justice and Environmental Integrity Project, *Cementing a Toxic Legacy? How the Environmental Protection Agency Has Failed to Control Mercury Pollution from Cement Kilns,* July 2008.) In summer 2010, the EPA announced its intent to begin regulating mercury emissions from cement plants. (U.S. EPA, "EPA Sets First National Limits to Reduce Mercury and Other Toxic Emissions from Cement Plants," press release, 9 August 2010.)

225 **Beneficial reuse of coal ash:** Public Employees for Environmental Responsibility, "RE: Docket Number EPA-HQ-RCRA-2009-0640, Notice of Proposed Rule for Hazardous and Solid Waste Management System,

Identification and Listing of Special Wastes; Disposal of Coal Combustion Residuals from Electric Utilities," letter to the U.S. EPA, 18 August 2010 [Appendix A]; U.S. EPA, *Cost and Impacts of Wasting Cement Kiln Dust or Replacing Fly Ash to Reduce Mercury Emissions*, December 2006.

225 **Tuna fish story:** S. Steingraber, "The Environmental Life of Children," in R. Cavoukian and S. Olfman (eds.), *Child Honoring: How to Turn This World Around*, (Westport, CT: Praeger, 2006), 107–115.

226 **FDA advice on tuna fish:** This consumption advisory has not been updated since 2004. On average, canned tuna contains 0.17 parts per million of methylmercury. Canned albacore ("white" or "solid") tuna averages 0.35 parts per million, as do fresh and frozen tuna fillets. In general "light" tuna has lower levels of mercury, but the variability between mercury levels in individual cans is so high that some cans of albacore contain less mercury than some cans of light tuna and vice versa. (U.S. Food and Drug Administration, *Draft Risk and Benefit Report: Section II, Exposure to Methylmercury in the United States*, Jan. 2009.) Because of this great variability in mercury levels among individual fish, tuna consumption guidelines for children and women of reproductive age are based on uncertainties. The Mercury Policy Project, in a letter signed by a group of thirty scientists, physicians, and consumer advocates, has urged the FDA to strengthen its fish consumption advisory for mercury and improve consumer warnings. www.mercurypolicy.org.

CHAPTER NINE

233 **The hypothalamus as the governor of puberty:** F.J.P. Ebling, "The Neuroendocrine Timing of Puberty," *Reproduction* 129 (2005): 675–83; M.M. Grumbach and D.M. Styne, "Puberty: Ontogeny, Neuroendocrinology, Physiology, and Disorders," in P.R. Larsen et al., eds., *Williams' Textbook of Endocrinology*, 10th ed. (Philadelphia: Saunders, 2003), 1115–1286; S.M. Hughes and A.C. Gore "How the Brain Controls Puberty, and Implications for Sex and Ethnic Differences," *Family and Community Health* 30 (2007, Suppl. 1): S112–14; A.S. Parent et al., "The Timing of Normal Puberty and the Age Limits of Sexual Precocity: Variations Around the World, Secular Trends, and Changes After Migration," *Endocrine Reviews* 24 (2003): 668–93.

233 **When these sleeping neurons wake up:** In girls, the activation of the hypothalamus pulse generator occurs about halfway through the first decade of life. The physical results, however, do not manifest for several more years (Frank Biro, personal communication).

234 **The hypothalamus as the receiver of messages:** H.M. Dungan et al., "Minireview: Kisspeptin Neurons as Central Processors in the Regulation of Gonadotropin-releasing Hormone Secretion," *Endocrinology* 147 (2006): 1154–58; A.C. Gore, "Gonadotropin-releasing Hormone (GNRH) Neurons: Gene Expression and Neuroanatomical Studies," *Progress in Brain Research* 141 (2002): 195–210; M. Tena-Sempere, "KiSS-1 and Reproduction: Focus on Its Role in the Metabolic Regulation of Fertility," *Neuroendocrinology* 83 (2006): 275–81; E. Terasawa, "Role of GABA in the Mechanism of the Onset of Puberty in Non-Human Primates," *International Review of Neurobiology* 71 (2005): 113–129.

234 **Infant puberty:** Ebling, "The Neuroendocrine Timing of Puberty"; A. Papathanasiou and C. Hadjiathanasiou "Precocious Puberty," *Pediatric*

Endocrinology Reviews 3 (2006, suppl. 1): 182–87; C.L. Sisk and J.L. Zehr, "Pubertal Hormones Organize the Adolescent Brain and Behavior," *Frontiers in Neuroendocrinology* 26 (2005): 163–174; K.E. Whitlock et al., "Development of GnRH Cells: Setting the Stage for Puberty," *Molecular and Cellular Endocrinology* 254–55 (2006): 39–50; C.W. Worthman, "Evolutionary Perspectives on the Onset of Puberty," in W.R. Trevathan, ed., *Evolutionary Medicine* (New York: Oxford, 1999).

235 **Endocrine disruptors identified so far:** For a thoughtful overview, E. Diamanit-Kandarakis et al., "Endocrine-Disrupting Chemicals: An Endocrine Society Scientific Statement," *Endocrine Reviews* 30 (2009): 293–342.

235 **Endocrine disruptors are not screened or regulated:** Safer Chemicals, Healthy Families, "The Health Case for Reforming the Toxic Substances Control Act," Jan. 2010.

236 **Bisphenol A:** K.S. Betts, "Body of Proof—Biomonitoring Data Reveal Widespread Bisphenol A Exposures," *Environmental Health Perspectives* 118 (2010): A353; E.C. Dodds and W. Lawson, "Synthetic Oestrogenic Agents Without the Phenanthrene Nucleus," *Nature* 137 (1936): 996; J.A. Taylor et al., "Similarity of Bisphenol A Pharmacokinetics in Rhesus Monkeys and Mice: Relevance for Human Exposure," *Environmental Health Perspectives*, Sept. 2010 [epub ahead of print]; L.N. Vandenberg et al., "Urinary, Circulating, and Tissue Biomonitoring Studies Indicate Widespread Exposure to Bisphenol A," *Environmental Health Perspectives* 118 (2010): 1055–70.

Animal studies link prenatal exposure to BPA with a multitude of chronic, hormonally mediated disorders, including early puberty, breast and prostate cancer, infertility, diabetes, and heart disease. F.S. vom Saal et al., "Chapel Hill Bisphenol A Expert Panel Consensus Statement: Integration of Mechanisms, Effects in Animals and Potential Impact to Human Health at Current Exposure Levels," *Reproductive Toxicology* 24 (2007): 131–38.

237 **Canada and bisphenol A:** Health Canada, "Government of Canada Takes Action on Another Chemical of Concern: Bisphenol A," press release, 18 April 2008; C. Mitchell, "Canada First Country to List BPA as Toxic," *Food Safety News*, 21 Sept. 2010.

237 **Properties of phthalates:** H.M. Koch and A.M. Calafat, "Human Body Burdens of Chemicals Used in Plastic Manufacture," *Philosophical Transactions of the Royal Society B* 364 (2009): 2063–78; J.D. Meeker et al., "Phthalates and other Additives in Plastics: Human Exposure and Associated Health Outcomes," *Philosophical Transactions of the Royal Society B* 364 (2009): 2097–113; L. Schierow, *Phthalates in Plastics and Possible Health Effects: CRS Report for Congress* (Congressional Research Service, RL34572, July 2008).

238 **Atrazine:** The precise molecular mechanism by which atrazine influences aromatase production is not yet understood, however, and appears to vary between cell types. W. Fan et al., "Atrazine-Induced Aromatase Expression is SF-1 Dependent: Implications for Endocrine Disruption in Wildlife and Reproductive Cancers in Humans," *Environmental Health Perspectives* 115 (2007): 720–27; J.A. McLachlaan et al., "Endocrine Disruptors and Female Reproductive Health," *Best Practice and Research. Clinical Endocrinology and Metabolism* 20 (2006): 63–75.

239 **EPA special review of atrazine:** EPA, "EPA Begins New Scientific Evaluation of Atrazine," press release, 7 October 2009.

239 **Atrazine banned in the EU in 2006:** J.B. Sass and A. Colangelo, "European Union Bans Atrazine, while the United States Negotiates Continued Use," *International Journal of Occupational and Environmental Health* 12 (2006): 260–67.

239 **At one such conference:** "Summit on Environmental Challenges to Reproductive Health and Fertility," convened by the University of California, San Francisco and the Collaborative on Health and the Environment, Jan. 28–30, 2007.

240 **My commissioned report on the falling age of puberty:** S. Steingraber, *The Falling Age of Puberty in U.S. Girls: What We Know, What We Need to Know*, Breast Cancer Fund, 2007.

241 **Description of development of male reproductive tract:** J.C. Achermann and I.A. Hughes, "Disorders of Sex Development," in H.M. Kronenberg et al. (eds.), *Williams Textbook of Endocrinology*, 11th ed. (Philadelphia: Saunders, 2008), pp. 783–848.

242 **Testicular dysgenesis syndrome:** G-X Hu et al, "Phthalate-induced Dysgenesis Syndrome: Leydig Cell Influence," *Trends in Endocrinology and Metabolism* 20 (2009): 139–45; J. Luoma, *Challenged Conceptions: Environmental Chemicals and Fertility* (Stanford University School of Medicine and Collaborative on Health and the Environment, 2005); K.M. Main et al., "Cryptorchidism as Part of the Testicular Dysgenesis Syndrome: The Environmental Connection," *Endocrine Development* 14 (2009): 167–73; S. Sathyanarayana, "Phthalates and Children's Health," *Current Problems in Pediatric and Adolescent Health Care* 38 (2008): 34–49; UCSF Program on Reproductive Health and the Environment, *Shaping Our Legacy: Reproductive Health and the Environment*, (University of California, San Francisco, Sept. 2008).

See also, K.M. Main et al., "Flame Retardants in Placenta and Breast Milk and Cryptorchidism in Newborn Boys," *Environmental Health Perspectives* May 2007 (115): 1519–26; Y. Xia et al., "Exposure to Combustion Byproducts Linked to Male Infertility," *Human Reproduction* 24 (2009): 1067–74; and *Vallombrosa Consensus Statement on Environmental Contaminants and Human Fertility Compromise*, Oct. 2005.

243 **Men from rural Missouri have fewer moving sperm than men from Minnesota:** S.H. Swan et al., "Semen Quality in Relation to Biomarkers of Pesticide Exposure," *Environmental Health Perspectives* 111 (2003): 1478–84; S.H. Swan et al., "Geographic Differences in Semen Quality of Fertile U.S. Males," *Environmental Health Perspectives* 111 (2003): 414–20. See also, J. Luoma, *Challenged Conceptions: Understanding Environmental Contaminants and Human Fertility*, Stanford University School of Medicine and Collaborative on Health and the Environment, October 2005.

243: **PCBs in Sweden and Taiwan:** Y.-L. Guo et al., "Growth Abnormalities in the Population Exposed in Utero and Early Postnatally to Polychlorinated Biphenyls and Dibenzofurans," *Environmental Health Perspectives* 103 (1995 S6): 117–22; L. Hardell et al., "In Utero Exposure to Persistent Organic Pollutants in Relation to Testicular Cancer Risk," *International Journal of Andrology* 29 (2006): 228–34.

244 **Anomalies in wildlife:** M.R. Milnes et al., "Contaminant-induced Feminization and Demasculization of Nonmammalian Vertebrate Males in

Aquatic Environments, *Environmental Research* 100 (2006): 3–17; J.M. Schwartz et al., *Shaping Our Legacy: Reproductive Health and the Environment—A Report on the Summit on Environmental Challenges to Reproductive Health and Fertility*, September 2008.

244 **How phthalates interfere with testicular development:** C.E. Talsness et al., "Components of Plastic: Experimental Studies in Animals and Relevance for Human Health," *Philosophical Transactions of the Royal Society B* 364 (2009): 2079–96.

245 **Anogenital distance and endocrine disruption:** M.H. Hsieh et al., "Associations Among Hypospadias, Cryptorchidism, Anogenital Distance, and Endocrine Disruption," *Current Urology Reports* 9 (2008): 137–42; S.H. Swan, "Environmental Phthalate Exposure in Relation to Reproductive Outcomes and Other Health Endpoints in Humans," *Environmental Research* 108 (2008): 177–184. See also, Schwartz, "Shaping Our Legacy."

248 **Far fewer data available in boys:** S.Y. Euling et al., "Examination of U.S. Puberty-Timing Data from 1940 to 1994 for Secular Trends: Panel Findings," *Pediatrics* 121 (2008): S172–91.

249 **Early blooming girls have more aggressive breast cancers:** C.C. Orgeas et al., "The Influence of Menstrual Risk Factors on Tumor Characteristics and Survival in Postmenopausal Breast Cancer," *Breast Cancer Research* 10 (2008): R107.

249 **Boys treated as leaders:** Grumbach and Styne, "Puberty: Ontogeny, Neuroendocrinology, Physiology, and Disorders."

249 **Early-maturing girls:** I review the data in *The Falling Age of Puberty in U.S. Girls* and "Girls Gone Grown-Up," in S. Olfman (ed.), *The Sexualization of Childhood* (Westport, CT: Praeger, 2009), 51–62. See also, M.S. Golub et al., "Public Health Implications of Altered Pubertal Timing," *Pediatrics* 121 (2008): S218–230.

250 **1999 decision to redefine precocious puberty:** P. Kaplowitz and S.E. Oberfeld, "Reexamination of the Age Limit for When Puberty Is Precocious in Girls in the United States: Implications for Evaluation and Treatment," Drug and Therapeutics and Executive Committees of the Lawson Wilkins Pediatric Endocrine Society, *Pediatrics* 104 (1999): 936–41.

250 **Large-scale study that finds few problems:** W. Copeland et al., "Outcomes of Early Puberty Timing in Young Women: A Prospective Population-Based Study," *American Journal of Psychiatry* 167 (2010): 1218–25.

251 **Puberty is a plastic trait in all mammals:** Ebling, "The Neuroendocrine Timing of Puberty."

252 **Procession of events:** Adrenarchy always precedes gonadarchy, although the physical manifestation of gonadarchy (breast development) usually appears sooner than those of adrenarchy (pubic hair). Frank Biro, personal communication.

253 **Loss of brain plasticity in puberty:** Frank Biro (personal communication); Grumbach and Styne, "Puberty: Ongeny, Neuroendocrinology, Physiology, and Disorders"; Sisk and Zehr, "Pubertal Hormones Organize Adolescent Brain and Behavior"; A.J. Yun et al., "Pineal Attrition, Loss of Cognitive Plasticity, and Onset of Puberty During the Teen Years: Is It a Modern Maladaptation Exposed by Evolutionary Displacement?" *Medical Hypotheses* 63 (2004): 939–50.

254 **Sixteen as average menarchal age in mid-nineteenth century:** Reliable

historical data in the United States go back only to 1900, but in Europe, data extend back several decades further. J.M. Tanner and P.B. Eveleth, "Variability Between Populations in Growth and Development at Puberty," in S.R. Berenberg, ed., *Puberty, Biologic and Psychosocial Components* (Leiden, Netherlands: H.E. Stenfert Kroese, 1975), pp. 256–73.

254 **Pubertal trends in girls over the past century:** S.E. Anderson and A. Must, "Interpreting the Continued Decline in the Average Age at Menarche: Results from Two Nationally Representative Surveys of U.S. Girls Studied 10 Years Apart," *Journal of Pediatrics* 147 (2005): 753–60; F.M. Biro et al., "Pubertal Assessment Method and Baseline Characteristics in a Mixed Longitudinal Study of Girls," *Pediatrics* 126 (2010): e583–90; S.Y. Euling et al., "Examination of U.S. Puberty-Timing Data from 1940 to 1994 for Secular Trends: Panel Findings," *Pediatrics* 121 (2008): S172–S191; D.S. Freedman et al., "The Relation of Menarchal Age to Obesity in Childhood and Adulthood: The Bogalusa Heart Study," *BMC Pediatrics* 3 (2003): 3; W.R. Harlan et al., "Secondary Sex Characteristics of Girls 12–17 Years of Age: the U.S. Health Examination Survey," *Journal of Pediatrics* 96 (1980): 1074–78; M.E. Herman-Giddens, "The Decline in the Age of Menarche in the United States: Should We Be Concerned?" *Journal of Adolescent Health* 40 (2007): 201–03; M.E. Herman-Giddens, "Recent Data on Pubertal Milestones in United States Children: The Secular Trend Toward Earlier Development," *International Journal of Andrology* 29 (2006): 241–46; Tanner and Eveleth, "Variability between Populations in Growth and Development at Puberty"; E.C. Walvoord, "The Timing of Puberty: Is It Changing? Does It Matter?" *Journal of Adolescent Health* 47 (2010): 433–39.

254 **Pediatricians who studied British orphan girls:** W.A. Marshall and J.M. Tanner, "Puberty," in F. Falkner and J.M. Tanner, eds., *Human Growth* (New York: Plenum, 1986): 171–209; W.A. Marshall and J.M. Tanner, "Variations in the Pattern of Pubertal Changes in Girls," *Archives of Disease in Childhood* 44 (1969): 291–303.

254 **Astonishing results published in 1997:** Herman-Giddens, "The Decline in the Age of Menarche."

255 **Danish study:** L. Aksglaede et al., "Recent Decline in Age at Breast Development: The Copenhagen Puberty Study," *Pediatrics* 123 (2009): e932–39; J. Raloff, "Report of Earlier, Longer Puberty in Girls," *Science News,* 5 May 2009.

255 **2010 U.S. study:** Biro, "Pubertal Assessment Method and Baseline Characteristics in a Mixed Longitudinal Study of Girls."

256 **Heredity clearly plays a role in pubertal timing:** I. Banerjee and P. Clayton et al., "The Genetic Basis for the Timing of Human Puberty," *Journal of Neuroendocrinology* 19 (2007): 831–38.

256 **U.S. black girls versus black girls in Cameroon, Kenya, and South Africa:** Grumbach and Styne, "Puberty: Ontogeny, Neuroendocrinology, Physiology, and Disorders"; Parent, "The Timing of Normal Puberty and the Age Limits of Sexual Precocity."

256 **Black girls a century ago had later puberties than white girls:** M.A. McDowell et al., "Has Age At Menarche Changed? Results from the National Health and Nutrition Examination Survey (NHANES) 1994–2004," *Journal of Adolescent Health* 40 (2007): 227–31. See also, Herman-Giddens, "The Decline in the Age of Menarche in the United States."

257 **Low birth weight:** C.M. Villanueva et al., "Atrazine in Municipal Drinking Water and Risk of Low Birth Weight, Preterm Delivery, and Small for Gestational Age Status," *Occupational and Environmental Health* 62 (2005): 400–05.

257 **Obesity plays a role and may explain racial disparities:** S.E. Anderson et al., "Relative Weight and Race Influence Average Age and Menarche: Results from Two Nationally Representative Surveys of U.S. Girls Studied 25 Years Apart," *Pediatrics* 111 (2003): 844–50; Anderson and Must, "Interpreting the Continued Decline in the Average Age at Menarche"; F.M. Biro, "Influence of Obesity on Timing of Puberty," *International Journal of Andrology* 29 (2006): 272–77; K. Foxhall, "Beginning to Begin: Reports from the Battle on Obesity," *American Journal of Public Health* 96 (2006): 2106–2112; R.T. Kimbro et al., "Racial and Ethnic Differentials in Children's Overweight and Obesity Among 3-Year-Olds," *American Journal of Public Health* 97 (2007): 298–305; D.H. Lee et al., "Association Between Serum Concentrations of Persistent Organic Pollutants and Insulin Resistance Among Nondiabetic Adults," *Diabetes Care* 30 (2007): 622–28; C.L. Ogden et al., "Prevalence and Trends in Overweight Among U.S. Children and Adolescents, 1999–2000," *JAMA* 288 (2002): 1728–32.

257 **Chinese study:** B. Cromer and C.M. Gordon, "Early Pubertal Development in Chinese Girls," *Pediatrics* 124 (2009): 799–801; H.-M. Ma et al., "Onset of Breast and Pubic Hair Development and Menses in Urban Chinese Girls," *Pediatrics* 124 (2009): e269–77. Nevertheless, China shows the highest increase in obesity rates in the world (Frank Biro, personal communication).

258 **Falling age of puberty may be linked to increasing insulin resistance:** P. Kaplowitz, "Pubertal Development in Girls: Secular Trends," *Current Opinion in Obstetrics and Gynecology* 18 (2006): 487–91; A.H. Slyper, "The Pubertal Timing Controversy in the USA, and a Review of Possible Causative Factors for the Advance in Timing of Onset of Puberty," *Clinical Endocrinology* 65 (2006): 1–8.

258 **Physical inactivity as a risk factor:** Grumbach and Styne, "Puberty: Ontogeny, Neuroendocrinology, Physiology, and Disorders"; Parent, "The Timing of Normal Puberty and the Age Limits of Sexual Precocity."

258 **Psychosocial stress:** M.A. Bellis et al., "Adults at 12? Trend in Puberty and their Public Health Consequences," *Journal of Epidemiology and Community Health* 60 (2006): 910–11; B.J. Ellis and J. Garber, "Psychosocial Antecedents of Variation in Girls' Pubertal Timing: Maternal Depression, Stepfather Presence, and Marital and Family Stress," *Child Development* 71 (2000): 485–501; P. Kaplowitz, *Early Puberty in Girls: The Essential Guide to Coping with This Common Problem* (New York: Ballantine, 2004); L.S. Zabin et al., "Childhood Sexual Abuse and Early Menarche: The Direction of Their Relationship and Its Implications," *Journal of Adolescent Health* 36 (2005): 393–400.

258 **Father absence and early puberty:** J. Deardorff et al., "Father Absence, Body Mass Index, and Pubertal Timing in Girls: Differential Effects by Family Income and Ethnicity," *Journal of Adolescent Health* September 2010 [epub ahead of print]. See also, Ellis and Garber, "Psychosocial Antecedents in Girls' Pubertal Timing"; M. MacLeod, "Her Father's Daughter," *New Scientist*, 10 February 2007, pp. 38–41; D. Maestripieri

et al., "Father Absence, Menarche and Interest in Infants Among Adolescent Girls," *Developmental Science* 7 (2004): 560–66.

258 **Siblings and household crowding:** S. Hoier, "Father Absence and Age at Menarche: A Test of Four Evolutionary Models," *Human Nature* 14 (2003): 209–34; R. Matchock and E.J. Susman, "Family Composition and Menarcheal Age: Anti-Inbreeding Strategies," *American Journal of Human Biology* 18 (2006): 481–91.

258 **Environmental endocrine disruption and pubertal timing:** G.M. Buck Louis et al., "Environmental Factors and Puberty Timing: Expert Panel Research Needs," *Pediatrics* 212 (2008): S192–207; S.Y. Euling et al., "Role of Environmental Factors in the Timing of Puberty," *Pediatrics* 121 (2008, S3): S167–71.

259 **PBBs in Michigan:** H.M. Blanck et al., "Age at Menarche and Tanner Stage in Girls Exposed in Utero and Postnatally to Polybrominated Biphenyl," *Epidemiology* 11 (2000): 641–47.

259 **Hormone levels in the Dutch study:** Aksglaede, "Recent Decline in Age at Breast Development."

260 **Environmental endocrine disruption and animal studies:** M. Fernandez et al., "Neonatal Exposure to Bisphenol A Alters Reproductive Parameters and Gonadotropin-Releasing Hormone Signaling in Female Rats," *Environmental Health Perspectives* 117 (2009): 757–62. Emerging evidence suggests that BPA exposures in males can trigger changes in gene expression (epigenetic alterations) that are not only permanent but heritable and so can influence future generations. See, for example, G.S. Prins et al., "Developmental Exposures to Bisphenol A Increases Prostate Cancer Susceptibility in Adult Rats: Epigenetic Mode of Action Is Implicated," *Fertility and Sterility* 89 (2008, S2): e41.

260 **Steps that parents can take:** Email alert from Zero Breast Cancer, received 11 August 2010.

261 **Bisphenol A in air:** F. Pingqing and K. Kawamura, "Ubiquity of Bisphenol A in the Atmosphere," *Environmental Pollution* 158 (2010): 3138–43.

CHAPTER TEN

264 **Urban sprawl and children's health:** Committee on Environmental Health, American Academy of Pediatrics, "The Built Environment: Designing Communities to Promote Physical Activity in Children," *Pediatrics* 123 (2009): 1591–98.

268 **Geology of Cayuga Lake and Taughannock Falls gorges:** D. Tall, *From Where We Stand: Recovering a Sense of Place* (New York: Knopf, 1993); outreach materials, Museum of the Earth, Ithaca, New York: www.museumoftheearth.org.

269 **Description of Marcellus Shale:** "Geology of the Marcellus Shale: A Paleontological Perspective on a Modern Resource," Museum of the Earth, Paleontological Research Institution, Ithaca, NY. www.museumof theearth.org.

270 **Description of hydrofracking:** M. Fischetti, "The Drillers Are Coming," *Scientific American* 303 (2010): 82–85; D.M. Kargbo et al., "Natural Gas Plays in the Marcellus Shale: Challenges and Potential Opportunities," *Environmental Science & Technology* 44 (2010): 5679–84. See also ProPublica's series of investigative stories, "Buried Secrets: Gas Drilling's Environmental Threat," www.propublica.org.

271 **Like a bat to a windshield:** A. Chung, "Quebec Between a Rock and a Hard Place on Gas from Shale," *Toronto Star*, 25 July 2010.

271 **Infrastructure and water requirements for fracking:** R.A. Kerr, "Natural Gas from Shale Bursts Onto the Scene," *Science* 328 (2010): 1624–26; New York State Department of Environmental Conservation, *Draft Supplemental Generic Environmental Impact Statement on the Oil, Bas and Solution Mining Regulatory Program, Well Permit Issuance for Horizontal Drilling and High-Volume Hydraulic Fracturing to Develop the Marcellus Shale and Other Low-Permeable Gas Reservoirs*, Albany, NY, Sept. 2009; U.S. Department of Energy, *Modern Shale Gas Development in the United States: A Primer*, April 2009; M. Zoback et al., *Addressing the Environmental Risks from Shale Gas Development*, Worldwatch Institute, July 2010.

271 **Between 34,000 and 95,000 wells envisioned for New York state:** These numbers are estimates based on assumptions about how much of the shale will be tapped over what period of time. 77,000 wells assumes that 17 New York State counties are drilled and that the shale is 70 percent developed over fifty years at a density of eight wells per square mile. T. Engelder, "Marcellus 2008 Report Card on the Breakout Year for Gas Production in the Appalachian Basin," *Fort Worth Basin Oil and Gas Magazine*, Aug. 2009, 18–22, and Anthony Ingraffea, Ph.D., personal communication.

271 **Speed of gas drilling in Pennsylvania:** M. Fischetti, "The Drillers Are Coming," *Scientific American* 303 (2010): 82–85; T. Williams, "Gas Pains," *Audubon Magazine*, May–June 2010.

273 **Forest fragmentation:** T. Williams, "Gas Pains," *Audubon Magazine*, May–June 2010.

273 **Impact of fracking on air quality:** Ozone levels have increased in gas-drilling areas of Texas, Wyoming, and Colorado. In the Dallas-Fort Worth area, natural gas wells are estimated to contribute more air pollution than all the cars and trucks in the area. A. Armendariz, *Emissions from National Gas Production in the Barnett Shale Area and Opportunities for Cost-Effective Improvements*, Environmental Defense Fund, Jan. 2009; M. Fischetti, "The Drillers Are Coming"; J. Kirkland, "Fears of Pervasive Air Pollution Stir up Politics in Texas Shale Gas Country," *New York Times*, 2 August 2010.

273 **Ozone from Hanesville Shale drilling:** "Ozone Impacts of Natural Gas Development in the Hanseville Shale," *Environmental Science and Technology*, 2010, in press.

273 **Accidents:** Kargbo et al., "Natural Gas Plays in the Marcellus Shale"; A. Worden, "Gas Spews from N.W. PA Well Rupture," *Philadelphia Inquirer*, 5 June 2010.

274 **Making water disappear:** The Chesapeake Energy Corporation argues that the water vapor generated during methane combustion compensates for the water lost to the hydrologic cycle when it is buried within the cracks of broken bedrock (M.E. Mantell, "Deep Shale Natural Gas and Water Use, Part Two: Abundant, Affordable, and Still Water Efficient," presentation at the Groundwater Protection Council Annual Forum, 28 September 2010). Water vapor is indeed produced when natural gas burns. However, regardless of how much water vapor is created by methane combustion far from the wellhead, millions of gallons of water permanently removed from a stream or small aquifer can signifi-

cantly deplete local water availability and so alter the ecology of a bioregion ("Water Withdrawals for Development of Marcellus Shale Gas in Pennsylvania," Penn State College of Agricultural Sciences, cooperative extension fact sheet, 2010). Moreover, not all natural gas is burned; a considerable amount is used as a feedstock for synthetic materials, such as fertilizer and plastics.

274 **Groundwater depletion:** Y. Wadacta, "Global Depletion of Groundwater Resources," *Geophysical Research Letters*, 37 (2010) L20402.

275 **Can fracking contaminate groundwater?** Food and Water Watch, *Not So Fast, Natural Gas: Why Accelerating Risky Drilling Threatens America's Water*, Washington, D.C., July 2010; A. Lustgarten, "EPA Launches National Study of Hydraulic Fracturing," *ProPublica*, 18 March 2010; J. Manuel, "EPA Tackles Fracking," *Environmental Health Perspectives* 118 (2010): a199.

276 **Documented cases of water contamination:** A. Lustgarten and ProPublica, "Drill for Natural Gas, Pollute the Water," *Scientific American*, 17 Nov. 2008; J.L. Schnoor, "Regulate, Baby, Regulate!" *Environmental Science and Technology* 44 (2010): 6524–25.

276 **Pavillion, Wyoming:** ATSDR, *Evaluation of Contaminants in Private Residential Well Water, Pavillion, Wyoming, Fremont County*, August 2010; A. Lustgarten, "Feds Warn Residents Near Wyoming Gas Drilling Sites Not to Drink Their Water," *ProPublica*, 1 Sept. 2010.

277 **Regulatory exemptions:** R.L. Kosnick, *The Oil and Gas Industry's Exclusions and Exemptions to Major Environmental Statutes*, Oil and Gas Accountability Project, 2007. In 2005, the Energy Policy Act exempted hydraulic fracturing from federal regulation, which allows the chemical formulation of the fracking fluids to remain a trade secret. There is no requirement for disclosure. Called the Halliburton loophole after the energy company whose former CEO, Vice President Dick Cheney, supported the exemption. Kargbo, "Natural Gas Plays in the Marcellus Shale." See also T. Colborn et al., "Natural Gas Operations from a Public Health Perspective," *International Journal of Human and Ecological Risk Assessment*, 2011, in press. www.endocrinedisruption.com.

278 **Chemicals in fracking fluid:** The Endocrine Disruption Exchange, "Chemicals Used in Natural Gas Fracturing Operations: Pennsylvania," April 2009. www.endocrinedisruption.com.

278 **Chemicals in flowback water:** A.W. Gaudlip et al., "Marcellus Shale Water Management Challenge in Pennsylvania," Society of Petroleum Engineers 119898, paper presented for presentation at the SPE Shale Gas Production Conference, Forth Worth, Texas, 16–18 Nov. 2008; A. Lustgarten, "Is New York's Marcellus Shale Too Hot to Handle?" *ProPublica*, 9 Nov. 2010. Drill cuttings also require disposal and are sometimes radioactive. Kargbo, "Natural Gas Plays in the Marcellus Shale." See also Colborn, "Natural Gas Operations from a Public Health Perspective."

279 **Quote on unclear disposal options for waste water:** Kargbo, "Natural Gas Plays in the Marcellus Shale."

279 **Possible migration of chemicals through shale fissure and old abandoned wells:** John Gephart, Ph.D., personal communication.

279 **Fracking's economic story:** J. Kirkland, "Big Money Drives Up the Betting on the Marcellus Shale," *New York Times*, 9 July 2010.

280 **Gas as the bridge fuel:** J. Kirkland, "Natural Gas Could Serve as the 'Bridge' Fuel to Low-Carbon Future," *Scientific American*, 25 June 2010. E. Moniz et al., *The Future of Natural Gas*, MIT Energy Initiative, Massachusetts Institute of Technology, June 2010.

280 **Gas has half the carbon dioxide of coal:** Kargbo, "Natural Gas Plays in the Marcellus Shale."

280 **First-pass lifecycle analysis:** R.W. Howarth, *Preliminary Assessment of the Greenhouse Gas Emissions from Natural Gas Obtained by Hydraulic Fracturing*, Cornell University, Nov. 2010 draft.

282 **Quote on green energy in twenty years:** M.Z. Jacobson and M.A. Delucchi, "A Path to Sustainable Energy by 2030," *Scientific American* 301 (2009): 58–65.

283 **Land already leased in my county:** Marcellus Accountability Project for Tompkins County, http://tcgasmap.org.

283 **"The shale army has arrived":** Bill Gwozd, vice president of gas services for Ziff Energy Group in Calgary, quoted in "New York Environmental Groups Worry About Shale Gas Drillings Effect on Water," *The News*, New Glasgow, Nova Scotia, 26 Feb. 2010. www.ngnews.ca.

Acknowledgments

A favorite book in our household, *The Strange Case of Dr. Jekyll and Mr. Hyde*, was, according to legend, penned by Robert Louis Stevenson over a course of four to six days. With the assistance of cocaine.

This was not my model for authorship. Instead, *Raising Elijah* unfolded over a period of eight years. *Long-awaited* would be the appropriate euphemism. (The actual drafting process took place over a period of three months. With the assistance of coffee.)

Thus, I had ample opportunity to audition the evolving ideas of this book, at, for example, lecture events. Afterwards, audience members sometimes offered me their comments and suggestions; a number of these have proved invaluable. One soft-spoken man—in an auditorium lobby somewhere in the Midwest—introduced me to the term "well-informed futility syndrome." Whoever you are, thank you. And my heartfelt appreciation to the Jodi Solomon Speakers Bureau for bringing me to podiums across the nation. Public speaking is an awesome privilege.

Some of this book's themes, stories, and arguments, were debuted, in different versions, in *Orion Magazine*. For this good fortune, I thank my editors there, H. Emerson Blake and Jennifer Sahn. My investigation of the explosion of Formosa Plastics in Chapter 4 expands on "The Pirates of Illiopolis," which appeared in *Orion* in spring 2005. My discussion of methylmercury in Chapter 8 incorporates elements of "Tune of the Tuna Fish," which appeared in *Orion* in winter 2006.

Other chapters of this book were informed by various reports that I was commissioned to write: *The Falling Age of Puberty in U.S. Girls:*

What We Know, What We Need to Know (Breast Cancer Fund, 2007); *Update on the Environmental Health Impacts of Polyvinyl Chloride (PVC) as a Building Material* (Healthy Building Network, 2004); and *The Ecology of Pizza* (Organic Trade Association, 2003). My investigation of arsenic-treated wood in Chapter 2 began with a story for *Rachel's Environment and Health News*. Chapters 3 and 4 draw from two essays, "The Organic Manifesto of a Biologist Mother" (for Organic Valley) and "But I Am a Child Who Does" (for the Center for Eco-literacy), and a video lecture project produced by Stonyfield Farm. Thanks to Gary Hirshberg and Alice Markowitz.

Child psychologist Sharna Olfman invited me to contribute two chapters to the book series she edits, *Childhood in America*, with Praeger Publishers. And thus, earlier versions of some of the material contained in chapters 3 and 9 first appeared in *Child Honoring: How to Turn This World Around* (coedited with Raffi Cavoukian, 2006) and *The Sexualization of Childhood* (2009).

For financial support during the research and writing process, I thank the Winslow Foundation, Jenifer Altman Foundation, and the Ceres Trust. For making possible my scholarship, I am grateful to the Division of Interdisciplinary and International Studies at Ithaca College and its dean, Tanya Saunders. Special thanks to Adelaide Gomer and Judy Kern.

Various researchers at the U.S. Department of Agriculture offered me their assistance. I acknowledge especially agricultural statistician Theresa Varner. I'm also grateful to the interlibrary loan department at Ithaca College and all the wonderful librarians at the Ulysses Philomathic Library.

Was I still filling up the Crock-Pot each night while piecing together the sentences of this book? No, I was not. Paul Martin and Evangeline Sarat, my CSA farmers at Sweetland Farm, kept me in blackberries, kale, squash, and eggs, and Margaret Shepard at the Trumansburg Farmers Market ensured that I would never run out of heirloom garlic, but it was my mother-in-law, age 87, who flew in from Ohio to chop, blanch, freeze, bake, and serve while I toiled over one chapter or another. And it was my daughter, age 11, who signed up for summer cooking lessons so she could keep the salads and pesto coming while her mother missed one dinner bell after another for the purposes of writing a book that features her younger brother. Meanwhile, my own mother in Illinois signed on as a reader, freely offering the authority of her memory of some of the past events narrated here, along with her good proofreading skills.

Many colleagues in science, medicine, education, and policymaking also contributed their expert knowledge to this project by reading parts or all of the manuscript. They are Ellie Biddle, Ellis Hollow Nursery School; Frank Biro, M.D., Division of Adolescent Medicine, Cincinnati Children's Hospital Medical Center; Judy Brady; Terra Brockman, The Land Connection; Laurine Brown, Ph.D., Illinois Wesleyan University; Theo Colborn, Ph.D., TEDX (The Endocrine Disruption Exchange); Antonia Demas, Ph.D., Food Studies Institute; Lise Eliot, Ph.D., Department of Neuroscience, Chicago Medical School; John Gephart, Ph.D.; Monica Hargraves, Ph.D., Cornell University Office for Research on Evaluation; Robert L. Hendren, D.O., Department of Psychiatry, University of California, San Francisco; Martha Herbert, M.D., Ph.D., Center for Child and Adolescent Development, Harvard Medical School; Anthony Ingraffea, Ph.D., School of Civil and Environmental Engineering, Cornell University; Colleen Kearns, Department of Ecology and Evolutionary Biology, Cornell University; Elise Miller, Collaborative on Health and Environment; Nancy Myers, Science and Environmental Health Network; Frederica Perera, Dr.PH., Center for Children's Environmental Health, Columbia University; Ivette Perfecto, Ph.D., School of Natural Resources, University of Michigan; Jennifer Phillips, Nose in a Book Publishing; Sandy Podulka, Cornell Laboratory of Ornithology; William Podulka, Ph.D., Marcellus Accountability Project; Carolyn Raffensperger, Science and Environmental Health Network; Peter Sandman, Ph.D.; Ted Schettler, M.D., M.P.H., Science and Environmental Health Network; Amy Seidl, Ph.D., University of Vermont; Helen Slottje, senior attorney, Community Environmental Defense Council; Shanna Swan, Ph.D., Center for Reproductive Epidemiology, University of Rochester; Karl Tupper, Pesticide Action Network; Bill Walsh, Healthy Building Network; and Wendy Wolfe, Ph.D., Division of Nutritional Sciences, Cornell University. A special thanks to Robert Armstrong.

Enormous generosity came from three readers—Chanda Chevannes of the People's Picture Company; Tom Lent of Healthy Building Network; and Carmi Orenstein of BreastCancer.Org—whose insights and guidance on this project were transformative.

For their good comments and criticisms, I am indebted to all of the above. Responsibility for accuracy, of course, rests entirely with me.

For their patience and enthusiasm, I'm grateful to my literary agent, Charlotte Sheedy, who championed this project from the start, and my long-time editor, Merloyd Lawrence, who has supported my writing life, in ways big and small, since the days when my only housemate was a Wheaten terrier with a fear of dogs.

Jeff, Faith, and Elijah: I'm finally finished. Let's make a bowl of popcorn and lie under the skylight. You are my sun and my moon and my stars. Thank you.

Index

activists, for climate change, 173, 179
ADHD (attention-deficit/hyperactivity
  disorder)
  CDC statistics on, 204
  EPA response to pesticides, 211
  linked to pesticide exposure,
    xiv–xv, 65, 208
  as neurodevelopment disorder,
    202
  prevalence of, 204
adrenal glands
  in babies born vaginally, 141
  hastening onset of puberty in
    girls, 257
  in lung and immune system
    development, 144
  in process of puberty in girls, 252
adult-onset asthma, 143
adults
  becoming heroes of climate crisis,
    182
  differences between children vs.,
    27–28, 30
  toxicology studies done for, 28
agriculture. *See also* organic foods,
    pesticides, *and names of
    individual crops*
  antibiotics for food animals, 71
  community supported agriculture,
    60–62, 78, 104
  compost as fertilizer, 77, 104
  dairy, 99–101
  economics of, 99–101
  ecosystem services and, 108–109

fertilizer, 104–105
  fossil fuel dependency of, 102,
    194
  green manure and, 104
  hidden costs of conventional,
    90–91
  institutional neglect of organic,
    86–87
  nitrogen fixation and, 104–105
  northeast wheat, 109
  organic farming feeding world,
    question of, 101–104
  ozone layer, effects on, 90
  resilience and evolution of
    organic, 106
  trends in organic, 86
  yields, conventional vs. organic,
    105–107
air pollution
  asthma and, xiv, 144–145,
    150–52, 153, 155, 157–159,
    161–62
  cars vs. lawn mowers, 187
  Clean Air Act, 156–157
  from coal, xiv, 153, 156, 158,
    160, 162, 222–224, 227, 280,
    287
  diesel exhaust, 12, 150, 152, 155,
    240, 272, 278
  effect on children's development,
    xiv–xv, 120, 144, 162, 172,
    222–223, 257
  effect on family health, 155–156
  effect on heart, 155

air pollution *(continued)*
  effect on respiration, 140,
    152–153
  from fertilizers, 185
  formaldehyde as, 145, 150
  and incineration, 158
  indoor, 120, 144–146, 155, 222
  from lawn mowers, 187–188
  linked to preterm birth, xiv, 11–13
  ozone (smog), 13, 75, 150, 152,
    157, 160–161, 172, 185, 188,
    223, 273
  pesticide drift, 69–70
  phthalates as, 11–12, 120,
    144–145, 156
  from PVC manufacturing, 119,
    122–123
  from slickwater hydrofracking, 273
  ultrafine particles, 150, 155
  from vehicles, 12, 151–152, 271
air travel, following 9/11, 23–24
allergies
  amendments to Clean Air Act for,
    156
  in atopic asthma, 143
  environmental agents augmenting,
    144
  from fossil fuels, 162
  from phthalate exposure, 120
alligators, reproductive tract
    anomalies in, 244
all-purpose flour, 91
American Academy of Pediatrics,
    on climate change, 172
amniotic membrane rupture, from
    barometric pressure, 12–13
amphibians
  demasculinization of, 244
  migration of, 2–3
  sensitivity to pesticides of, 70–71
animal studies
  amphibian sensitivity to
    pesticides, 70–71
  anomalies in alligators, panthers,
    polar bears, 244
  pubertal timing and, 260
anogenital distance, endocrine
    disruption and, 245–246
antibiotics, used in food animals,
    71–72
apples, price of Ding Dongs vs.,
    89–90
Argonne National Laboratory report,
    132
Armstrong family, 24–26
aromatase, 238

arsenic
  2001 report on drinking water
    containing, 38
  development of CCA wood, 34–36
  geology of, 32
  glucocorticoids and, 32–33
  government allows CCA wood
    playgrounds, 40–42, 47–48
  harm from CCA wood, 36–38
  lifetime risk of cancer from
    exposure to, 38–39
  methylation and, 33
  as most toxic element, 32, 47
  organic vs. inorganic, 32
  in pressure-treated wood, 47
  special vulnerabilities of children
    to, 33–34
  toxicology of, 31–32
asbestos, backing asphalt tile,
    147–149
Asperger's syndrome, 204
asphalt tile, 147–149
asthma
  air pollution exposure and,
    149–153, 155
  demographics of, 151
  development of lungs, 140–141
  diesel exhaust and, 150–151
  early-life experiences contributing
    to, 144
  effect of Ground Zero cloud on,
    22
  effects that are not, 153
  environmental agents augmenting
    allergic reactions, 144
  global warming-induced air
    pollution and, 161
  indoor air pollution and, 145, 152
  linked to toxic chemical exposures,
    xiv
  living near busy roads and, 152
  pesticides and, 151
  phthalates inciting inflammatory
    responses, 144
  respiration and emotions, 140
  statistics, 149
  surface area of lung and, 139
  taxonomy of, 143
  time trends, 150
Atlanta, asthma and ozone levels in,
    157
atmosphere
  causes of preterm labor within,
    10–12
  environmental crisis of heat-
    trapping gases in, xiii

atopic asthma, 143–144
atrazine
    as endocrine disruptor, 238–239
    links to shorter pregnancy and
        lower birth weights, 257
    sensitivity of amphibians to,
        70–71
attention-deficit/hyperactivity
        disorder. *See* ADHD (attention-
        deficit/hyperactivity disorder)
autism
    2010 review of causes of, 206
    California cases, 205
    chlorpyrifos linked to, 93
    core features of, 204–205
    lifetime costs of, 202–203
    as neurodevelopment disorder,
        202
    prevalence of, 205
    thimerosal and, 206
    toxic chemical exposure linked
        to, xv

barometric pressure, onset of labor
        linked to, 12–14
Bates College, and composting, 186
bats
    rabies from. *See* rabies
    with white-nose syndrome,
        194–195
bedrock
    arsenic found in, 32
    contaminated from PCBs, 4
    extracting natural gas by fractur-
        ing shale, xvii, 269, 271, 281
    Marcellus Shale, 3, 269–270,
        272
behavioral patterns, children vs.
        adults, 27–28
Bell Telephone, using CCA-treated
        wood, 34
benzene, 11, 21, 188, 276, 278
Binghamton, New York, 2010 train
        derailment in, 132
biometeorology, onset of labor and,
        12–13
birth defects
    in alligators, polar bears, panthers,
        242–243
    environmental threats to repro-
        ductive health, 239
    testicular dysgenesis syndrome,
        242–246
    in wheat-growing areas, 92–93
birth weight, and onset of puberty in
        girls, 257

bisphenol A
    early puberty and, 260
    as endocrine disruptor, 235–237
    statistics on exposure to, 236
black girls
    average menarchal age in U.S. for,
        254, 256
    early breast development in, xv,
        254–255
    early onset puberty in, 250, 256,
        257–258
Black Plague, cheese-making during,
        99
bladder cancer
    arsenic linked to, 32, 38
    author's survival of, 39
body size
    of children vs. adults, 27–28
    risk of early onset puberty and,
        257–258
boys
    biological vulnerability of, 22–23,
        204, 205, 237
    impact of Ground Zero cloud on
        baby, 22
    pubertal timing, 248–249
    testicular dysgenesis syndrome in,
        242–246
Bradshaw, Bradford, 128, 135
brain development
    chlorpyrifos linked to cognitive
        defects, 93
    digital media, unknown effect of,
        219–221
    effect of air pollution on,
        222–223
    effect of organophosphate pesti-
        cides on, 217–218
    how it occurs, 214–215
    methylmercury and, 223–225
    neurotoxicants and. *See* neuro-
        toxicants
    organophosphates interfering
        with, 209
    polycyclic aromatic hydrocarbons
        interfering with, 222–223
    during puberty, 253
    stages of, 215–218
bread. *See* wheat farming
bread flour, cake flour vs., 91
breast cancer
    aromatase inhibitors for, 238
    early sexual maturation and risk
        of, xv, 249, 251
    failure to breastfeed increasing
        risk of premenopausal, 17

breast cancer *(continued)*
   garlic consumption may lower
      risk of, 98
   steps parents can take, 260
breast development
   age at onset over last century,
      254–256
   bisphenol A and early, 260
   Danish study of age of onset,
      255, 257
   in chubby girls, 257
   procession of events in puberty,
      252
breastfeeding
   benefits to child's health, 16
   benefits to mother's future
      health, 17
   effect of 9/11 terrorist attack
      on, 19
   working mothers and, 18
bridge fuel, natural gas as, 280
Brockman, Terra, 103
Brown, Laurine, 66
burns
   children with, 117–118
   flammability of PVC, 121
   trickiness of, 116–117

Caesarean birth, 1, 141
cake flours, bread flours vs., 91
California
   air pollution and child asthma in,
      151–153
   comprehensive pesticide registry
      for, 96
   methyl iodide classed as
      carcinogen in, 76
   organophosphates affecting
      cognition of children, 209
   pesticide exposure of farmworkers'
      children, 69, 210
   pesticides in fresh-market
      tomatoes, 75, 96
   pesticides in strawberries, 75, 78
   San Francisco compost pick-up,
      186
   study in autism, 205
Canada
   declaring bisphenol A as toxic
      substance, 237
   fracking for natural gas in, 281
cancer
   arsenic linked to, 38–39, 42
   breast. *See* breast cancer
   methyl bromide linked to,
      75–76

methyl iodide as carcinogen, 76
   prevalence of testicular, 243
   vinyl chloride and vinyl acetate as
      carcinogens, 121
Carson, Rachel, xiii
Cayuga Lake, geology of, 268–269
CCA (chromated copper arsenate)
   wood
   1993 ban on arsenical pesticides,
      51–52
   2001 report on arsenic in drinking
      water, 38
   2003 draft of EPA risk assessment,
      52–53
   2004 refinement of mathematical
      model, 53
   2007 challenge, 53
   2008 final risk assessment, 53
   2010 study in New Orleans, 53
   amount of arsenic in, 35, 47
   arsenic leaching from, 41–42
   cancer patients created by, 38–39,
      52–54
   consequences of disposing, 37
   exempted from hazardous waste
      rules, 37
   fact sheets in retail stores, 36
   Hauserman's reporting on, 38
   high risk of decks/play equipment,
      40–44
   in home building trades, 35
   in landfills, 48–49
   lifetime risk from exposure to
      arsenic, 38–39
   parental debate on, 44–46
   PBS story on playgrounds, 58
   popularity of decks made
      with, 35
   production increases 1970–1995,
      37
   results of sealant study, 53
   safety tips for homeowners, 55
   signs of harm, 36
   Sonti Kamesam and history
      of, 34
   Stillwell's discovery, 37–38
   well-informed futility regarding,
      47–50
cement plants, mercury emissions
   from, 224–225
Centers for Disease Control
   blood mercury levels in
      reproductive-age women,
      223–224
   exposure to phthalates, 120, 236
   prevalence of ADHD, 204

prevalence of autism disorders, 205

rabies protocols, 168

statistics on bisphenol A levels, 236

cheese-making
during Black Plague, 99
during economic downturn, 101

chemical pollutants
in detonation of Formosa Plastics Plant, 123–124
effect on immune system/lung development, 144
effect on onset and tempo of labor, 10–12
effects of endocrine disruptors. *See* endocrine disruptors
effects of neurotoxicants. *See* neurotoxicants
effects of PVC. *See* PVC (polyvinyl chloride or vinyl)
environmental crisis of, xiii
in fracking fluid, 278–279
health concerns after Formosa Plastics Plant explosion, 128–132
keeping children safe from, 135–136
and national security, 121–122, 131–133
our bodies and brains altered by, 212
proneness to detonation, 120
regulations governing, 28–29, 36–37, 140, 156–157, 251, 261, 277
risk of early puberty from, 260–261
terrorism and report on infrastructure of, 132
transporting, 132–133

Chemical Security Act, 119

The Chemical Security Act of 2003, 119

chemical toxicity, testing for, 29–30

chickens, organically raised, 86

children
with asthma. *See* asthma
books on climate change for, 177–178
with burns, 117–118
climate change and health of, 171
communicating about climate change with, 174–177
differences between adults and, 27–28, 30
diseases linked to toxic chemical exposures, xiv–xv
effects of fast-food branding on eating habits of, 67–68
farmworker, 69–70
fed organic foods, 62–63
hand-to-mouth interactions per hour, 30
hazards of PVC to, 119–120
injured by lawn mowers, 188
literature on climate change for, 177–178
as main victims of environmental crisis, xiii–xiv
Roman street, 200
toxicology studies not representative of, 28–30

China
air pollution measures and results in, 157, 222–223
most commercial-grade arsenic from, 32
study on age at onset of breast development, 257

chlor-alkali plants, 223–224

chlorine gas, manufacturing PVC, 119, 121

chlorpyrifos
effect on brain development, 217–218
interference with child's neural pathways, 151
link to cognitive defects and autism, 93

chromated copper arsenate. *See* CCA (chromated copper arsenate) wood

citizen advocacy, PCB contamination, 4

civil rights approach, to climate change, 179

Clean Air Act
analyzing proposed amendments to, 156–157
awaiting legislation on heat-trapping gases under, 160
fracking exemption from, 277
monitoring only tiny portion of total, 150

climate change. *See also* global warming
activists at the barricades, 173
air pollution exacerbated by, 160–163
barriers to household energy efficiency, 181

climate change *(continued)*
  as biggest health threat to children, 171–172
  children's literature on, 177–178
  civil rights approach to, 179
  clothes dryers, 191–193
  critics of American lawns, 190
  curtailing fossil fuel emissions, 172
  as de-creation, 176
  personal responses to list for how to green your house, 181
  emissions from residences, 180
  Eremozoic Era, 176
  food and yard waste, 185
  and intergenerational equity, 163
  lawnmowers and, 187–188
  living on planet with, 211–212
  mammals threatened with extinction, 176
  methane in landfills, 185
  ocean, effects on, xiii, 163–165
  policies ending subsidies for fossil fuels, 178
  rapid 80-percent reduction, 181
  social silence surrounding, 173–174
  systems thinking about, 186–187
  white-nose syndrome in bats and, 194
clothes dryers, 191–194
coal
  air pollution from power plants, 150, 153
  ash in cement and consumer products, 225
  arsenic release when mining, 32
  carbon dioxide of natural gas vs., 280–281
  China closes polluting power plant, 157, 222–223
  creating mercury contamination in fish, 223–224
  development of CCA wood to protect miners of, 34
  global temperature increasing from, 161
  mercury poisoning from, 223–225, 227
  mountaintop removal, 281
  as nonrenewable resource, 165
  PCBs synthesized from, 3
  preterm birth linked to, xiv
  public involvement in climate crisis and, 178–179
  PVC synthesized in China from, 121, 134

Code of Federal Regulations, 1992, 37
cognitive defects. *See* brain development
Cold War, and children, 178
colon cancer, and garlic consumption, 98
communal bereavement, and Ground Zero, 23
composting, 185–186
consumptive water use, fracking, 275
convenience, trace chemical exposure as price of, 55
conventionally grown foods
  destroying earth's ozone layer, 75–79
  field tomatoes, 96
  hidden costs of, 90
  television marketing, 67–68
  vulnerabilities of children to pesticides in, 62–65
co-op food stores, benefits of, 72–73
costs
  of asthma, 149
  of autism over lifetime, 202
  of child neurodevelopment disorders, 202–203
  of conventionally grown food, 90
  of conventional vs. organic flour, 94
  of conventional vs. organic tomato paste/tomatoes, 97
  of developmental disabilities, 202
  of mercury-induced intelligence loss, 224
cows, organically raised, 86, 100
Crock-Pots, 80
cryptorchidism, 242–243
CSA farms, 60–62, 78
CSB. *See* U.S. Chemical Safety and Hazard Investigation Board.
cytokines, 153

dairy
  cheese-making during Black Plague, 99
  economic pressures on farmers, 101
  lack of bailout for organic, 100–101
  plummeting costs in 2009 economic collapse, 99–100
databases, pulled from Web, 121
debridement, burn treatment, 117
de-creation, climate change as story of, 176

DEHP (di(2-ethylhexyl)phthalate)
  creating inflammatory responses, 144
  as endocrine disruptor, 120
  preterm labor caused by, 11
  human exposure to, 120
Demas, Antonia, 66
denial, from well-informed futility, 47
depression, in early-maturing girls, 249–250
DES (diethylstilbestrol), 236–237
detonation
  chemicals prone to, 120, 121
  of Formosa Plastics Plant, 123–124
developmental disorders
  brain development and. *See* brain development
  trends in learning and, 203
di(2-ethylhexyl)phthalate. *See* DEHP (di(2-ethylhexyl)phthalate)
diabetes
  agricultural policies and, 90
  air pollution contributing to, 155
  arsenic linked to, 32–33, 54
  breastfeeding reducing risk of, 16–17
  food pricing and, 90
  insulin resistance in early puberty and, 258
diesel exhaust
  asthma linked to, 150–152, 155–156, 240
  disruption to hormonal messages, 12
  role in fracturing deep bedrock, 271, 273, 276, 278
  not monitored and regulated, 150
diet, of children vs. adults, 27–28
diethylstilbestrol (DES), 236–237
digital media, effect of on brain development, 219–221
Ding Dongs, price of apples vs., 89–90
dioxin
  burning of PVC dispersing, 118
  Formosa Plastics Plant explosion dispersing, 130–131
  World Trade Center disaster dispersing, 19–21
diversity, in organic farming, 93–94, 98–99
DNA damage
  impact of Ground Zero cloud on children, 22
  methyl iodide causing, 76–77

drinking water
  2001 report on arsenic in, 38
  ethylene dichloride in, 121
  exposure to atrazine through, 236
  fracking contaminating, 276–277
  fracking depleting, 274–275
  with herbicides in wheat-growing areas, 93
durum wheat, 91, 93
dyscalculia, 202
dysfunctional families, learning disabilities in, 208
dyslexia, as neurodevelopment disorder, 202

eating habits
  developing healthy, 66–67
  effects of fast-food branding on, 67–68
  healthy foods for healthy, 78
  lack of TV influencing healthy, 67–68
  USDA food plan, 84–85
  working parents and 78–81
ecological impacts, from slickwater hydrofracking, 272–273
economic externalities, 90
economics. *See also* costs
  of fracking, 279–281
  impact on dairy farmers of 2009 collapse, 99–100
ecosystem services, 108–109
electricity, used by clothes dryers, 193
Eliot, Lise, 214–215
emotions, and respiration, 140
endocrine disruptors
  animal studies on, 260
  anogenital distance and, 245–246
  anomalies in alligators, panthers, polar bears, 244
  atrazine, 238–239
  bisphenol A, 236–237
  DEHP, 120
  early maturity in girls and, 252
  neurotoxicants, 235
  phthalates, 237–238
  pubertal timing and, 258–259
energy efficiency, of organic oil production, 95
environmental agents, augmenting allergic reactions, 144
environmental health, disregarding policies for children, 28–29

EPA (Environmental Protection
Agency)
1978 review of arsenic pesticides,
35
1988 decision on arsenic pesticides, 36
2001 PCB contamination study, 4
2002 revoking of new CCA
for residential settings,
42, 49–50
2003 draft of arsenic risk assessment, 52–53
2005 results of CCA sealant study,
53
2008 final risk assessment for
CCA wood, 44, 53
response to organophosphate
pesticides, 211
reviewing of atrazine, 239
vulnerabilities of children to
pesticides, 63–64
Eremozoic Era, 176
estrogen
atrazine tricking body into making
more, 238–239
bisphenol A as mimic of,
236–237
endocrine disruption and animal
studies, 244, 260
procession of events in puberty,
252–253
ethylene dichloride, manufacturing
PVC, 121
European Respiratory Society, 161
European Union, atrazine ban
in, 239
exercise
lowering risk of early puberty,
258, 260
using reel mowers for, 189–190

fallout shelter construction, 135–136
family life
going from one child to two,
9–10, 16
single life vs., 8–9
farms, leaking pollutants, 93
farmworkers
hidden costs of pesticide poisoning, 90
pesticide poisoning in children of,
69–70, 209
fatalism, about global warming,
177
father absence, and early puberty,
258

fertilizer
calculating nitrogen for wheat, 93
in cost of conventionally grown
foods, 90
from natural gas, 102, 104–105,
185
nitrogen as, 102
for organic wheat production, 94
using composting for, 185–186
field tomatoes, conventional vs.
organic, 96–97
The Firebird (children's book), 196
fish
advisories, 226
mercury contamination in,
223–224
flour. *See* wheat farming
Food Quality Protection Act of 1996,
63–64
forest fragmentation, in hydrofracking, 272–273
formaldehyde, and child asthma,
145
Formosa Plastics Plant disaster
CSB report on cause of, 123–124
detonation, 123
enormity of, 125
health and environmental
concerns after, 128–130
history of, 123
nearby residents on night of,
125–127
toxic emissions from, 123
transporting hazardous materials
from, 132
fossil fuel emissions
80-percent cut required to prevent
calamity, 181
acting now to curtail, 172
changing policies, 178–179
economic opposition to removing,
172–173
global warming effects from,
162–164
harming children's cognitive
development, 222–223
from individual residences, 180
from lawnmowers, 187–189, 191
mercury poisoning from fish,
223–224
fracking. *See* slickwater hydrofracking
freeze tolerance, of wood frogs, 2
Fresh from the Vegetarian Slow Cooker
(Robertson), 80
fresh-market field tomatoes, 96
Fresno, pesticides on tomatoes in, 96

garages, lawnmower emissions in attached, 188
gardening
    composting food and yard waste, 185–186
    saving fossil fuels by, 183–184
    saving genetic diversity of seeds through, 184–185
garlic, 97–98
GE (General Electric), dumping PCBs into Hudson River, 3–4
Gelobter, Michel, 162–163
genetics
    preserving in diversity of seeds, 184–185
    role in autism, 205–206
geology
    of arsenic, 32
    of Cayuga Lake, 268–269
    creating healthy rural county of Woodford, 103–104
    Marcellus shale, 269–270
girls
    early-maturing, psychosocial problems of, 249–251
    early maturity, as natural process, 251–252
    early-maturity, causes of, 256–260
    early-maturity in, 248–249
    procession of events in puberty, 252–253
    trends in puberty over last century, 254–256
glial cells, in brain development, 216–218
global warming. *See also* climate change
    air pollution induced by, 160–163
    children's books on climate change, 177–178
    effects on ocean, 163–165
    talking to children about, 174–177
glucocorticoids
    guiding lung development, 144
    interference of arsenic with, 32–33
    produced during birth process, 141
gonad-stimulating neurons
    during infant puberty, 235
    overview of, 233–234
    procession of events in puberty, 252–253
grass. *See* lawnmowers
green energy within 20 years, 282
greenhouse gas emissions, 160–162, 185–186

green manure, organic farming, 94, 104
groundwater
    effects of buried pressure-treated lumber on, 49
    fracking contaminating, 275–277
    fracking depleting, 274–275
Ground Zero cloud effects, 21–23
gubernaculum, 240–241
Guinness Book of World Records, tomatoes in, 77

Haber process, 104
Halliburton, 280, 282
Hancock, Randy, 125, 128
handloading, 53
hand-to-mouth interactions, in children, 30
Hansen, James, 179
hard wheat, 91
Hauserman, Julie, 38
hazardous materials, risks of transporting, 132
hazardous waste rules, 1992 exemption of CCA from, 37
health ranking study, 2010, 103
heart
    air pollution effects on, 155–156, 273
    arsenic's potential for destroying, 32
    breastfeeding protecting mother's, 17
heredity, role in pubertal timing, 256
heroes for climate change, 182–184
high specificity, of rabies, 172
Holt's Diseases of Infancy and Childhood, 206–207
Homeland Security Newswire, 133
hormones
    endocrine disruption and animal studies, 260
    in male reproductive tract development, 241–242
    of neuroendocrine system, 233–234
    pesticide influence on children's, 64–65
    reproductive, measured in Danish study, 259
    sensitivity of amphibians to pesticides, 70–71
household crowding, and early puberty, 258
households. *See* residences

How to Grow World Record Tomatoes (Wilber), 77
hydrofracking. *See* slickwater hydro-fracking
hypospadias, prevalence of, 242
hypothalamus
  as governor of adolescent puberty, 233–234
  procession of events in puberty, 252

Illiopolis
  enormity of devastation at, 125
  evacuation of, 125–127
  Formosa Plastics Plant accident, 123–125
  Formosa Plastics Plant emissions, 122–123
  health and environmental concerns after explosion, 128–131
immune system
  air pollution altering functioning of, 155–156
  air pollution exposure in pregnancy and, 152
  creating inflammatory response in, 143–144, 153
  garlic stimulating, 98
  influencing onset of labor in pregnancy, 10
indoor air pollution
  child asthma linked, 144–145, 152
  improving during green renovations, 146
industries, early onset of labor linked to, 12
infancy
  benefits of breastfeeding vs. formula, 16–17
  heightening inflammatory response in, 144
infant puberty, 234–235
inflammatory response
  in atopic asthma, 143–144
  cytokines and, 153
  phthalates inciting, 144
insulin-like factor 3, 237, 244
insulin resistance, and early puberty in girls, 258
intellectual disability. *See* mental retardation
Italy, lack of clothes dryers in, 191

Jones, Zach, 173

Kamesam, Sonti, 34
Kansas wheat production, 91
Kimbrell, Andrew, 90
kisspeptin, 234

labels, reading food, 72–73
labor, agricultural
  ecosystem services as unpaid labor, 108
  exposure to pesticides of farm-worker children, 69–70
  organic foods requiring more, 85, 100
labor, onset of human
  air pollutants, 12
  barometric pressure, 12–14
  epidemiological studies of environmental influences on, 10–11
  obscure origins of, 10
  phthalates and early, 11–12
  preterm births. *See* preterm births
laboratory testing, models not repre-sentative of children, 28–29
Lancet study of 2006, neurotoxicants, 201–202
landfills
  CCA wood in, 48–49
  food and yard waste producing methane in, 185
laundry. *See* clothes dryers
lawnmowers
  benefits of reel mowers, 189–191
  children injured by, 188
  emissions from, 187–189
  lead batteries in, 189
  ozone layer and, 188
lawns, critics of American, 190
lead
  ability to maim children's brains, 207
  causing preterm labor, 11
  drop in mental retardation from policies on, 203
  in lawnmower batteries, 188
  prenatal exposure to, 208
learning disabilities
  costs of, 202–203
  dysfunctional families, 208
  factors creating risk of, 208–209
  as neurodevelopment disorder, 202
  prevalence of, 204
  toxic chemical exposure linked to, xiv–xv
  trends in, 203

LEED (Leadership in Energy and Environmental Design) rating, 119
leopard frogs, 18
leptin, 234
Leydig cells, 244
linoleum, genuine, 120
literature, children's
on climate change, 177–178, 181
The Firebird, 196
heroes of climate crisis in, 182
Lovejoy, Elijah, xi–xii, xvii, 4–5
low birth weight, and onset of puberty in girls, 257
lungs
air pollution stunting development of, 153
development of, 140–141
environmental chemicals altering architecture of, 144
surface area of, 139

male birth rate, effect of Ground Zero cloud on, 22–23
male reproductive tract
description of, 241–242
vulnerability to damage from phthalates, 237
manure, in organic farming, 94, 104
Marcellus Shale
air pollution from fracking, 273
contamination of groundwater from fracking, 275–279
economic story of fracking, 279–281
forest fragmentation from fracking, 272–273
geology of, 269–270
hydrofracking at, 270–271
potential of accidents at, 273–274
water disappearing from fracking, 274–275
McDonald's branding, 68–69
McKibben, Bill, 179, 182, 211
medication
for childhood ADHD, 204
managing asthma with, 153, 156
melanoma, from destruction of earth's ozone, 75–79
melatonin, 234
memory, PCB affecting development of, 208
menstruation (menarche)
age of breast budding and onset of, 254–255
black girls in U.S., 256

history of average age of onset, 254
procession of events in puberty, 252–253
mental retardation (intellectual disability)
costs of, 202
diagnosing autistic children with, 205
linked to toxic chemical exposures, xiv–xv
as neurodevelopment disorder, 201–202
regulations on lead and, 202–203
mercury
emissions from cement plants, 224–225
poisoning from fish, 223–224
thimerosal for flu shots based on, 206
methane in landfills, 185–186
methylation, arsenic and, 33
methyl bromide
destruction of earth's ozone from, 75–76
methyl iodide alternative to, 76–77
as neurotoxicant, 75
methyl iodide, 76
methylmercury, 223–225
Michigan, PBBs in, 259
midwives, on weather and onset of labor, 12–13
migration stage, of brain development, 215–218
migratory birds, hydrofracking threatening, 272–273
milk. *See* dairy
Missouri, sperm counts in men from rural, 243
Montessori, Maria, 200
morbidities of childhood, from toxic chemicals, xv
motor vehicles
air pollution from. *See* fossil fuel emissions
carbon dioxide emissions, 180
reducing dependency on, 180
time children spend in outdoor play vs., 221
mouth breathing, in children vs. adults, 28
mouthing frequencies
in children, 32, 40–41
estimating arsenic exposure of children in playgrounds, 41

myelination stage, of brain development, 216–217

Narrow Bridge Farm, 59–60, 67
National Agricultural Statistics
    Service Web site, 92
National Children's Study of 2000,
    29, 65
national pesticide registry, 92
natural gas. *See also* slickwater hydro-
    fracking
    accidents and, 273–274
    carbon dioxide of, 280–281
    chemicals in fracking fluid for,
        278–279
    depletion and contamination of
        groundwater, 274–277
    economic story of fracking,
        279–281
    hydrofracking at Marcellus Shale,
        270–271
    as industrial feedstock, 272
    in Louisiana, 273
    New York State moratorium,
        271–272
    properties of, 272
    speed of drilling in Pennsylvania
        for, 271
    in Texas, 273
    in Wyoming, 276–277
neurodevelopmental disorders
    costs of, 202–203
    from neurotoxicants. *See* neuro-
        toxicants
    overview of, 201–202
    pandemic of, 201–202
    trends in, 203–204
neuroendocrine system, 233–234
neurogenesis stage, of brain develop-
    ment, 215–216
neurons
    sexual maturation and gonad-
        stimulating, 233–234
    stages of brain development,
        215–218
neurotoxicants
    acting in concert, 208
    brain development and, 214–215
    endocrine disruptors and, 235
    exposure to air pollution and,
        222–223
    methyl bromide, 75
    need for organophosphate
        regulation, 209–211
    neurodevelopment disorders and,
        201–202

overview of, 201
pesticides and ADHD, 209
stages of brain development,
    215–218
timing of exposures, 208
New Orleans study on CCA wood in
    play structures, 53
nitrogen
    calculating for wheat, 93
    as fertilizer, 102
nonrenewable resources, 164
northeast wheat farming, 109
Norway, removing arsenic-treated
    wood, 48
No Yuck Rule, 66
nursery school
    abandoning CCA playground at,
        43–44
    parental debate on children using
        CCA playground, 44–46
    testing for arsenic leaching at,
        41–42

obesity
    hastening onset of puberty in
        girls, 257
    rise of, 151
ocean, global warming effects on,
    163–165
olive oil, 94
organic arsenic, vs. inorganic, 32
organic farming, 101–105
organic foods
    benefits of co-op food stores,
        72–73
    costs of milk and, 99–100
    CSA farms growing, 60–62
    developing healthy eating habits,
        66–69
    garlic, 98–99
    higher cost of, 85
    institutional neglect of farming
        for, 86–87
    olive oil, 94–95
    organic farming feeding the
        world, 101–104
    pesticide residues in, 62–63
    resilience and evolution of fields,
        106
    retail growth figures for, 85–86
    seasonal memories of, 73–75
    tomatoes, 97
    trends in agriculture, 86
    vulnerabilities of children to
        pesticides and, 63–65
    wheat, 93–94

organophosphate pesticides
asthma and exposure to,
150–151
in children eating organic vs.
conventional foods, 62–63
chlorpyrifos in wheat linked to
cognitive deficits, 93
disorders in children with higher
exposure, 65
effect on neurodevelopment of
farmworker children, 210
exposure per child, 211
lack of regulation for abolishing,
209–211
vulnerabilities of children to,
63–65
in wheat, 94–95
outdoor play, vs. time spent in
vehicles, 221
ovulation, in puberty, 252
oxygen cycle, phytoplankton and,
164–165
ozone layer
Clean Air Act and, 157
lawnmowers and, 188
methyl bromide as destroyer of,
75–76

Paarlberg, Robert, 101–102, 104
PAHs (polycyclic aromatic hydrocarbons), 201, 222–223
panthers, reproductive tract
anomalies in, 244
Pavillion, Wyoming, 276
PBBs (polybrominated biphenyls),
in Michigan, 259
PCBs (polychlorinated biphenyls)
contamination
affecting memory centers of
brain, 208
causing preterm labor, 11
citizen advocacy about, 4
EPA investigation, 4
health effects of, 3
history of in Hudson River, 3–4
in Michigan, 259
seeping through shale bedrock, 4
testicular cancer in Sweden and
France linked to, 243
thyroid hormone in pregnancy
and, 4
pediatric disorders, xiv–xv
Pediatrics for Parents, 55
Pennsylvania, gas drilling in, 271
perineum, anogenital distance and,
244–245

peroxisome proliferator-activated
receptors, in preterm labor,
11–12
pesticides. *See also* organophosphate
pesticides
in California tomatoes, 96
in children of farmworkers, 69–70
in children with ADHD, 209
effects of methyl bromide, 75–79
in field tomatoes, 96
foods including costs of, 90
on garlic, 98
keep pace with population growth
using, 102
lack of regulation for abolishing,
209–211
linked to asthma, 150–151
linked to preterm labor, 11
linked to reproductive anomalies
in alligators, 244
no national registry for, 92
in olive oil, 94–95
organic farming prohibiting, 62
sensitivity of amphibians to, 70
special vulnerabilities of children
to, 63–64
in wheat, 92–93
wood treated with. *See* CCA
(chromated copper arsenate)
wood
pH of ocean's surface, 163
photosynthesis, 164–165
phthalates. *See also* PVC
banned from children's toys, 145
CDC statistics on exposure to,
236
as endocrine disruptor, 237–238
hazards of PVC flooring, 120
inciting inflammatory responses,
144
interfering with testicular development, 244–246
linked to child asthma, 144–145
linked to preterm labor, 11–12
physical environment, onset and
tempo of labor, 10–11
phytoplankton, 164–165. *See also*
plankton
pizza study, 87–89, 110
plankton, ocean's, xiii, 163–165
plastic, manufacturing using
bisphenol A, 236–237
playgrounds, pressure-treated wood
and, 40–48. *See also* CCA
(chromated copper arsenate)
wood

polar bears, xiii, 174–175
policies
    agricultural, 77–78, 87, 89–91
    changing carbon emission,
        178–179
    chemical, xv, 28–29, 133–134,
        201, 240, 261
    ending subsidies for fossil fuels,
        178
    requirements to prevent calamity,
        181
politics, adults becoming heroes of
    climate crisis, 182
polybrominated biphenyls (PBBs).
    See PBBs, (polybrominated
    biphenyls)
polychlorinated biphenyls. See PCBs
    (polychlorinated biphenyl)
    contamination
polycyclic aromatic hydrocarbons
    (PAHs), 201, 222–223
polyvinyl chloride or vinyl. See PVC
    (polyvinyl chloride or vinyl)
PONI enzymes, and organophos-
    phates, 215
potato study, 106
poverty
    and dysfunctional families, 208
precocious puberty, 250–251
pregnancy
    air pollution exposure in, 152
    development of lungs during,
        140–141
    experiences heightening inflam-
        matory response in, 144
    exposure to methylmercury
        during, 223–224
    stages of brain development in,
        215–216
    vulnerability for autism in early,
        206
pressure-treated wood. See CCA
    (chromated copper arsenate)
    wood
preterm births
    disruption of lung development
        in, 141
    hastening onset of puberty in
        girls, 257
    impact of Ground Zero cloud on,
        22
    link to lower barometric pressure,
        13
    link to toxic chemical exposures,
        xiv, 10–12
    rise of, 151

processed tomatoes, 96
Product Safety Improvement Act of
    2008, 145
progesterone, timing of birth
    and, 12
Project Laundry List, 194
psychosocial problems, in early-
    maturing girls, 249–250, 258
puberty
    1990 debate on precocious,
        250–251
    adolescent, 234
    age at onset of, 248–251
    brain development in, 253
    drivers of early onset of, 256–260
    endocrine disruptors and,
        236–239
    endocrine disruptors and early,
        252
    environmental threats to repro-
        ductive health, 239–241
    hypothalamus as governor of,
        233–234
    infant, 234–235
    linked to toxic chemical exposures,
        xv
    loss of brain plasticity in, 253
    menarchal age in 1850s, 254
    natural process of early maturity
        in girls, 251–252
    procession of events in,
        252–253
    report on falling age of,
        240–241
    testicular dysgenesis syndrome,
        242–243
    trends in girls over last century,
        254
PVC (polyvinyl chloride or vinyl).
    See also phthalates
    avoiding items made with,
        133–134
    briefing paper on, 118–119
    contributing to atmospheric
        mercury, 224
    detonation of. See Formosa
        Plastics Plant disaster
    improving indoor air quality by
        removing, 146–147
    making vinyl using, 120–122
    manufactured using chlorine gas,
        119
    manufacturing plants as leaky
        places, 122
    World Trade Center disaster and,
        19–20

rabies
bringing same approach to climate
change, 171–172
concerted efforts to prevent,
168–170, 171
differences between climate
change and, 172–173
statistics on, 168, 170–171
reel mowers, 189–191
Reference man, 28–29
regulation
abolishing organophosphate
pesticides, 209–211
exemptions for fracking,
277–278
success of lead-paint, 203,
210–211
reproductive tracts
anomalies in alligators, panthers,
polar bears, 244
anomalies in male, 242–243
male, 241–242
residences
barriers to energy efficiency,
181
clothes dryers, 191–193
emissions contributed from,
180
handling of food and yard waste,
185
lawn, critics of, 190
lawnmowers, 187–188
respiration
asthma and. *See* asthma
and emotions, 140
process of, 139–140
Ringgold, Faith, 4–5
Robert Wood Johnson Foundation,
2010 health ranking
study, 103
Roman street children, and Maria
Montessori, 200

Sandman, Peter, 47
San Francisco compost pick-up,
186
school days, asthmatic children and
missed, 149
school spending, 201
Sculpture Space, 111
sealant study results, for CCA wood,
53
Secchi disk, used to gauge global
warming, 164–165
security lines, air travel post-9/11,
23

September 11, 2001. *See* World Trade
Center collapse
sex
controlling olive flies with, 95
effect of disasters on gender ratio,
22–23
performing tomato, 96
sexual maturation
boys. *See* boys
girls. *See* girls
heredity setting pace of, 256
reactive state of female, 251
shale army has arrived, 283
shale bedrock
drilling for natural gas.
See Marcellus Shale
PCBs seeping through, 4
shell-bearing species, global warming
effects on, 163–165
siblings, lowering risk of early
puberty, 258
Silent Spring (Carson), xiii
single life, family life vs., 8–10
skin cancer (melanoma), from
destruction of earth's ozone,
75–79
slickwater hydrofracking. *See also*
natural gas
chemicals in fluid for, 278
ecological impacts of, 272–277
economic story of, 279–280
exempt from regulations,
277–278
infrastructure and water requirements for, 271
overview of, 270–271
speed of in Pennsylvania, 271
social silence
children's literature on climate
change, 177–178
surrounding climate change,
173–174
talking to children about climate
change, 174–176
soft wheat, 91
species extinction, xiv, 176
sperm
in male reproductive tract,
241–242
prevalence of lowered counts of,
243
spring wheat, 91, 93
Stanford study, 102, 161
Stillwell, David, 37
Stochastic Human Exposure and
Dose Simulation model, 53

Stowe, Harriet Beecher, xi
strawberries
    methyl bromide used in, 76–77
    seasonal memories of organic,
        73–75
stress, in development of immune
        system/lungs, 144
Subra, Wilma, 122
sunburn in whales, 75
Sweden, PCBs in, 243
Switzerland, air pollution measures,
        157
synaptic pruning stage, of brain
        development, 216–217
synaptogenesis stage, of brain
        development, 216–217
systems thinking
    applied to clothes drying, 192
    applied to mowing lawns,
        187–188
    overview of, 186–187

taxonomy, of asthma, 143
television sets in homes, 67–68
terrorism, and chemical infrastructure
        report, 132. *See also* World
        Trade Center collapse
Terrorism and the Chemical Infra-
        structure report, 2006, 133
testicles
    cryptorchidism (undescended),
        242–243
    in male reproductive tract,
        241–242
    phthalates interfering with
        development of, 244–246
testicular cancer, 243
testicular dysgenesis syndrome
    anogenital distance and,
        245–246
    anomalies in alligators, panthers,
        polar bears, 244
    description of, 242–243
    origins of, 243
testosterone
    anomalies in alligators, panthers,
        polar bears, 244
    male reproductive tract develop-
        ment, 241–242
    prevalence of lowered levels of,
        243
    shortened anogenital distance
        indicating inadequate, 245
thimerosal, link to autism, 206
Thomas Road amphibian migration,
        2–3

thyroid hormone
    endocrine disruptors and, 235
    PCBs during pregnancy and, 4
tobacco smoke, prenatal exposure to,
        208
tomatoes
    field vs. fresh-market, 96
    in *Guinness Book of World Records*,
        77
    history of cultivation, and sex, 95
    methyl bromide used for, 76
    pesticides used on California, 96
    pricing own, 87
    seasonal memories of organic,
        73–75
    vulnerabilities of, 96
Toxic Chemicals Safety Act of 2010, 29
toxicology
    of arsenic, 31–32
    limitations of regulatory system,
        29–30
    of methyl iodide, 76
    of pesticides in organic foods, 62
    small amount of chemicals tested,
        29
    studies not representative of
        children, 28–30
Toxics Release Inventory, 122
Toxic Substances Control Act of
        1976, 29
traffic, timing of birth linked to
        intensity of, 12
traffic-related air pollution, and
        asthma, 151–152
transverse lie fetal position, 1
tumor-suppressor genes, arsenic
        suppressing, 33
tuna
    fish advisories, 226
    mercury exposure from, 224
2.4-D herbicide, wheat-growing
        areas, 92

Uncle Tom's Cabin (Stowe), xi
University of Michigan study, organic
        farming, 105
U.N. Ozone Secretariat, petition to,
        76
Up in the Air (Kirn), 23
urban sprawl, children's health and,
        264–268
U.S. Chemical Safety and Hazard
        Investigation Board (CSB)
    2007 final report, 124
    overview of, 123
    results, 131

USDA (U.S. Department of Agriculture)
   Food Plan, 84
   progress of organic agriculture, 86
   wheat yield for northeast U.S., 109
U.S. Fish and Wildlife Service quote, 195
USGS, on mercury in freshwater fish, 224
U.S. study, age at onset of puberty, 255
Utah steel mine, and asthma, 157
uterine quiescence, causing preterm labor, 11

vaccine–autism links, 206
The Victory Garden Cookbook (Morash), 80
vinyl. *See* PVC (polyvinyl chloride or vinyl)
vinyl acetate, 121
vinyl chloride, 121, 123–124

weather, and onset of labor, 12–13
weight
   early puberty and, 260
   hastening onset of puberty in girls, 257
   rise of obesity, 151

well-informed futility syndrome
   creating despair and denial, 47
   defined, xvi, 46
   use of CCA wood and, 47–50
whales, with blistering sunburns, 75
wheat farming
   2.4-D and birth defects from, 92
   history of, 91
   nitrogen calculation for, 93
   in northeast, 109
   pesticides used on, 92–93
   production in, 91
white-nose syndrome, bats with, 194–195
Wiebe, Gerhart, 46
Wilber, Charles, 77
winter wheat, 91, 93
Wisconsin study, organic farming, 105
Wolfe, Wendy, 66
wood frogs, freeze tolerance of, 2
World Trade Center collapse
   air travel following, 23–24
   effect on children, 22–23
   effect on nursing mothers and infants, 19
   effect on pregnant women, 20
   effect on sex ratio, 22–23
   enormity of devastation, 21, 125
   event of, 18–19

Zooplankton, 163–164

ABOUT THE AUTHOR

Biologist and poet Sandra Steingraber, Ph.D., lives with her family in the Finger Lakes region of upstate New York. She is the author of *Having Faith: An Ecologist's Journey to Motherhood*; *Post-Diagnosis*, a volume of poetry; and *Living Downstream: An Ecologist's Personal Investigation of Cancer and the Environment*. Released in a second edition in 2010, *Living Downstream* is now the subject of an award-winning documentary film produced by the People's Picture Company of Toronto.

A passionate public speaker, Steingraber lectures in many universities and medical schools. She has provided Congressional briefings, testified before the President's Cancer Panel, and spoken in the European Parliament and Commission. For her research and writing, she has received the Rachel Carson Leadership Award from Chatham College, the Environmental Health Champion Award from Physicians for Social Responsibility, Los Angeles, and, from Healthy Child, Healthy World, the Mom on a Mission Award for Prevention. In 2010, Steingraber was named one of "25 visionaries who are changing your world" by *Utne Reader*.

A scholar in residence at Ithaca College, she is a columnist and contributing editor at *Orion Magazine*. Please visit her Web site: www.steingraber.com.